LABORATORY LOTUS® A Complete Guide To Instrument Interfacing

Louis M. Mezei

Prentice Hall
Englewood Cliffs, New Jersey 07632

Library of Congress Cataloging-in-Publication Data

MEZEI, LOUIS M.
Laboratory Lotus : a complete guide to instrument interfacing / Louis M. Mezei.
p. cm.
Bibliography: p.
Includes index.
ISBN 0-13-519885-2
1. Scientific apparatus and instruments—Data processing.
2. Research—Data processing. 3. Lotus 1-2-3 (Computer program)
I. Title.
Q180.55.E65M49 1989
507'.8—dc19 88-26755

Editorial/production supervision and
interior design: BARBARA MARTTINE
Cover design: DIANE SAXE
Manufacturing buyer: MARY ANN GLORIANDE

The publisher offers discounts on this book when ordered in bulk quantities. For more information, write:

Special Sales/College Marketing
College Technical and Reference Division
Prentice Hall
Englewood Cliffs, New Jersey 07632

Printed in the United States of America

10 9 8 7 6 5 4 3 2

ISBN 0-13-519885-2

Prentice-Hall International (UK) Limited, *London*
Prentice-Hall of Australia Pty. Limited, *Sydney*
Prentice-Hall Canada Inc., *Toronto*
Prentice-Hall Hispanoamericana, S.A., *Mexico*
Prentice-Hall of India Private Limited, *New Delhi*
Prentice-Hall of Japan, Inc., *Tokyo*
Simon & Schuster Asia Pte. Ltd., *Singapore*
Editora Prentice-Hall do Brasil, Ltda., *Rio de Janeiro*

To
Becky, Heather, and Collette
. . . We're Goin' to Disneyland® . . .

Contents

PREFACE xiii

1 INTRODUCTION 1

How This Book Is Organized 1
Your Keys to Success 3
Disclaimer 3
Assumptions about Your Abilities 3
Assumptions about Your Software and Personal Computer 4
A Simple Explanation of Lotus 7
Components of ''Modern'' Instruments 8
Components of ''Older'' Instruments and Transducers 9
Instrument Communications Categories 9
What's Next? 10

2 PRELIMINARY INSTRUMENT AND COMPUTER SETUP 11

Plan Ahead 11
Brief Description of the Example Instrument 11
Getting Ready to Communicate 14
Assembling the Instrument 14

Communications Fundamentals 15
Transmission Rate: Baud 16
Character Bits, Parity, and Stop Bits 17
Setting Communications Parameters 18
Observing the Speed Limit 20
Making the Connection 21
Plug It in! 24
Making the DOS MODE.com Program Accessible 25
What's Next? 26

3 SPREADSHEET TEMPLATES AND COMPUTER INITIALIZATION 27

Brief Description of the Example Instrument 27
Spreadsheet Templates 28
Planning Your Template 29
Overview of Creating Templates 30
How to Create the Example Template 30
Formatting the Cells 34
Backing up Your Work 35
Macro Programming 35
Creating Macro Programs 36
Macro Execution 37
Auto-Executing Macros 37
Differences in Lotus 1-2-3® and Symphony® Auto-Executing Macros 38
Creating an Initialization Macro for Lotus 1-2-3® 39
Creating an Initialization Macro for Symphony® 40
Giving a Second Name to \0 Macros 42
A Few Comments on Commenting 43
What's in a Name? 44
What's Next? 44

4 THE CORE PROGRAM 45

Some Alternatives 46
The Structure of a Program 47
On with the Example! 48
Where to Start 49
Getting Acquainted with Creating Macros 51
Initializing Using Auto-Executing Macros 52
Overview of the "Main" Program 52

Opening Communications Ports 53
Setting the Initial Cell-Pointer Position 54
Instrument Initialization 54
Initializing the ACRO-931 55
Looping to Prompt an Instrument for Data 56
Receiving and Parsing Data from an Instrument 57
Cutting a Piece of the Pie 58
How to Determine a Parsing Formula 59
Explanation of the Parsing Function in the Example Program 60
Forcing Recalculations 61
Moving Data into the Template 62
The Grand Finale 63
The ''Scratch-pad'' 63
Nesting an @Function into a {PUT} Command 64
Test It! 65
What's Next? 65

5 PROGRAM ENHANCEMENTS 66

Brief Overview of the Enhancements 67
Where to Start 71
Program Portability 73
Starting Menus from Auto-Executing Macros 74
Menus 75
Calling a Subroutine 75
Other Programming Techniques for Menus 76
Quitting a Menu 77
Obtaining Information from the User 77
Providing Status Reports 78
Eliminating Screen ''Flicker'' 79
Getting the Date and Time 79
Determining Elapsed Times 80
Using the {WAIT} Command to Generate Delays 80
Printing a Report with a Macro 82
Saving a File with a Macro 83
Programming for Maximum Performance 84
Making Your Job a Little Easier 85
Ways to Get More Efficiency from {PUT} and {WRITELN} 85
Crank up the Speed! 86
Test It! 87
Analysis and Documentation—The Easy Way 87
Back up Several Copies 88
What's Next? 88

6 ADDING YOUR OWN RS-232 INTERFACE 89

The Association of Analog and Digital Signals 89
Analog Instruments and Experiments 90
Benefits of A/D Converters 91
Basic Data Acquisition Concepts 92
How to Choose an A/D Converter 93
The ACRO-900™ System from ACROSYSTEMS® 96
The ACRO-400™ 97
The Nelson Analytical Series 900 98
The Model WB-40 Interfacing Microcomputer 98
The Model WB-31 Real World Interface 99
Others to Try 100
Setting up the Interface 101
Tapping into an Instrument's Meter 102
Brief Description of an A/D Example Experiment 106
Obtaining A/D Data 107
Running in the Background 111
Retrieving Data from Background Mode 117
Taking the Snapshot 117
Other Ways to Take a Snapshot 122
Looking for Intangibles 127
RS-232 and Plug-in Board Snapshots 127
Monitoring and Controlling Processes 128
What's Next? 128

7 TOTAL AUTOMATION: CONTROLLING INSTRUMENTS THAT HAVE MOTORS 129

Brief Description of the Example Instrument 129
Installation 131
Making the Connection 132
Setting the Communications Parameters 132
Description of an Assay Template 133
Creating the Template for this Example 137
Test the Template 138
Establish Communications 138
The Macro Program 139
The Auto-Executing Macros 139
Instrument Initialization Macros 139
Description of the Example Initialization Macro 141
Testing for Successful Initialization 142
Timing During Initialization 142

Standardization Macros 143
Linear Regression Macros 145
Macros for Unknown Samples 147
Help Windows 147
Finding the Cell-Pointer 149
Testing Error Codes 151
Strings that Contain Multiple Pieces of Information 151
Parsing Variable Length Strings 152
Exponents in Strings 154
Program Timing 154
The "Scratch-pad" and String Commands 154
Creating the Program 155
Creating Windows 157
Test It! 158
What's Next? 158

8 GETTING DATA FROM INSTRUMENTS THAT YOU CANNOT PROMPT 159

Brief Description of the Example Instrument 160
Installation 161
Making the Connection 162
Setting the Communications Parameters 164
Report Templates 165
Creating the Template 166
The Macro Program 168
Starting with a Slow Baud Rate 168
"Linked" Menus 169
Self-Modified Range Specifications 169
Centering Data on the Display 169
Data Capture Macros for Non-Prompted Instruments 171
Handling Errors 172
Waiting for the Data 174
The Importance of Trap Loops 175
Testing for the Presence of Data 175
Resetting Error-Handling Protocols 176
Speed Considerations While Designing Data Collection Macros 177
Parsing Multiple Data Points from Strings 178
Extracting and Filing Just the Data Portion of a Spreadsheet 180
Other Utilities 181
The Scratch-Pad 183
On-Screen Prompts 183

Creating the Program 184
Printing Wide Reports in Compressed Mode 186
Crank up the Speed! 187
What's Next? 187

9 UNCONVENTIONAL EQUIPMENT: LOTUS MEASURE™ 189

Brief Description of Lotus Measure 193
Installing and Starting Lotus Measure 194
Installing Lotus Measure for Lotus 1-2-3 195
Installing Lotus Measure for Lotus Symphony 196
Starting Lotus 1-2-3 with Measure 197
Starting Lotus Symphony with Measure 197
Lotus Measure RS-232 Module 197
A Measure RS-232 Example: The Mettler Program 202
Other Measure RS-232 Commands 204
Taking a Snap shot of METTLER 206
Expanding the METTLER Program to Fit Your Needs 207
The MBC16 Module: Plug-in Boards 207
Configuring A/D Plug-in Boards 208
Voltage Configuration Formulas 213
Installing an A/D Plug-in Board 214
Configuring Measure's MBC16 Module 215
Testing MBC16 Settings 219
Programming for MBC16 219
Component Oriented Differences 222
Using Multiple Input Channels with the MBC16 Module 225
IEEE-488/GPIB/HP-IB 226
GPIB Addresses 229
The Role of the Cable in GPIB Control 231
Configuring GPIB Plug-in Boards 232
Configuring Instruments for GPIB 234
Gathering the Pertinent Information 235
Configuring the Measure NAT488/GPIB and HP488/HP-IB Modules 235
A Simple GPIB Example Program 238
GPIB Auto-Executing Macros 240
Initializing the GPIB System 240
Turning on the DMA Channel 241
Setting up a Mechanism for System Status Checks 241
Resetting Instruments to Their Power-on States 243
Writing to GPIB Instruments 244
Reading Data from GPIB Instruments 245

Testing {STATUSON} Codes 247
Polling and Servicing Instruments 249
Waiting for SRQs/Monitoring Several Instruments 252
Other GPIB Macro Commands 254
What's Next? 254

10 GETTING DATA FROM AN INSTRUMENT'S DIGITAL DISPLAY 255

Binary Coded Decimal 255
Other Forms of BCD 257
Digital Input/Output Interfaces 258
How to Choose a Digital I/O Interface 259
Brief Description of the Example Instrument and Interface 259
Connecting the Instrument to the Digital I/O Interface 259
The Macro Program 261
Obtaining BCD Data and Converting to Decimal 261
Counting in Hexadecimal 265
What's Next? 267

11 SUMMARY 268

Thirty Easy Steps to Successful Integration 268
Programming Tips 270
Communications Tips 277
Worksheet Protection: Design Tips 278
Communications Troubleshooting Guide 279
Program Troubleshooting Tools 281
What's Next? 283

APPENDICES

A DETERMINING AND BUILDING RS-232 CABLES 284

Brief Description of the Auto-Cabler 286
Rules for Using the Auto-Cabler 286
Using the Auto-Cabler 286
How to Get a Cable 288
Making an Adapter 289
Making a Cable 290

B TABLE OF ASCII CHARACTERS 293

C RS-232 PINOUT SPECIFICATIONS 297

D BINARY CODED DECIMAL AND HEXADECIMAL TABLES 299

E ARITHMETIC, RELATIONAL, AND LOGICAL OPERATORS 301

F MACRO COMMAND SUMMARY 304

System Commands 304
Interaction Commands 305
Program Flow Commands 305
Cell Commands 306
File Commands 307
Menu Commands 308
Macro Commands for Special Keys 308

G @FUNCTION SUMMARY 309

Mathematical @Functions 309
Statistical @Functions 310
String @Functions 311
Special @Functions 312
Logical @Functions 313
Date and Time @Functions 313

INDEX 315

Preface
How I Got Hooked on the Lotus® Spreadsheets and Programming Language

Have you ever noticed how scientific software manufacturers corral you into a one-inch square box and then charge you $2000 for the box?

Have you ever noticed how commercial software manufacturers try to make you do it their way, or not at all?

Have you ever bought a piece of equipment and couldn't find software to support it?

Have you ever purchased a piece of equipment that was billed as "RS-232 compatible" only to find it meant that there was a lifeless connector at the rear of the instrument?

Do you own a piece of equipment that was born before the computer revolution and doesn't even have an RS-232 connector?

Do you run experiments that collect data with sensors?

Well, at one time or another I was able to say yes to each of the above. That's what prompted me to look into spreadsheet programs to see if they could answer my needs. At first, I was awed at the number of things that I thought I would need to learn. What's worse, all of the books written on spreadsheets were written from the business standpoint. After buying about 16 books on Lotus® spreadsheets and trying to weed through such foreign topics as "rate of return", "sales forecasts", "commission schedules", and the like, I became dismayed. But, I was able to see a glimmer of hope. Lotus 1-2-3 and Symphony seemed to have all of the features that I needed to reduce my data and graph it.

More importantly, the Lotus programming language seemed to have commands that I could use to control instruments. All that I needed to do was to direct the information in my books away from a business focus to a scientific standpoint and try to find a way to use the features in a way that would address the kinds of programming and spreadsheet activities that a scientist could benefit from.

My incentives were numerous. I was buying expensive software that was designed by people who had never worked in a laboratory environment. The software was never a good "fit" for how things are really done in laboratories. I had doubts that software designers had ever caught onto the fact that every experiment performed in a laboratory setting is different. This is especially true in research laboratories. In addition, data reduction tends to be highly variable. In fact, most researchers will look at the data from many different angles . . . the number of angles varying with the inverse of the quality of the data.

To try to fit more potential user requirements, many manufacturers' programs will store the data in a form that can be imported into a Lotus spreadsheet. From this, it is proclaimed, you can do your own data reduction.

However, this results in a slow and awkward process in which you must capture the data using their software, save the data, exit their program, get into the spreadsheet, import the data, move the data to appropriate cells and then do your data reduction and/or graphics. If you are like me, I am anxious to see my data; this method eliminates the advantage of immediate viewing of the data and, quite frankly, is a lot of work. It also seems absurd to buy a piece of expensive software just to capture the data from an instrument, only to import the data into a spreadsheet. It seemed easier to me to find a way to get it directly into the spreadsheet.

I also had other reasons for wanting to integrate my data collection instruments with a spreadsheet. Every piece of software that I bought controlled just one instrument. This constraint meant that I had to keep buying (and paying for) more software. It also meant that I had to weed through tons of manuals and try to remember many different software formats.

I kept running into another problem as well. I could not obtain software for some of my older instruments. Some of the instruments did not have RS-232 connectors. Some of them did, however, have chart recorder outputs or displays. I even had experiments that did not collect data with an instrument, but had sensors.

These are the reasons why I considered a spreadsheet program. As I said earlier, Lotus looked promising; so I purchased Lotus Measure.™ Aside from the expense, I was very disappointed with the manual. Measure also took a lot of programming to get it to work with the spreadsheet and my instruments.

Fortunately, I had done a considerable amount of C programming prior to working with Lotus; so I decided to try to apply some C tricks to Lotus programming to see if they would work. They did. I was able to find a method to talk to instruments directly from the spreadsheet, **WITHOUT Lotus Measure.** This finding exploded into a cost effective way to beat the problems with purchased software. In fact, it solved all of my data acquisition, reduction, and graphics problems.

Lotus is a very rich environment that you too can use to solve your data acquisition, reduction, and graphics problems. You do not need to learn the hundreds of Lotus program commands and special-purpose formulas to use it effectively. You also do not need to spend a lot of time learning how to connect an instrument to a personal computer. Early in this book, I will teach you the quick and easy way to figure out cabling and how to set communication parameters. Your time will be better spent designing and carrying out experiments.

From robots to doughnuts (and everything in between), if your central mission is the collection and reduction of data, this book can help you. Armed with this book, you can control just about everything from a robot that automates the most sophisticated genetic engineering project to a texture analyzer that measures the adhesion, bloom strength, crispness, and tackiness of your doughnuts to make sure that they are done or have not become stale. And that can be very important at coffee break time.

This book will also give you a rational explanation of Lotus Measure. There are some instances in which you MUST use Measure. For example, if you have a plug-in data acquisition board or are using IEEE-488, you need to use Measure.

After you have your applications up and running, you can impress your colleagues with show-stoppers that you have created with your newly found skills. That is, after reading this book you will be able to write *your own* $2000 piece of software in a couple of hours.

I will teach you all of the electronics, robotics, automation, cabling, computer jargon, and programming skills that you will need to know to fully interface your instrument. In the process, I will show you how to use Lotus commands to do some real programming.

Once you have learned how to capture the data from your instruments and place it in your spreadsheet, you can make your data come alive with rich graphics and sophisticated data reduction. You can confidently share your reports with your colleagues. In turn, your reports can be much more enjoyable when they contain Fast Fourier Transforms, linear and curvilinear regressions, integrations, differentiations, and very hot graphics.

LOUIS M. MEZEI

1

Introduction

This book will teach you everything you need to know to be able to merge scientific instruments, your personal computer and either a Lotus 1-2-3® or a Lotus Symphony® spreadsheet to produce a single, powerful unit that can meet your exact automation, data reduction, and graphics needs.

The process of creating this all-encompassing system is known as "integration" and the cabling, the communication parameters, and the part of your program that deals specifically with the communications that take place with the instrument are referred to as the "interface". To avoid confusion, both "integrating" and "interfacing" will be used in reference to the communications between the spreadsheet program and your instrument throughout the remainder of the book.

HOW THIS BOOK IS ORGANIZED

This book is progressive in format, but contains a certain degree of redundancy (when necessary). The remainder of this chapter is devoted to stating any assumptions I've made, and provides you with all the information you need to complete the objectives of the remaining chapters successfully.

The major objective of Chapters 2 through 5 is to show you how to plan and organize the integration of your instruments with your personal computer, Lotus 1-2-3, and/or Symphony. These chapters point out ways to avoid some of the common pitfalls found in RS-232 communications and Lotus. They also show you how to avoid problems arising from quirks in Lotus' operation, and what to do if you fall into a trap.

Chapters 2 through 5 use a single, comprehensive, example instrument. They follow the order that you would normally follow when you create an application:

- Chapter 2 shows you how to set up the communications parameters inside an instrument, how to determine which cable to use, and how to connect an instrument to your personal computer.
- Chapter 3 shows you how to set up your spreadsheet to receive data from an instrument and how to build a simple program that will automatically configure your personal computer to the proper communications parameters.
- Chapter 4 shows you the basic programming that is necessary to communicate with an instrument.
- Chapter 5 shows you how to add a number of very useful features to your program.

The objective of Chapters 6 through 10 is to make you comfortable with interfacing a variety of instruments. The chapters are designed to make the process of integrating instruments as easy as possible for those of you with no prior programming and/or interfacing experience.

- Chapter 6 shows you how to convert analog voltages to the digitial signals that computers can understand. This is an important chapter if you are using sensors to monitor experiments. The chapter also shows you how to get data from an instrument that has a chart recorder output and an instrument that has a meter as a display.
- Chapter 7 shows you how to control a robotic instrument. It includes a discussion of how to command an instrument to move its motors to perform an assay and return the data to your spreadsheet.
- Chapter 8 shows you how to receive data from instruments and/or barcode readers that transmit their data without being prompted for it from the computer.
- Chapter 9 shows you how to use Lotus Measure™ for instruments that utilize RS-232. It also thoroughly discusses plug-in circuit boards and IEEE-488 communications.
- Chapter 10 shows you how to get data from instruments that have digital displays, but not RS-232 connectors. It also contains thorough discussions of Binary, Binary Coded Decimal (BCD) and Hexadecimal numbering systems.

Chapter 11 is a summary chapter. It contains a step-by-step list of all the tasks needed to integrate any instrument. It also gives you numerous tips on programming and communications. Additionally, it gives you time-proven methods for pinpointing the cause of a program or communications problem.

Appendix A contains *very simple* methods of determining the configuration of the cable for your application and how to build the cable.

YOUR KEYS TO SUCCESS

If you work through this entire book, you should know how to integrate a wide variety of instruments with Lotus. The real secret to becoming an expert, however, is to actually work through the examples contained within this book. You will be able to get a much better feel for the structure of a program if you type the examples into a spreadsheet. Hands-on experience is the best way for you to learn. Therefore, if you have an instrument in the category being described in a chapter, you should try to interface the instrument at the time that you read through the chapter.

After you have an idea of what each example program can do, you should work with any features that you don't fully understand. Do so by making variations of the program to see the effects of your changes. Place a copy of the @functions that are within macro commands into cells of the spreadsheet and see how they update as the program proceeds. Frequently, after you try using a specific Lotus command or @function and see it in action, the explanation in the book will become clearer.

One thing that you will discover when reading through this book is that there are often many solutions to a single problem. There often isn't a single, best way to achieve a programming goal with software. I will usually present only one or two ways to address each programming challenge. However, with a programming language as versatile as this one, there are usually *many* ways to implement almost any task. Some ways are better than others, depending on the circumstances and the application. Sometimes you will be surprised to find that what you thought was a very inefficient method, is actually the most efficient. Do not be afraid to investigate several alternative programming schemes to accomplish a particular task and then choose the best.

DISCLAIMER

Lotus will give you no guarantees and neither can I. However, I can assure you that I have taken great pains to make sure that all of the programs that I present to you in this book work. I have tried all of them in both Lotus 1-2-3 and Symphony on an IBM XT, AT and Model 80. However, the examples are just examples. Although all of the examples can be used directly in your laboratory, they are meant to teach you certain specific concepts about how to interface instruments and are not necessarily the most efficient way to program for your particular situation.

ASSUMPTIONS ABOUT YOUR ABILITIES

I wrote this book for the person who has access to a copy of Lotus 1-2-3 and/or Symphony and has at least a beginner's knowledge of how to use it. A basic discussion of Lotus spreadsheets and their components can be found in a subsequent section of this chapter.

I also assume that you have a real desire to learn how to connect a scientific instrument to a computer and to learn some basic programming techniques.

ASSUMPTIONS ABOUT YOUR SOFTWARE AND PERSONAL COMPUTER

Software

I developed and tested all of the examples in this book using Lotus 1-2-3 (Releases 2.0 and 2.01) and Symphony (Versions 1.1, 1.2, and 2.0). Because the Lotus macro commands and @functions are fully established, there should not be any changes in how they are invoked or how they perform (other than additions required to support the multiple layering of spreadsheets). For this reason, the information and examples given in this book should be generally applicable to all future releases of the two Lotus programs.

If you haven't purchased a Lotus spreadsheet program yet, I would recommend Symphony. Lotus 1-2-3 contains a spreadsheet, graphics, and a crude database. Symphony also contains these three environments plus word processing, telecommunications, and windowing capabilities. Additionally, it can run DOS programs directly from the spreadsheet.

Although 1-2-3 and Symphony have many features in common, especially in the spreadsheet area, Symphony is much more powerful than 1-2-3. Lotus Development has enhanced the spreadsheet and graphics functions for Symphony and has rewritten the database functions to make this part of the program more powerful and easier to use. Symphony can easily keep track of samples and/or results for a single lab or several different labs.

Personal Computer Equipment

The computers that I used were all IBM based. I have a Personal System/2 Model 80, an AT, and an XT. Nearly all of the programs worked on all three computers. The XT proved too slow in a couple of instances, so I used an Orchid Technology PCturbo 286e plug-in accelerator board to boost the speed of the XT. (The PCturbo 286e is available from Orchid Technology, 45365 Northport Loop West, Fremont, CA, 94538.)

Drawing from my experience, you should be able to use any IBM personal computer or close compatible that supports Lotus. Get one with a speed at least as fast as an eight megahertz AT. The difference in price between an AT and an XT is inconsequential, and the AT's speed will make it much easier to automate a wider variety of instruments.

Hard drive. A hard drive is a necessity. Do not try to use either Lotus program without a hard drive. The current cost of hard drives will make it one of your best investments. The difference in price between a 10 and a 20 megabyte hard drive is so insignificant that I recommend you buy at least a 20 as you will soon surpass 10 megabytes.

RS-232 communications port. The last necessity is at least one serial RS-232 communications port. If you have two, however, two serial ports can automate two instruments. And, if you will be buying a serial printer, the second serial port is an absolute must because the printer will occupy the first one.

Graphics equipment. If you plan on using graphics, make sure that your monitor and printer can handle graphics.

Expanded memory. If you are going to be dealing with a *very large* amount of data, you could run out of computer memory. You may want to consider adding expanded memory to your computer. Lotus, Intel®, and Microsoft® Corporations have jointly developed a specification that allows programs to use memory above the standard 640K maximum. Memory boards conforming to this specification are referred to as *E*xpanded *M*emory *S*pecification (EMS) boards. Lotus 1-2-3, Release 2 and Symphony can take advantage of the memory on expanded memory boards. These boards are marketed by Intel, Quadram, AST, and many others. You can use up to four megabytes of EMS memory from a Lotus Spreadsheet.

If expanded memory is present in your computer, Lotus will use the EMS memory to store formulas, floating point (decimal) numbers, and labels. Integers are stored in standard memory. The following table outlines the type of memory (CONVentional or EMS) where various types of information reside.

	1-2-3 2.0/2.01	Symphony 1.1/1.2	Symphony 2.0
INTEGERS	CONV	CONV	CONV
DECIMAL NUMBERS	EMS	EMS	EMS
LABELS (TEXT IN CELLS)	EMS	EMS	EMS
MACRO PROGRAMS	EMS	EMS	EMS
FORMULAS	EMS	EMS	EMS
@FUNCTIONS	EMS	EMS	EMS
RANGE NAMES	CONV	CONV	EMS
BLANK (ERASED) CELLS	CONV	CONV	CONV
ADD-IN APPLICATION PROGRAMS	CONV	CONV	CONV
SETTINGS SHEETS	CONV	CONV	EMS
LOTUS 1-2-3/SYMPHONY	CONV	CONV	CONV
PRINT SETTINGS	CONV	CONV	EMS
LOTUS TEXT (MESSAGES)	CONV	CONV	EMS
MEMORY RESIDENT PROGRAMS	CONV	CONV	CONV

As you can see, there are subtle, but very important, differences in how the various Lotus programs utilize expanded memory.

EMS memory is not nearly as fast as conventional memory. Thus, the use of EMS memory can profoundly affect the speed of your program and can even affect the communications between your instrument and spreadsheet. This difference in speed is especially evident in Symphony, Release 2 where your programs will execute approximately 35% slower if you are using expanded memory. This difference in speed is sometimes significant enough to have a considerable influence on the types of things that you will need to do in order to successfully establish communications with your instruments.

Lotus claims that you can utilize up to four megabytes of expanded memory. However, this statement is misleading. For each "segment" ("packet") of four active cells that are stored in expanded memory, Lotus allocates about 16 bytes of conventional memory (i.e., *about* 4 bytes of conventional memory are consumed for each EMS cell). This conventional memory is used for control purposes and establishes a link to the expanded memory location where each cell's information is located. The amount of unused conventional memory, therefore, will place limitations on the number of cells of data that your Lotus program can support (no matter how much memory each cell takes up in expanded memory). The following table will give some *rough* guidelines. It assumes that you have a computer with 640K of conventional memory and at least 640K of EMS memory, storage of floating point (decimal) numbers and no memory-resident, print spooler, add-in application, etc., programs present:

	1-2-3 2.0/2.01	Symphony 1.1/1.2	Symphony 2.0
# CELLS WITHOUT EMS	34500	24500	23000
# CELLS WITH EMS	84000	68500	74000

Thus, EMS memory can help you collect massive amounts of data, but be forewarned that EMS memory is not nearly as fast as standard memory and it will slow down your spreadsheet and programs.

Unless you anticipate collecting more than 23000 data points in your experiments, I strongly recommend disabling the expanded memory. Your spreadsheet will run about 25% to 35% faster for nearly all operations. You can disable your expanded memory by removing the command lines that were placed in the CONFIG.SYS file to specify the EMS device driver (i.e., the ".SYS" file) and then rebooting your computer.

There is one last EMS quirk that you need to be aware of. If you are using Lotus with a co-processor accelerator board, such as the Orchid Technologies PC-turbo 286e, Lotus spreadsheets WILL NOT be able to access the EMS memory in accelerator mode. If you have added one of these boards to an old IBM PC/XT to try to improve its speed, you will need to choose between EMS memory and speed.

A SIMPLE EXPLANATION OF LOTUS

Lotus 1-2-3 and Symphony both have over 600 commands. This magnitude tends to overwhelm and discourage new users. What new users fail to understand is that they need to know only a handful of the commands to get started.

Consider all of the programming commands available as if they were tools in a toolbox. To perform a certain task, look up the name of the tool and use it. The more commands that you are aware of, the more tools you have at your disposal. Half the battle of programming is knowing what is possible. To get a grasp of the Lotus tools available, skim the manuals accompanying the software, browse through Appendices E through G at the end of this book, and/or look over some of the books listed as references at the end of this chapter.

A number of the most powerful commands are not documented in the Lotus manuals and many of the reference books listed. This book uses many of these undocumented features of the Lotus programming language. It also describes many ways to use the commands in programs . . . in both conventional and unconventional manners.

Macro Commands and @Functions

A set of instructions in a format and language that Lotus can understand is called a Lotus ''Macro'' program. The language that is used is called the ''Lotus Command Language'' and the instructions are called ''macro commands''.

There are six broad categories of Lotus macro commands. They are

- System commands, which allow you to control the screen display and the computer's speaker.
- Interaction commands, which allow you to create interactive macros that pause for the user to enter data from the keyboard.
- Program flow commands, which let you include branching and looping in your program.
- Cell commands, which transfer data between specified cells and/or change the values of cells.
- File commands, which work with data in DOS files.
- Menu commands, which allow you to design and manage menus.

Special-purpose Lotus formulas are called ''@functions''. Lotus contains numerous @functions. Each @function can extend your calculating power beyond simple arithmetic and text handling operations. There are several broad categories of Lotus @functions: mathematical, text, logical, financial, statistical, database, and date/time.

In general, macro programs use both macro commands and @functions to carry out the kinds of processes that your program implements to get your instrument to perform data collection chores. Your goal is to plan and write efficient macro programs. Efficient programs:

- Get specific work accomplished in a specific order.
- Specify exactly what you want your instrument to do and when.
- Get data into the cells of your spreadsheet.
- Carve out small pieces of data from larger pieces.

Planning ahead is perhaps the most important concept of efficient programming. Knowing what you need to do, what you can do, and when you need to do it is over 90% of the data collection game.

The attachment of your instrument to your personal computer is just as important to your system as the program that you write. Getting the correct cable to attach your instrument with your personal computer is a major part of creating your application. Both the instrument and the personal computer *must* be talking and listening at exactly the same speed and with words that have letters of the same size and meaning. Otherwise, mass confusion will occur. The next sections will explain the instrument to equipment interface in more detail.

COMPONENTS OF "MODERN" INSTRUMENTS

Most state-of-the-art scientific instruments can be viewed as having three main components. The first component is a power source. This component converts the alternating current (AC) from the wall outlet to direct current (DC) and provides a steady and unchanging voltage to each of the other components of the system.

The second component contains all of the electrical and mechanical elements of the instrument. These elements actually perform the work necessary to carry out the measurements that you want to make.

The third component is the *C*entral *P*rocessing *U*nit (CPU). The CPU is a computer within the instrument that contains the circuits that control and coordinate all of the functions of the instrument. If the instrument is listed as having "RS-232 compatibility", then the CPU will also have the ability to "talk" to a personal computer. That is, the CPU of the instrument will be able to transmit data from its RS-232 connector (also referred to as a "port").

Some instruments can also receive instructions from a personal computer and execute them. This capability gives the instrument the ability to perform at a very high level of automation and flexibility.

A substantial portion of this book shows how to interact with an instrument that has a CPU and RS-232 compatibility. That is, much of this book is aimed at teaching how to send commands to instruments and receive data back.

You can also get data from older (non-computer compatible) instruments and from experiments that produce voltage changes. The following section discusses the basic layout of these instruments and experiments.

COMPONENTS OF "OLDER" INSTRUMENTS AND TRANSDUCERS

Many laboratories have very high quality instruments that were manufactured prior to the computer revolution. These instruments do not have CPU components; nor do they have RS-232 connectors from which to send data. Most of these instruments, however, do have either a chart recorder output or a meter. Either of these sources of electrical signal can express a voltage level that is proportional to the event being measured. The voltage level that is obtained represents a continuously changing quantity and is referred to as an "analog" output. In other words, an analog signal can take on any value between the lower limit and upper limit of the range of acceptable values. In contrast, digital signals used by computers switch suddenly between two very different voltage levels to give either a binary zero or a binary one.

Likewise, many experiments are performed using simple sensors to collect the data. Most signals which are acquired are either from a natural electrical souce (e.g., biopotentials from an electrocardiogram, voltages from a thermocouple, electromotive force of two redox half-cells, etc.) or from transducers.

A transducer is a sensor that produces an electrical output that is proportional to the physical or chemical property that is being measured. Transducers can be used to measure pressure, flow rate, temperature, force, light, inductance, position, velocity, acceleration, strain, or even light intensity.

To automate an experiment which produces an analog signal you first need to "digitize" the signal before a personal computer can use the data. Digitizing the data results in a number that corresponds to the analog voltage level. The device that performs this conversion is called an "analog-to-digital" (A/D) converter. After the analog signal is digitized, you can receive the information from the converter and place it into your spreadsheet via your personal computer.

One additional category of "older" instruments exists that deserves brief mention. In this category, the instrument has a digital display, but no RS-232 capability. If your instrument has a digital display, you can obtain the information directly from the display with a device called a "Digital Interface".

INSTRUMENT COMMUNICATIONS CATEGORIES

As you can deduce from the previous two sections, the world of instruments is relatively small. From a programming point of view, only four different categories of instruments use RS-232 communications:

- Instruments that transmit data only after being asked
- Instruments that have motors (robotics)
- Instruments that sit idle and transmit data after a user presses a key
- Instruments that transmit data continuously

Instruments that do not have RS-232 capability are

- Instruments that have chart recorder or voltage outputs
- Sensors and transducers
- Instruments that have analog (continuous scale) meter displays
- Instruments that have digital displays
- Instruments that use IEEE-488 communications
- Circuit boards that plug into the slots of your computer

WHAT'S NEXT?

The next chapter will show you how to set your instrument to the proper communication parameters and how to determine the type of cable to use when connecting the instrument to your personal computer. These are the first tasks that you will need to perform when you begin your integration process.

BEGINNER USERS' REFERENCES

ADAMIS, EDDIE, *Command Performance: Lotus 1-2-3.* Bellevue, WA.: Microsoft Press, Div. of Microsoft Corp., 1986.

MILLER, STEVEN, *Lotus Magazine: The Good Ideas Book.* Reading, MA.: Addison-Wesley Publishing Co., Inc., 1988.

The staff of Lotus Books, *The Lotus Guide To Learning Symphony Macros.* Reading, MA.: Addison-Wesley Publishing Co., Inc., 1986.

The staff of Lotus Books, *The Lotus Guide to Learning Symphony Command Language.* Reading, MA.: Addison-Wesley Publishing Co., Inc., 1985.

STARK, ROBIN, *Encyclopedia of Lotus 1-2-3.* Blue Ridge Summit, PA: Tab Books, Inc., 1987.

ADVANCED USERS' REFERENCES

FENN, DARIEN, *Symphony Macros And The Command Language.* Indianapolis, IN.: Que Corporation, 1987.

ZUCKERMAN, MARCIA, Dyrud, Anne, and Posner, John, *The Lotus Guide to 1-2-3 Advanced Macro Commands.* Reading, MA.: Addison-Wesley Publishing Co., Inc., 1986.

CAMPBELL, MARY, *1-2-3 Power User's Guide.* Berkeley, CA.: Osborne McGraw-Hill, Inc., 1988.

2

Preliminary Instrument and Computer Setup

This chapter will explain how to configure an instrument's communications parameters and how to connect the instrument to a personal computer. These are the first steps that you will need to perform when you want to unite *your* instrument with *your* personal computer and *your* Lotus spreadsheet.

An outline of all the steps in the process of creating an instrument application is given in Figure 2–1. This same outline will appear in Chapters 3 through 5. In each chapter, the steps that will be discussed in the chapter will be marked by arrows.

PLAN AHEAD

The tasks in this chapter may require you to purchase a computer cable, locate a copy of your particular instrument's User's Manual, and/or contact an instrument manufacturer for more information. Any of these pre-requisite steps could produce a delay in the time that you could actually start creating your application. So, plan on initiating the tasks outlined in this chapter well in advance of the time that you want to begin working on the integration process.

BRIEF DESCRIPTION OF THE EXAMPLE INSTRUMENT

I have chosen the ACROSYSTEMS® Model ACRO-900™ "Stand-Alone Interface System" as the example for Chapters 2 through 5.

The ACRO-900 is available from ACROSYSTEMS, Inc., 66 Cherry Hill Dr.,

```
                STEPS TO A SUCCESSFUL APPLICATION

->Assemble The Instrument
->Test Instrument In Manual Mode
->Determine Wiring Configuration
->Get Cable
->Connect Cable To Computer And Instrument
->Secure Cable
->Determine How To Set Communication Parameters
->Set Module Addresses
->Reset Instrument
->Add PATH Command To AUTOEXEC.BAT File
  Add MODE Command to AUTOEXEC.BAT File
  Design Template And Macro Program
  Start The Spreadsheet
  Type Template Into Spreadsheet
  "Activate" Template
  Test The Template
  Create And "Activate" The Auto-executing Macros
  (Symphony): Specify The Name Of The Auto-executing Macro
  Use The DOS MODE Program To Set Communication Parameters
  Type In A "Bare Bones" Program
  "Activate" The Test Program
  Try To Establish Minimal Communications
  Evaluate The Test Program
  Expand Test Program To Retrieve Data
  Add Amenities And Menus To The Program
  "Activate/Re-activate" Macro Program
  (Symphony): Create Windows
  Create A Table Of Range Names
  Save The File
  Evaluate The Program
  Increase Communications Speeds
  Final Test The Program
  Save/Backup Several Copies Of The File
  Save Several Copies In Separate Safe Places
```

Figure 2-1

P.O. Box 487, Beverly, MA. 01915. This system is the same as the OM-900, which is available from Omega Engineering, Inc., P.O. Box 2669, Stamford, CT 06906.

This example is only intended to show you how to integrate a personal computer with an instrument. Your personal computer and instrument will surely be different. However, the steps and concepts will be the same.

The ACRO-900 system is an excellent example because it is simple to use and it requires all of elements that you will need to interface just about any instrument to a Lotus spreadsheet. In future chapters, I will show you how easy it is to modify the setup and programming steps used for the ACRO-900 for other example instruments. Each example instrument, by the way, is typical of the other categories of instrument-personal computer- spreadsheet arrangements. That is, collectively the example instruments of the next few chapters will represent all of the commonly required cable arrangements, methods of configuring communications parameters, methods of communicating, degrees of automation, and programming.

You should therefore think about how your instrument exhibits each of these

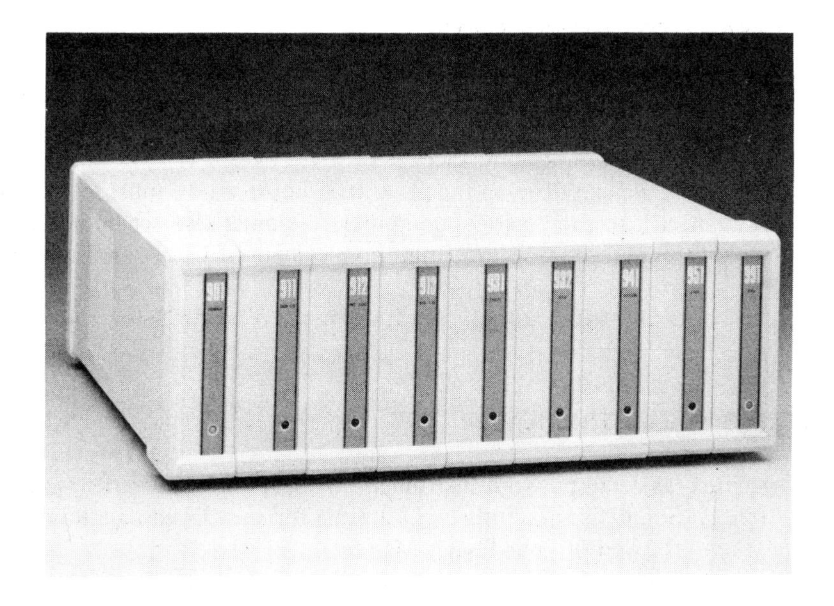

properties and requirements. You can then implement the concepts that are appropriate when you begin to integrate your system.

Below is a brief description of the ACRO-900. Think about your configuration in the same context as you read it.

The ACRO-900 or OM-900 package includes a power supply, a microprocessor (CPU) module, and module(s) to meet your specific data acquisition and/or control needs.

Some of the modules that are available include a thermocouple interface, an RTD (Resistance Temperature Detector) interface, a strain gauge input, a linear/rotation position transducer (LVDT/RVDT) converter and several analog-to-digital converters.

As I mentioned in Chapter 1, the analog-to-digital converter modules are particularly useful if you have an older instrument that does not have an RS-232 interface. I have an analog-to-digital converter in my lab that I connect to the chart recorder outputs and meters of some of my older instruments. The converter takes the analog (voltage) signals, changes them into digital signals, and I retrieve the information from the RS-232 output of the module. A detailed description of this technique is given in Chapter 9.

Other modules are also available. Some of them are called General Purpose Input/Output (GPIO) modules and are available in several formats. These modules are ideally suited for monitoring an experiment, using the spreadsheet to store the data and make decisions, and then sending an output signal back to the experiment to effect a change. The control signals can be either analog or digital. Thus, you can control the experiment as it proceeds.

GETTING READY TO COMMUNICATE

In the next few chapters, we will begin to develop some programs to send commands to the ACRO-900 and receive temperature data back from one of its modules, the OM-931 Thermocouple Interface.

Before we can develop the programs, however, we must set up the instrument and connect it to the personal computer. We must also set up a mechanism in the personal computer that will facilitate convenient setting of communications parameters.

These are tasks that you, too, will need to perform for your instrument.

ASSEMBLING THE INSTRUMENT

Your first task is to put your instrument together and make sure that it is performing correctly. Some manufacturers will do this for you. Others require that you follow instructions given in your User's Manual.

I cannot emphasize too strongly how important it is for you to test your instrument. If at all possible, make sure that it works in a manual mode before trying to control it from your personal computer. Adding communications from the instrument to a personal computer before the instrument is operating properly complicates matters and rapidly becomes difficult to track down the causes of problems.

Some instruments may have individual components that you need to assemble as a unit (i.e., they are modular). Modular instruments also usually require you to give each module within the instrument an "address".

An "address" is usually just a memory location that the instrument's CPU uses to find and activate the module that it needs to interact with. Each module MUST have a unique address number or the entire system will get confused.

Using addresses is a desirable feature in any system because you can switch back and forth between the modules within the system to make various kinds of measurements. For example, in the ACRO-900 system, you can alternate between a thermocouple measurement and an analog measurement by just sending out module addresses or numbers. You can also use the address system to switch between several thermocouple modules to take measurements from a large number of probes.

Let me describe briefly how you would put together the ACRO-900 because it illustrates some important points that are common to many other instruments (especially those that are modular). The CPU, power supply, and modules of the ACRO-900 system are joined together by placing each of the modules next to each other, pushing inwards and then backwards. This motion automatically snaps the modules into place.

Next, the modules are connected together electrically. This connection allows information to flow between the modules and the CPU. The connection is accomplished by lifting the front panel doors and pushing a flat cable containing many wires (called a "ribbon" cable) onto the end connectors of each of the modules in the system. This cabling interconnects all of the modules in the system and is referred to as a "bus".

A "bus" is a group of parallel connections that is used to transfer signals back and forth between modules and the CPU. Buses are very common in instruments, but are usually part of the internal circuit boards and are not removable. Some of the lines on the bus are used in the address system, some are used to regulate data flow, and some carry actual program data.

Needless to say, any time that you make bus type connections, it is critically important that the connections be made correctly. Do not bend any of the pins when you push the connectors together. Perhaps more importantly, make certain that the cable is oriented correctly. Pin one usually has a colored wire and/or a mark on the connector. This mark will allow you to determine which way the connector needs to be oriented when you place it onto the module's pin 1 (which will have similar markings).

Since the ACRO-900 is a system that utilizes addresses, you need to carry out one more very important task: You need to set the address for each of the modules. The "address" of an ACRO-900 is a number (between 1 and 15) and is set by a rotary switch on each module. If you have a system that uses addresses, jot them down for each of the modules. They will be important later on, when we begin programming.

COMMUNICATIONS FUNDAMENTALS

Your next step is to set up the communications parameters in your instrument.

All instruments and personal computers that communicate through RS-232 connectors do so by transmitting one data bit at a time down a single wire of the cable. For this reason, the transmission is referred to as "serial" transmission.

A data bit (*B*inary Dig*it*) is the most basic element of computer information. It consists of either a 0 or a 1. The receiving unit assembles a specific quantity of these bits into a number that represents one of the 128 numerals, punctuation marks, letters, or special control characters.

The conversion that takes place is based on a standard, called *A*merican *S*tandard *C*ode for *I*nformation *I*nterchange (ASCII, pronounced "askee"). An ASCII code is a seven-bit binary code whose decimal values are between 0 and 127 and is used to represent alphanumeric characters.

For example, an alphabetic "A" is represented as a 65, a "1" is actually a 49, and a space is a 32. The ASCII standard is almost universally used as "the" system for representing characters in computers. It is an important standard and I will refer to the system often throughout this book. The following table, an abstract of Appendix B, will give you a sampling of some ASCII conversions:

Decimal	Binary	ASCII
32	0100000	space
33	0100001	!
43	0101011	+
45	0101101	−
46	0101110	.
48	0110000	0
49	0110001	1
50	0110010	2
51	0110011	3
52	0110100	4
53	0110101	5
54	0110110	6
55	0110111	7
56	0111000	8
57	0111001	9
65	1000001	A
66	1000010	B
67	1000011	C
68	1000100	D
69	1000101	E
97	1100001	a
98	1100010	b
99	1100011	c
100	1100100	d

As you can see, integers and decimal numbers are actually transmitted as ASCII *characters,* with each digit of the number being represented by an individual *ASCII decimal number.* For example, a 527 is really transmitted as a 53 50 55; NOT THE NUMBER 527.

An organized and connected sequence of characters is referred to as a *string.* Thus, the characters 53 50 55 form a string that contains a 5, a 2, and a 7 (NOT the number 527). Likewise, characters 65, 66, and 67 form the string ABC. ASCII characters, strings, and the way that numbers are transmitted are all important programming elements that you will need to consider in just about every program that you develop.

TRANSMISSION RATE: BAUD

When an ASCII character is sent, each bit is transmitted at a certain speed. This speed is called the Baud rate and is stated in bits per second. The common Baud rates are 110, 150, 300, 1200, 2400, 4800, 9600, and 19200 bits per second. A Baud rate of 1200 translates to a speed of roughly 120 characters per second. As a rule of thumb, dividing the Baud rate by ten usually gives an approximation of the number of characters that can be transmitted per second or the number of English words that can be transmitted per minute.

It is critically important that the Baud rate of the personal computer and the instrument be identical. If they are not, your program will not work. This requirement may be simple, but believe it or not, about 40% of the problems that you are going to have are going to be due to incorrect Baud rates in either the instrument or the personal computer. Fortunately, Baud rate problems are among the easiest to correct.

CHARACTER BITS, PARITY, AND STOP BITS

There are some other parameters that are critically important to the communications between your personal computer and the instrument. Like the Baud rate, they must be set correctly and identically in both the instrument and the personal computer. They are

- The number of character bits. This term refers to the number of bits that are transmitted to make up one ASCII character. This number is usually either 7 or 8.
- The parity. This term refers to a check that is made to ensure that the data coming across the cable retains its integrity. This method is actually very simple. When a string of characters are sent from a communications port, the sending unit makes sure that the bit count is even (or odd, depending on how the system has been configured). In order to ensure this, a "parity bit" is appended at the end of each character. The extra bit is either 0 or 1 to make the total number of 1's even for even parity (or odd for odd parity). When the characters arrive at the port of the receiving unit, the receiving unit will make a check on the number of "1" bits to ensure that the number is still even (or odd, depending on how the system has been configured). No matter how many characters are received, the number of one bits must still be of the same "parity" that was sent. If it is not, then the unit knows that some of the data was either lost or some interference added some extraneous data.

  The following are some examples to clarify the concept of the parity bit. The first represents eight bits traveling across an RS-232 line. The parity bit is added to give even parity:

  PARITY BIT
  ↓
  0 1 0 0 1 1 0 1 number of 1 bits = 4 (even)
  1 1 1 0 0 1 1 1 number of 1 bits = 6 (even)

  The second example is the parity bit that is added if odd parity checking is used:

PARITY BIT
↓
1 1 0 0 1 1 0 1 number of 1 bits = 5 (odd)
0 1 1 0 0 1 0 0 number of 1 bits = 3 (odd)

The parity check can also be "none", which means no checking of the parity bit.

- The number of stop bits. This term refers to the number of bits that are added to a character to indicate the end of the character. There are usually 1 or 2 stop bits.

The reason that the number of character bits, parity, and number of stop bits are critically important is simple: The unit that receives the data expects each character to be made up of a certain, specific total number of bits. The separation of one character that is received from the next character is found by counting off this number of bits. If the number of character bits plus parity bit plus stop bits is different than expected, the data will be garbled. Furthermore, when parity checking occurs, the wrong bit will be used and the computer will turn up numerous parity errors.

It will be easy for you to understand the reason for garbled data and parity errors when you begin to add up the actual total number of bits that are transmitted per character. For example, suppose you set the number of character bits to seven, the parity to even, and the number of stop bits to one. The total transmission for the character is 10 (all transmissions start with one start bit). This example is quite different than a situation of eight character bits, even parity, and two stop bits. In this case, 12 bits are transmitted per character!

SETTING COMMUNICATIONS PARAMETERS

Manufacturers use the four methods outlined below to allow you to set the Baud rate and communications parameters on their instruments. Your instrument will use one of them.

1. Perhaps the oldest method is to use "jumpers". A jumper is a small square block of plastic that has a metal conductor encased within it. When it is placed over two adjacent pins on a circuit board, it closes the circuit between the pins. Some instruments use a series of these small pins to set communications parameters. (See Figure 2-2.) For example, a block of several sets of adjacent pins is often used for each of the possible Baud rates. If your instrument uses jumpers, look at the discussion in Chapter 9.
2. The newest method is to set communications parameters through the instrument's keyboard or keypad. In this case, the information on the parameters is electronically "written" onto a computer chip, which remembers it. If your instrument uses its keyboard to set the parameters, look at the discussion in Chapter 8.

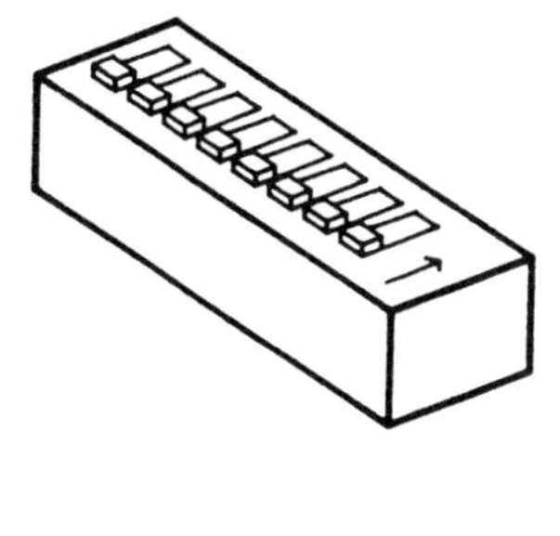

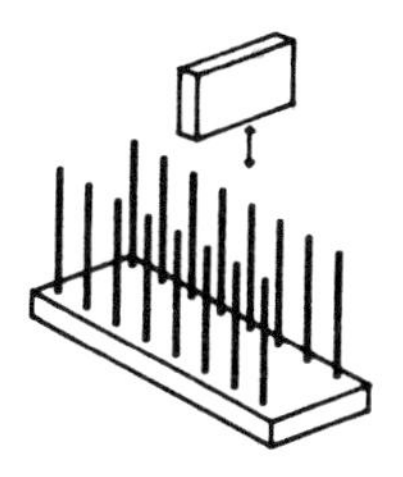

Figure 2–2

3. The most common method is to use DIP (*D*ual *I*n-line *P*ackage) switches. A DIP switch is a series of small on-off switches mounted in a rectangular block on the circuit board. (See Figure 2–2.) The ON and OFF pattern of the switches specifies the communications parameters to be used. The ACRO-900 uses this method and is described further below.
4. The last, and least common, method is not to allow you to change the parameters at all. This method allows you very little flexibility in your programming. The Perkin-Elmer LS-2B (described in Chapter 7) is an example.

To find out how to set the Baud rate and communications parameters for your instrument, first look in your User's Manual. If your manual does not contain sufficient information, look at the circuit boards inside your instrument. Circuit boards will often have the information printed right on them. Another place to look is at the rear of the instrument. If none of these places contain the information, call the manufacturer and ask for assistance.

No matter which method you use to set your instrument's communications parameters, you MUST turn the instrument off and then on again each time that you change the settings, *before* using the instrument. This is critically important because most instruments check the communications settings only when they start up and ignore them thereafter. If you change the settings without a powerdown/powerup sequence, the old configurations will probably be the ones that the instrument will use.

As previously stated, the ACRO-900 is an example of an apparatus that uses DIP switches. To set the Baud rate on an ACRO-900, go to the back of the CPU module and lift a plastic flip cover. Under this cover are two banks of DIP switches. (On other instruments these switches are often located inside the unit on one of the circuit boards.)

To set the switches correctly, I looked into the ACRO-900 manual to find out how to set the Baud rate to 4800. I also looked to see how to set the other parameters that are important to communications.

The sample program has a *Baud rate* of 4800 bits per second, an even *parity,*

7 *character bits,* and 2 *stop bits.* The following is the position for the settings on the ACRO-900:

Switch 1	Position		Designation
1	RIGHT	→	OFF
2	LEFT	←	ON
3	RIGHT	→	OFF
4	RIGHT	→	OFF
5	RIGHT	→	OFF
6	RIGHT	→	OFF
7	RIGHT	→	OFF
8	LEFT	←	ON
Switch 2	**Position**		**Designation**
1	LEFT	←	ON
2	LEFT	←	ON
3	LEFT	←	ON
4	LEFT	←	ON
5	LEFT	←	ON
6	LEFT	←	ON
7	LEFT	←	ON
8	LEFT	←	ON

Take a few minutes now to look up the switch settings for your instrument, read the next section of this chapter, and then set the switches.

OBSERVING THE SPEED LIMIT

If you plan to control your instrument directly from the spreadsheet, then you must observe some data transmission speed limitations that exist. With this approach, Lotus 1-2-3 and Symphony obtain their data from DOS buffers. DOS uses a temporary memory storage area (called a "buffer") to compensate for the differences in the rate of data flow. This buffer is needed because the personal computer's microprocessor handles data at a different rate than it is received from the communications port. The size of the buffer is 512 bytes (i.e., 4096 bits, or about 400 characters). This buffer, therefore, serves to decrease the possibility of overrunning the processing capabilities of the microprocessor as it works under the direction of the spreadsheet's program.

The Baud rate that you use will become important if long strings of characters are transmitted to, or received from, the instrument. Thus, if the Baud rate is too fast and a string larger than 512 bytes enters the DOS buffer, it may overrun the buffer and cause a "System Error" message to be displayed by Lotus. This problem does not occur if the system has had sufficient time to process enough data in the DOS buffer to keep its size below 512 bytes before any new characters come in.

The occurrence of the problem also depends on how efficient your program is, the speed of your personal computer (IBM XT vs. AT vs. Model 80), and whether you are using expanded memory. Therefore, when you are initially setting up the Baud rate, you should use a setting of 300 bits per second. After you have your program written and working, you can start doubling it (300->600->1200->2400->4800->9600) until the program stops working. Then you should take the Baud rate just slower than the one that does not work. This setting will give you the greatest transmission rate (and thus efficiency) while still providing you with a margin of safety.

One helpful tactic that you can employ to get the fastest Baud rate possible is to improve the speed of your program by disabling any expanded memory. EMS memory is not nearly as fast as conventional memory. Thus, the use of EMS memory can profoundly affect the speed of your program and can thereby affect the communications between your instrument and spreadsheet. This speed difference is especially evident in Symphony, Release 2, where your programs will execute approximately 35% slower if you are using expanded memory. This difference in speed is sometimes significant enough to affect the Baud rate that you can use.

Unless you anticipate collecting more than 23000 data points in your experiments, I strongly recommend disabling the expanded memory. Your spreadsheet will run about 25% to 35% faster for nearly all operations. Disable your expanded memory by removing the command lines in your CONFIG.SYS file that specify the EMS device driver (i.e., the ".SYS" file) and then rebooting your computer.

MAKING THE CONNECTION

Your next task is to physically connect the instrument to the personal computer. This connection has proven to be a problem area for RS-232. According to the RS-232 standard (Appendix C), manufacturers may use either pin 2 or pin 3 in the connector to transmit data. If the manufacturer uses pin 2 to transmit data, it must use pin 3 to receive data, and vice versa.

Other pins may be used for "flow control". The more common name for flow control is "handshaking". Handshaking is concerned with the methods that instruments and personal computers use to control the flow of data coming to them from another piece of equipment. Handshaking then, is an important method that prevents the overpowering of an instrument's and personal computer's ability to accept data. Handshaking creates a potential problem for RS-232.

In addition to the variability in pins 2 and 3, different manufacturers use different handshaking pins for different purposes. Because of this variability, you must consult your instrument's User's Manual to see if it can be of assistance in determining what kind of cable to use. If the manual is of little help, call the manufacturer. If neither is of any help, look at Appendix A at the end of this book to determine the kind of cable to use. Appendix A contains a very simple method to judge the configuration of the cable that best fits your instrument.

Before becoming too concerned about this connection problem you may want to try either a "straight-through" cable, a "null-modem" cable, or both. (See below for a description of these cables.) You have nothing to lose; even the worst mismatch won't harm either piece of equipment as long as both conform to the RS-232 standard.

To decide which of these two cables to try, consult the instrument's User's Manual and see if the instrument is listed as a DTE (Data Terminal Equipment) or a DCE (Data Communications Equipment). A DTE will, by definition, transmit its data on pin 2 of the connector and receive its data on pin 3. A DCE is just the opposite. It transmits on 3 and receives on 2.

IBM PCs, XTs, ATs, Personal System/2s and most compatibles are DTEs. Modems are DCEs. Some scientific instruments are DTEs and some are DCEs. Again, consult your instrument's User's Manual or the manufacturer to see which it is. (Many times one or the other will tell you what kind of cable to use to connect the instrument to a personal computer.)

For transmission to travel in the right direction and not become confused, only one of the devices can transmit on pin 2. The other must transmit on pin 3. You can think of this situation as a divided expressway with two one-way highways headed toward opposite directions. Imagine the havoc that a car would cause if it were traveling in the wrong direction!

If your instrument is a DCE and the personal computer is a DTE, you can often use a "straight-through" cable because one side will transmit on pin 2 and the other on pin 3. A straight-through cable is one in which every pin is connected to the like-numbered pin on the other end. (e.g., pin 1 to pin 1, pin 2 to pin 2, etc.) The following is a wiring diagram of a typical straight-through cable:

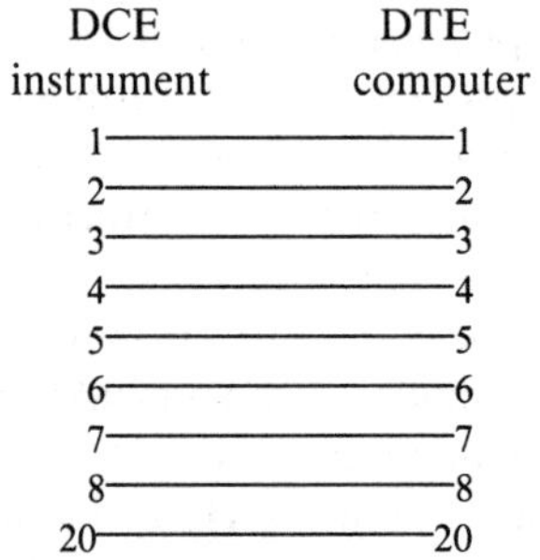

```
    DCE            DTE
 instrument      computer
     1──────────────1
     2──────────────2
     3──────────────3
     4──────────────4
     5──────────────5
     6──────────────6
     7──────────────7
     8──────────────8
    20──────────────20
```

If your instrument is a DTE and your personal computer is a DTE, you will need to use a "null-modem" cable.

A null-modem is a cable that will take pin 2 of the first side and connect it to pin 3 of the opposite side; likewise, it will take pin 3 of the first side and connect it to pin 2 of the opposite side. This kind of connection is often called a "cross" and will allow a DTE to talk to a DTE or a DCE to talk to a DCE.

Two other pins that are commonly "crossed" in a null-modem are pins 6 (Data Set Ready, DSR) and 20 (Data Terminal Ready, DTR). From a DTE's perspective, the Data Terminal Ready pin is a general purpose output generally used to signal the DCE that the DTE has been powered up and is ready to go. The Data Set Ready pin is a general purpose input that signals the DTE that the DCE has been powered up and is ready to go. Like the 2 "cross" 3 configuration, the 6 "cross" 20 is usually necessary in a null-modem to fool the two DTEs or two DCEs into thinking that the transmission line is available. The following is an example of the wiring that would occur in this type of null-modem:

DTE instrument	DTE computer
1	1
2	3
3	2
4	4
5	5
6	20
7	7
8	8
20	6

In addition to the changes in these four pins of a standard null-modem, several of the other pins may need to be either "crossed" or "jumped" (connected to another pin in the same connector).

Thus, there are many ways to build a null-modem cable. For example, pin 4 (Request To Send) and pin 5 (Clear To Send) are often "crossed". As in the case of the 6 "cross" 20 cable, the Request To Send is a general purpose output and the Clear To Send is a general purpose input from the DTE's perspective. Therefore, you may also need to "cross" the 4 and 5 pins when you use a null-modem. The following is an example of the wiring that would occur in this type of null-modem:

DTE instrument	DCE computer
1	1
2	3
3	2
4	5
5	4
6	20
7	7
8	8
20	6

To summarize the pin situation, pins 2, 4, and 20 are usually output pins for DTEs; pins 3, 5, and 6 are usually output pins for DCEs. If you want DTEs to talk to DTEs or DCEs to talk to DCEs, investigate using a null-modem cable.

I used a straight-through cable for the ACRO-900. Throughout this book I will provide you with the cables that I used for some of the other instruments that I interfaced. These applications should give you a better feel for the more common configurations. Additionally, the other example cables will illustrate the fact that many cables that you may need to use will be neither conventional straight-through nor null-modems.

Please refer to Appendix A for a simple (but thorough) discussion of cabling if you are confused or if you are having trouble deciding on the wiring to use for your instrument's cable. You should also refer to Appendix A for a discussion on how to prepare your own cable.

Before departing this topic, two important statements of caution need to be heeded. The first one is: Be very careful about trusting any information given in an instrument's User's Manual. In many cases, I have found the information to be wrong as often as it was correct.

The second one is: Sometimes different software programs require different cable formats. Some software programs manipulate (activate/deactivate) the handshake lines for successful management of the flow of data, while others do not have this ability. Further, programs that control the handshake lines usually require the lines to have a certain specific configuration for proper flow control or breakdowns in communications will occur. For example, using Lotus Measure to obtain data from an instrument will quite often need a different cable than the one required for obtaining the data directly into the spreadsheet using the techniques outlined in the next few chapters. This is because Lotus Measure controls the handshaking lines, while direct communication does not.

The cable that you use can be quite long. The maximum length depends on Baud rate, but even at 9600 Baud, the specifications of the RS-232 standard provide for a distance of about 250 feet with shielded cable. At 4800 Baud, you can have about 500 feet and at 1200 Baud, you can have about 1750 feet.

Cables are available in various configurations from most computer shops. Inmac® Corporation (2465 Augustine Drive, Santa Clara, CA 95054) offers a good selection of most cable types and also null-modem converters that can be fitted to the end of straight-through cables.

PLUG IT IN!

Connecting an instrument to the personal computer is easy. Just plug it in. The connectors that you use must be complimentary. For example, if the instrument has pins, then you need a female connector. The only way that you can go wrong is to attach the cable to the wrong connector at the back of the personal computer. If the personal computer has two serial communications ports, make sure that you use the correct port.

Another caution: Make sure that you do not connect the cable to the parallel port. A parallel port often looks like a serial port. On an IBM PC/XT, the parallel port usually has a female connector and is always 25 pin. In contrast, the serial ports are usually either 9 pin females or 25 pin males. If you are unsure of which port is serial, consult the Owner's Manual for your personal computer and/or video graphics board (a common site for a parallel port).

After you have plugged in the cable, I strongly recommend securing it with screws. It is frustrating to spend time trying to find a ''bug'' (error) in a program that won't communicate, only to find that the cable is out of its socket. This oversight is one of the most common communications problems and is also the easiest to avoid. Connectors can easily become disconnected from personal computers and instruments. Even loose connections can cause problems. All connectors should be secured with screws. This rule will save you hours of frantic work trying to pinpoint the cause of a seemingly ''dead'' system.

MAKING THE DOS MODE.com PROGRAM ACCESSIBLE

Your final step in preparation for programming is to set up a DOS file mechanism in your personal computer that will be used in the next chapter.

The communications rate and parameters of your personal computer must be set to the identical values that you configured your instrument to. If they are not, your program will not be able to decipher the data characters. Setting these parameters will be the subject of the next chapter. In that chapter, you will be using the DOS MODE.com program from a Lotus spreadsheet to set the Baud rate, parity, data bits, and stop bits of the personal computer. The DOS MODE.com program is known as an ''external command''. This term means that before you can use the file it must be accessible from the directory or subdirectory that contains your Lotus 1-2-3 or Symphony files.

You can make this file accessible in two different ways. The first way is to copy it into the directory or subdirectory containing your Lotus files using the DOS COPY command. To implement this method, place your DOS diskette into drive A. Next, make the drive containing your Lotus files current by typing "X:" (where "X" is the letter of drive, e.g. C:) and pressing return. Then, issue the CD (Change Directory) command to move to the directory of your hard drive that contains your Lotus spreadsheet. Finally, issue the COPY command to copy the MODE.com file into the subdirectory. For example,

```
copy a:MODE.com X:
```

Alternatively, if you have a subdirectory that contains the DOS MODE.com program on your hard disk, you can add the subdirectory's name to the DOS path.

When you enter the name of an executable file, DOS looks for the file in the current directory. The PATH command tells DOS where to look for the executable file if it is not in the current directory. To use the command, type ''PATH'' followed by the path(s) that you want DOS to follow in its search for the executable file. If you want to specify more than one path, separate the paths with semicolons.

For example, if your MODE.com file is in the \DOS subdirectory of drive C, your PATH command would look like the following

```
path c:\DOS
```

A convenient way to invoke the path command is to create a batch file named autoexec.bat in the root directory of your boot disk and include the path command in the file. When DOS boots, it looks for a file named autoexec.bat. If it finds one, it automatically executes the series of commands contained in the file.

To create a new autoexec.bat file (or update an existing one), use your favorite word processor to create the file in the main directory of your boot disk and type a line consisting of "PATH C:\DOS" (or the subdirectory containing the MODE.com file). Alternatively, you can use the DOS COPY command to create the batch file. Do so by typing CD\ and pressing return to make the main directory of the boot drive current. Then, type the following

```
COPY CON: AUTOEXEC.BAT        (NOTE: ''CON:'' means console)
PATH C:\DOS     (or drive:\directory containing MODE.com)
^Z                  (control Z)
<return>
```

After you type <control> Z and <return>, the disk drive will activate and the batch file "autoexec.bat" will be written to the disk. Press <control>, <alternate>, and <delete> simultaneously to reboot your personal computer and execute the new batch file.

Again, it is very important that you either copy the MODE.com file into the Lotus subdirectory or set the path. In Chapter 3 we will be using this file and if the file cannot be found, you will receive a Lotus error message.

WHAT'S NEXT?

The next chapter will show you how to make a spreadsheet template and how to create macro programs that will automatically set the personal computer's communications parameters (Baud rate, character bits, stop bits, parity, etc.). You need to complete these tasks before you start programming; they are necessary regardless of the instrument that you plan on interfacing with a personal computer and spreadsheet.

3

Spreadsheet Templates and Computer Initialization

The first step in developing a program that integrates your instrument and a Lotus spreadsheet is to prepare a structured framework in the spreadsheet. This framework will be used by your program and will serve as a repository for the data received from the instrument.

After this re-usable infrastructure has been developed, you will need to craft a mechanism to automatically configure your personal computer so that its communications parameters match your instrument's.

In this chapter, I am going to show you how to set up your spreadsheet template and how to automatically set your personal computer to the proper communications parameters. Figure 3–1 shows the location of these steps in the development process.

BRIEF DESCRIPTION OF THE EXAMPLE INSTRUMENT

In this chapter, I will continue with the ACROSYSTEMS® Interface example that I described in the previous chapter. The unit that I will specifically use is the ACRO-931 Thermocouple Input Module. The module has eight connectors on its back for thermocouples. These input connectors are called "channels".

The channels have been designed for direct connection of thermocouples formed by the junction of two wires of dissimilar metals. Specific combinations of these metals are referred to as "types" and are assigned the names B, E, J, K, T, R, or S. The module can be instructed to provide automatic cold junction compensation, linearization, and conversion to degrees Celsius (Centigrade), Fahrenheit, Kelvin, or Rankine.

```
                STEPS TO A SUCCESSFUL APPLICATION

  Assemble The Instrument
  Test Instrument In Manual Mode
  Determine Wiring Configuration
  Get Cable
  Connect Cable To Computer And Instrument
  Secure Cable
  Determine How To Set Communication Parameters
  Set Module Addresses
  Reset Instrument
  Add PATH Command To AUTOEXEC.BAT File
->Add MODE Command to AUTOEXEC.BAT File
->Design Template And Macro Program
->Start The Spreadsheet
->Type Template Into Spreadsheet
->"Activate" Template
->Test The Template
->Create And "Activate" The Auto-executing Macros
->(Symphony): Specify The Name Of The Auto-executing Macro
->Use The DOS MODE Program To Set Communication Parameters
  Type In A "Bare Bones" Program
  "Activate" The Test Program
  Try To Establish Minimal Communications
  Evaluate The Test Program
  Expand Test Program To Retrieve Data
  Add Amenities And Menus To The Program
  "Activate/Re-activate" Macro Program
  (Symphony): Create Windows
  Create A Table Of Range Names
  Save The File
  Evaluate The Program
  Increase Communications Speeds
  Final Test The Program
  Save/Backup Several Copies Of The File
  Save Several Copies In Separate Safe Places
```

Figure 3-1

The instructions that select the proper channel and/or the proper combination of calculations are strings of characters that you send out via the RS-232 port. The module is easy to use because it does not require extensive software development. A thorough description of how to control an instrument with character strings will be given in Chapter 4.

SPREADSHEET TEMPLATES

Before you start writing your program, you must create a spreadsheet template to accept the data.

A spreadsheet template is simply a portion of your spreadsheet that you set aside to receive raw data from your program, then automatically format it and/or perform calculations. The template usually contains only labels and formulas. You can view a template as a kind of re-usable "skeleton" into which your program will place the data. If you create a template before you start programming, you will find it much easier to create the program.

The template in this example is very simple. Subsequent chapters will contain more complicated (and useful) templates. Calculations need not be just simple averages, maximum values, or minimum values (as they are in the example). They can be formulae of immense complexity. Nearly all of the scientific formulae that you use can be placed into the template or the spreadsheet. For example, you can obtain logarithms, exponentials, exponents, trigonometric functions, standard deviations, and statistical functions. You can even set up the template for polynomial linear regression analyses, integrations and differentiations.

With the templates that you create for your data, you can change and experiment with the raw data. If you have formulae in the template, they will instantly recalculate and display the results of any changes you make. This capability is extraordinarily powerful when working with experimental data and provides you with the flexibility that you want when you examine the data.

PLANNING YOUR TEMPLATE

To make the best use of a spreadsheet's power, take the time to plan your template. Plan it with the intention that the template will serve as a report for your notebook records. You should also take into consideration that the template's primary purpose is to serve as a recipient of data from your instrument and so the template must work in harmony with the program that obtains the data. The time that you spend planning your template will be more than offset by the time that you will save when you begin programming. If you do not properly plan your worksheet, you may have to make changes that will require inserting rows, deleting columns, moving data to different cells, etc. Redesigning templates and rewriting programs take a lot of time and are prone to errors.

When you design the template, determine the location of each piece of data. Pay particular attention to whether the data would be better presented horizontally across the spreadsheet or vertically down. From this, you will be able to determine the locations for the macro programs. Coordinating the positions of the two is important because the template and its interactive program must form an efficient unit that exchange data with the greatest harmony and safety. Remember, a well planned template will lead to efficient programs.

You will also want to organize the areas of the template so the program can efficiently place the instrument's data into it. For example, you may want a section for standards, one for regression analysis, and one for unknown samples.

In addition, remember that experiments increase in size as projects proceed and new ideas are generated. If you plan your template for expansion from the start, you can reserve room on the spreadsheet for extra rows and columns that can be used to hold the extra data.

Sometimes, a little planning will permit you to design a template that will serve dual purposes. For example, you may be able to get day-to-day performance data for quality control purposes from the same template that holds your experimental data. Or, if you work in a centralized assay laboratory that performs sample analyses from several other laboratories, your need to issue individualized reports that contain only the data that a submitting lab is interested in can be met by laying out a template that can be individualized. (You can use the /Worksheet Column Hide command (Lotus 1-2-3) or the /Width Hide command (Symphony) to hide data from other submitting laboratories and just print the appropriate data for the lab in question.)

There are three styles guidelines to consider when designing a template. The first one is: The more formulae that you have in cells, the slower the spreadsheet will recalculate. If you have several hundred formulae in the template, every time the spreadsheet recalculates it will take a substantial amount of time. More importantly, a slow recalculation may affect the program's ability to receive data from the instrument. Therefore, if you need to place data summaries in the spreadsheet, you should have the program do as much of it as is practical.

The second style consideration is: Take the user's ability into account. Use headings and labels that are easily understood. Use as little jargon as feasible. If you use non-descriptive terms, it may confuse your audience. This confusion will lead to inefficient usage and errors that will waste time and may cause serious errors in the data (and, therefore, the interpretation of the experiment).

The third style consideration is: Do not place too many different types of experiments on one spreadsheet. Crowded layouts can make it cumbersome and confusing for the user. Try to keep the template as small and uncluttered as possible.

OVERVIEW OF CREATING TEMPLATES

Creating any template is very easy after you have planned it out. You just type in text headers and formulae, create named ranges, and specify formats for specific cells.

Since a template is nothing more than a completed "skeleton" without the data, you should test the template by manually entering characteristic data into it to make sure that it works correctly.

After all of the labels, formulae, and programs have been entered and the template is working correctly, store it. Store the spreadsheet without the data. With a "clean slate" template, you will not have to worry about deleting entries for each subsequent use.

You should also create copies of the spreadsheet on several floppy disks and store them in separate locations. These backup copies will be a safeguard against having to recreate the spreadsheet if the current one is damaged or lost.

HOW TO CREATE THE EXAMPLE TEMPLATE

Figure 3-2 shows a simple template that I created for the ACRO-931. If you recall, the ACRO-931 can handle eight thermocouple probes. By repeatedly taking a measurement from each of these probes, you can get temperature readings over time.

```
--------A--------B--------C--------D--------E--------F--------G--------H--------I--------J--------K--------L--------M--------N-----
 1                                         THERMOCOUPLE TEMPERATURE MEASUREMENTS
 2
 3                  COMMENTS:
 4                     DATE=                                                                       ROW SUMMARIES
 5                         1        2        3        4        5        6        7        8     AVG       MAX       MIN
 6           ELAPSED
 7 TIMEPOINT  TIME
 8                                                                                              ERR       ERR       ERR
 9                                                                                              ERR       ERR       ERR
10                                                                                              ERR       ERR       ERR
11                                                                                              ERR       ERR       ERR
12                                                                                              ERR       ERR       ERR
13                                                                                              ERR       ERR       ERR
14
15
16
17          COL AVG      ERR      ERR      ERR      ERR      ERR      ERR      ERR      ERR
18
19
20                           MAX           ERR          AVG           ERR
21                           MIN           ERR          RANGE         ERR
22                                                      +/-           ERR
23
24
```

Figure 3–2

This format fits nicely into a grid (matrix) wherein each row of data will contain a single timepoint reading for each of the eight probes.

Data from most instruments usually fit into convenient matrices. Take a moment to think about how you could organize the data from your instrument so it would fit into a matrix. Matrices are the cornerstones of good templates. Since a matrix is such a major part of a template, if you make an effort to organize the matrix that holds your data now, it will be easier to work with the data in the future.

Even if you are not using the ACRO-900 interface for the example, type in the template as you see it in Figure 3–2. Creating this template will teach you some important concepts that you can apply when you create *your* template.

The following instructions teach you how to create the template shown in Figure 3–2. The steps for your template will be essentially the same.

Using the arrow keys, move the cell-pointer (i.e., the reversed video, highlighted cell) to the appropriate cells and type in the text or number. Into cell C3 type "COMMENTS: and into C4 type "DATE=. Then, type "THERMOCOUPLE TEMPERATURE MEASUREMENTS into cell E1. The quotation marks (") at the beginning of the titles right-justify the text in the cells.

The remainder of the column titles are centered. The worksheet will automatically perform the centering task if you issue the following commands. In Symphony, issue the {Menu} Settings Label-Prefix Center command. You can issue the {Menu} command by either pressing slash (/) or the F10 key. Then, either move the menu-pointer (the reversed-video, highlighted rectangle) over the option using the arrow keys or type the first letter of the menu choice. For example, type S for Settings, L for Label-Prefix, C for center, and then Q for quit.

In Lotus 1-2-3, the equivalent command is /Worksheet Global Label-Prefix Center (/WGLC).

Into cell B6, type ELAPSED; into cell A7 type TIMEPOINT; and into cell B7 type TIME. Into cells C5 through J5, type in numerals 1 through 8 to represent the 8 probes. Use numerals instead of labels whenever you create column headings to have the columns in an appropriate form to plot the results. If labels are used, the Lotus graphics environment would treat them all as character strings, assign their values to zero and not give proper graphics displays. Finish the text headings on this line by typing AVG, MAX, and MIN into cells K5 through M5, respectively and ROW SUMMARIES into cell K4.

Into cells K8, L8, and M8, type the following Lotus @functions to calculate the average, maximum, and minimum values for the row:

```
K8: @AVG(C8..J8)
L8: @MAX(C8..J8)
M8: @MIN(C8..J8)
```

Use the copy command to copy these cells to row 9 through 13. Issue the / Copy command. When you are prompted for the "Cells To Copy From", press ESCape, use the arrow keys to move to cell K8, type a period (.), and use the arrow keys to move right to cell M8. This sequence will highlight all three cells. Press return. When prompted for the "Cells To Copy To", use the arrow keys to move to cell K9, type a period (.), and use the down arrow key to move to cell K13. Press return.

ERR will appear in all of the cells because the ranges on which their calculations depend do not have values as yet. This display will appear in cells of your template containing Lotus @functions whenever there are no data within the @function's specified range. These displays are normal. It will also occur in cells that depend on other cells whose @functions evaluate to ERR.

To set up the summaries at the bottom of the report range, type the following text into the specified cells:

```
B17      ''COL AVG
D20      ''MAX
D21      ''MIN
G20      ''AVG
G21      ''RANGE
G22      ''+/-
```

Into cell C17, type the formula @AVG(C8..C13). Use the copy command to copy the formula in cell C17 to columns D through J. Issue the / Copy command. When you are prompted for the cells to copy from, press ESCape, use the arrow keys to move to cell C17, and press return. When prompted for the cells to copy to, use the arrow keys to move to cell D17, type a period (.), and use the right arrow key to move to cell J17. Press return.

Notice again that ERR appears in each of these cells. It appears because the ranges on which their calculations depend do not contain values.

Finish the template by typing the following into the specified cells:

```
E20      @MAX(C8..J13)
E21      @MIN(C8..J13)
H20      @AVG(C8..J13)
H21      +E20-E21
H22      +H21/2
```

Now, take the time to see how the template works. Type some numbers into cells C8 through J13. The @functions will automatically update after each number is added. Figure 3-3 is an example that I prepared by typing in random numbers representative of the data that one might expect to collect.

Actually, by typing in representative data you will be performing an even more important task. You will be testing the template.

Testing a template *before* you start programming is very important. If the template does not perform correctly alone, you cannot expect it to perform correctly when it is interacting with a program.

As a general rule, it is best to test a template or a program as much as possible before going on to the next programming task. This action makes it easier to identify and track down errors. That is, if you wait until after you write your program, there will be many program lines and interdependencies in your template. These relationships will make it much more difficult to isolate errors. However, if you test

```
--------A--------B--------C--------D--------E--------F--------G--------H--------I--------J--------K--------L--------M--------N-----
 1                                          THERMOCOUPLE TEMPERATURE MEASUREMENTS
 2
 3                    COMMENTS:INITIAL TESTING ON S/N 1002
 4                       DATE= 12/25/87                                                                 ROW SUMMARIES
 5                              1        2        3        4        5        6        7        8  AVG      MAX      MIN
 6          ELAPSED
 7 TIMEPOINT  TIME
 8        0 00:00:00    19.55     19.5    19.35    20.49    19.55    19.35    19.55    19.74    19.64    20.49    19.35
 9        1 00:00:02     19.5    19.45    19.35    20.49     19.5    19.35    19.55    19.74    19.62    20.49    19.35
10        2 00:00:04    19.45     19.4     19.4    20.49    19.45    19.35    19.55    19.69    19.60    20.49    19.35
11        3 00:00:05     19.3    19.74    19.35    20.54    19.45     19.3     19.5    19.69    19.61    20.54    19.30
12        4 00:00:07     19.3    19.74    19.25    20.54     19.4     19.3    19.45    19.69    19.58    20.54    19.25
13        5 00:00:09     19.3    19.69    19.25    20.49    19.45     19.3     19.4    19.69    19.57    20.49    19.25
14
15
16
17          COL AVG     19.40    19.59    19.33    20.51    19.47    19.33    19.50    19.71
18
19
20                           MAX            20.54             AVG             19.60
21                           MIN            19.25             RANGE            1.29
22                                                            +/-              0.64
23
24
```

Figure 3-3

as you build, you can verify that each step works correctly before you add another layer of complexity.

When you are finished testing the template, erase the artificial data. In Symphony, issue the / Erase (/E) command. In Lotus 1-2-3, issue the / Range Erase (/RE) command. When prompted for the range to erase, press ESCape, arrow over to cell C8, type a period (.), arrow over to cell J13, and press return. These commands will return your spreadsheet to an empty template.

FORMATTING THE CELLS

Cell D4 will be used to display the current date. Because templates are created to be used repeatedly, it is important for you to have a date-stamp to show the date that an experiment was actually performed. (If several experiments are being performed on the same day, then you may also want to *time*-stamp the experiment.)

Lotus 1-2-3 and Symphony represent dates by integer values: Any given date is equal to the number of days that have elapsed since December 31, 1899. For example, December 25, 1987 has the value of 32136 and is 32136 days past December 31, 1899. This method of keeping track of dates allows for date arithmetic, which will be used subsequently. For now, it presents a problem because 32136 does not look much like a date.

Fortunately, Lotus 1-2-3 and Symphony allow you to format cells to recognizable displays. To assign a date format to a cell in Symphony, issue the {Menu} Format Date (/FD) command. To assign a date format in Lotus 1-2-3, issue the / Range Format Date (/RFD) command.

Both date format commands have a menu of format choices. Option 4 is common: MM/DD/YY. This format would display 32136 as 12/25/87. To choose option 4, either move the menu-pointer over the option or type 4. The program will then prompt you for the range of cells to format. If cell D4 is indicated, press return. If not, press the ESCape key, move the cell-pointer to cell D4, and press return.

Other formats are available. For example, menu option 1 gives the DD-MMM-YY format. This format would display the 32136 as 25-DEC-87. However, if you choose this option for this example template, the date will be displayed as a line of asterisks (*********) across the cell because the cell is not wide enough to display the entire date. You can correct this problem by widening the column to 10. To widen a column in Symphony, move the cell-pointer to the column and issue the {Menu} Width Set (/WS) command. In Lotus 1-2-3, issue the / Worksheet Column Set-Width command (/WCS). When the program prompts for a width, type 10 and press return.

Don't worry about typing the date into the cell for now, the program that you create later will automatically enter it. Remember, this is a template!

Finally, you will want to include significant figure considerations in your template. That is, if the data that comes from your instrument is good only to a certain number of significant figures, then you will not want to display the data past that number of digits.

For example, the data that comes out of the ACROSYSTEMS interface has two significant decimal digits. You can format the columns containing the @function summaries to two digits in Symphony by issuing the / Format Fixed (/FF) command. In Lotus 1-2-3 you issue the / Range Format Fixed (/RFF) command. When prompted for the number of digits, type 2 and then press return. When prompted for the range, type K8..M13, and press return.

Repeat this procedure to format ranges C17..J17 and D20..H22 to two digits.

Notice that we did not format the entire spreadsheet. The reason for not formatting the entire spreadsheet is: When a cell is formatted, Lotus allocates memory for the cell whether it contains data or not. Therefore, if you format several thousand cells, a great deal of memory will be used up, thus limiting the amount of data that you can place into your spreadsheet. You may run out of memory. Formatting large numbers of cells also tends to slow the recalculation of the spreadsheet.

BACKING UP YOUR WORK

I recommend storing a spreadsheet template (without data) as soon as you have all of the components in place. In fact, I recommend storing it often during creation. That way, if you have a power failure or a computer lockup, you will not lose the work you have already completed. If you have a large macro program and spreadsheet, you may want to store it under different names and on several floppy disks.

Before you store the template and begin to program the macros, take a minute to reset the label alignment settings in the spreadsheet. Recall that we set the label alignment to "center" for column headers. For macros, you will want to left-justify the text. As before, the worksheet will automatically perform the justification. In Symphony, issue the / Settings Label-Prefix Left command (/SLLQ). In Lotus 1-2-3, the equivalent command is / Worksheet Global Label-Prefix Left (/WGLL).

To store the file in Symphony, issue the {Services} File Save command (F9FS). In Lotus 1-2-3, the command is / File Save (/FS). When the program prompts for a name, type one that is descriptive. For this example, I chose the name "Thermo". Press return.

After the first time that you save a spreadsheet, the program will detect the file's presence. It will prompt you to confirm that you want to overwrite the old one. At this point, choose the Replace option for Lotus 1-2-3 or the Yes option for Symphony.

MACRO PROGRAMMING

After you have tested your template, your next step will be to write your macro program.

Lotus 1-2-3 and Symphony have a built-in programming language, called the Lotus Command Language. The Lotus Command Language allows you to automate virtually any procedure that you can perform manually through the keyboard. A program that uses the Lotus Command Language to automatically perform Lotus procedures is called a "macro".

In their simplest form, macros are time-savers. You can save time and keystrokes by creating a macro that "presses the keys" for you. That is, a simple macro contains the instructions that identify which keystrokes you would use to perform a specific task. These keystrokes are stored as labels (text) within cells of your spreadsheet. When you run a macro, Lotus 1-2-3 or Symphony reads and executes the keystrokes and commands stored in the macro. For example, you can use the following macro to "type" the word "hello" into two adjacent cells (~ means "press return" in the Lotus Command Language):

```
\R hello~
   {RIGHT}
   hello~
```

Similarly, you can create a macro to automatically use the menu to perform certain tasks. For example, use one of the following macros to erase cells A1 through B7:

For Lotus 1-2-3		For Symphony	
\R	/REA1..B7~	\R	/EA1..B7~

The Lotus Command Language has numerous programming commands that allow you to prompt the user for input, perform file operations, calculate formulae, assign values to cells, transfer data from one cell to another, control the display screen, move the cell-pointer, control the path of macro execution, display custom menus, trap errors, and perform a host of other useful functions. In fact, Lotus has more than 40 specialized macro commands available.

Thus, this macro language is like having a built-in BASIC programming language. It is just as easy, or easier, to use and holds many advantages over BASIC. These advantages will become readily apparent to you as you proceed through the remainder of this book.

CREATING MACRO PROGRAMS

You create a macro program by typing a list of commands horizontally down a single column of the spreadsheet. These instructions are entered directly into the cells just like any other data. Each instruction can be made up of one, or a combination of, the following

- keystrokes that you would press if you were entering data.
- keystrokes that you would use to select an item from a Lotus menu.
- keystrokes that you would use to respond to a prompt.
- commands that use one of the Lotus Command Language keywords.

You would then assign a name to the macro by placing the name in the cell to the left of the first cell of the macro and using the /Range Name Label Right command.

You can have as many macros in your spreadsheet as you like. You can also place them anywhere in the spreadsheet. Although the length of a cell entry is limited to 240 characters, the size of your macro is virtually unlimited because the macro can include as many cells in the column as it takes to accomplish the task that you need it to perform. The only limitation is the amount of available memory.

MACRO EXECUTION

To start a macro running, specify it by name. For example, in the simple macros given above, the macros were called by the name \R. To start a macro with this name, you would hold down the [ALT] key, press R, and then release both keys. (That is, you would *not* type the backslash (\)).

Lotus reads program lines starting with the cell to the right of the cell containing the name of the macro and proceeds downward. (That is, Lotus starts with the cell that was specified by the /Range Name Label Right command, which is the cell just to the right of the macro's name.) Lotus performs instructions in sequence. As each instruction is completed, Lotus automatically proceeds to the next cell below.

A macro terminates when it encounters an instruction to stop (the {QUIT} command), a cell entry that contains an instruction that is not recognized, or an empty cell. Most macros in this book terminate with an empty cell.

AUTO-EXECUTING MACROS

You can also program Lotus 1-2-3 and Symphony to start a macro automatically when the worksheet that contains the macro is retrieved. This special macro is the one that you will create in this chapter and is the first macro program that you would normally create whenever you start programming a new application.

A macro program that automatically executes when a spreadsheet is retrieved is called an "auto-executing macro". The example macro programs shown in this chapter are auto-executing macros that aid in configuring the communications port each time the spreadsheet is retrieved.

Auto-executing macros are also very commonly used to perform many other tasks that MUST be accomplished at the start of a Lotus session.

The characteristic that all of these tasks have in common is that they all need to be performed BEFORE the template or other macros in the spreadsheet can be used successfully. For example, auto-executing macros are frequently used to place starting values into key cells of the template. Similarly, auto-executing macros sometimes contain commands to set the cell-pointer to a starting position in the spreadsheet.

However, the most common and important use of auto-executing macros is to coordinate the process of creating an environment that can be used by your program. For example, before a Symphony add-in application program can be used, it must be attached. This task is a prime candidate for an auto-executing macro. Other prime candidates are tasks that are required to set up parameters in the instrument, start up user menus, start up the main program, etc.

These processes are referred to as "initialization" processes and will be explained in detail throughout the remainder of this book. The initialization macros in this chapter will contain both simple keystrokes and simple programming techniques. In addition, they will be useful macros because they will make sure that the communications port is configured correctly for the spreadsheet each time it is retrieved.

DIFFERENCES IN LOTUS 1-2-3® AND SYMPHONY® AUTO-EXECUTING MACROS

Lotus 1-2-3 and Symphony handle most macros identically. However, auto-executing macros are handled in slightly different manners by the two spreadsheets. For example, Lotus 1-2-3 requires that the auto-executing macro be given a special, unique name (\0). Symphony allows any legitimate macro name for the auto-executing macro as long as you tell Symphony the name.

Differences also exist in the ways that the two Lotus programs implement DOS capabilities. Lotus 1-2-3 has a "System" menu choice that is always ready to be executed, while you must attach a "DOS" application program to Symphony before you can use it.

Finally, a major difference lies in the ways that Lotus 1-2-3 and Symphony handle the interface between you and DOS. In Lotus 1-2-3, you issue the / System command. Lotus 1-2-3 responds by giving you a blank display with the familiar DOS prompt. While you are at the DOS prompt, you can issue DOS commands (like MODE COM1:4800,E,7,2˜, modified to your needs). You must then type EXIT and press return to exit back to Lotus 1-2-3.

Symphony represents substantial improvement in this area. It not only supports the Lotus 1-2-3 method of using DOS, but it allows you to execute DOS programs directly from the spreadsheet. Thus, you can completely automate the MODE-COM1: process without typing a DOS command and then EXITing.

Because of these differences in the two Lotus programs, creating auto-executing macros are separated by section. The next section describes the auto-executing macro for Lotus 1-2-3 and the following section describes the auto-executing macro for Symphony.

Again, this particular difference is a special case. Most of the macros that you create will translate directly back and forth between the two spreadsheets.

CREATING AN INITIALIZATION MACRO FOR LOTUS 1-2-3®

The way that Lotus 1-2-3 handles the / System command severely hampers your setting of the communications port from within the spreadsheet. This restriction makes the auto-executing macro ridiculously short. However, I am going to set up the auto-executing macro anyway because I will be adding to it in future chapters. Also, now is a logical time for you to learn how to create an auto-executing macro for Lotus 1-2-3.

Let me begin with an alternative to the initialization macro. You could automatically set the communications parameters at the time that your personal computer boots by adding a command to the autoexec.bat file that you created in Chapter 2. If you choose to set the communications port from the autoexec.bat file when your system boots, just add the MODE COM1:4800,E,7,2 command (modified to your own specific needs) to your autoexec.bat file below the PATH command.

Using this method to set the port is not the safest, however, because you would not be able to ensure that the port had not been reset by some other user or program prior to your using it from the spreadsheet. This method, however, eliminates the trouble of typing the command and EXITing each time you retrieve the spreadsheet.

Whether you set the port from the DOS autoexec.bat file or from the spreadsheet auto-executing macro, read through the rest of this section to familiarize yourself with Lotus 1-2-3 auto-executing macros. This knowledge will be needed later.

To make a macro an auto-executing one in Lotus 1-2-3, just give the macro the special name \0 (backslash zero). Assigning this name to the macro causes Lotus 1-2-3 to automatically execute it as soon as it retrieves the worksheet.

The auto-executing macro for the ACRO-900 is shown in Figure 3–4. To create the macro, start by using the arrow keys to move the cell-pointer to cell N3. Type '\0) (single quote, backslash, zero). Move the cell-pointer to cell 03. Type '/S in cell 03, making sure that you do not forget to type a single quote (') at the beginning of the command. If you forget the single quote, you will issue a command to immediately display the Lotus 1-2-3 main menu and then issue the System menu choice.

Before you use any macro, you must assign it a formal name and tell Lotus 1-2-3 where the macro begins. Assigning a formal name to the macro will "activate" the macro. To activate the macro, move the cell-pointer to cell N3, issue the /Range Name Label Right (/RNLR) command, and press return. Store the spreadsheet.

There is one alternative that will at least decrease the problem of having to remember the parameters for the MODE command and then typing EXIT. You can create a DOS batch file to do the typing for you. For example, you could create a batch file named "setup.bat". Then, when the spreadsheet issues the / System com-

```
---------N--------O--------P--------Q--------R--------S--------T--------U-----
1
2
3 \0       /S
4
5
6
7
```

Figure 3-4

mand, you could type SETUP, press return, and the personal computer would set the port up and automatically return you to the spreadsheet. "Setup" is certainly faster to type and easier to remember than MODE COM1:4800,E,7,2 <RETURN> EXIT <RETURN>.

Creating the "setup.bat" file is the same process as the one that you used when you created the autoexec.bat file in the last chapter. From the DOS prompt, make sure that your are in the subdirectory that contains Lotus 1-2-3. Use your word processor to create the file. Alternatively, you can use the DOS COPY command, as follows, to create the file:

```
COPY CON: SETUP.BAT
MODE COM1:4800,E,7,2
EXIT
^Z              (control Z)
<return>
```

After you type the <control> Z and <return>, the batch file "setup.bat" will be added to your Lotus 1-2-3 directory. As always, you should modify the text for this batch file to your particular instrument application.

CREATING AN INITIALIZATION MACRO FOR SYMPHONY®

The format of a Symphony auto-executing macro is shown in Figure 3-5. This macro is more elaborate than its Lotus 1-2-3 equivalent because it must check if the Lotus DOS.app application program has already been attached before it can invoke

```
---------N--------O--------P--------Q--------R--------S--------T--------U-----
1
2
3 AUTOEXEC {IF @ISERR(@APP("DOS",""))}{HOOKUP}
4          {SERVICES}AIDOS~MODE COM1:4800,E,7,2~
5
6 HOOKUP   {SERVICES}AADOS~Q
7
8
9
```

Figure 3-5

it. This macro also sets the communications parameters directly from the spreadsheet.

The two @functions in the auto-executing macro (@ISERR and @APP) work with each other to determine whether the DOS application program is attached. The @APP requires two pieces of information. The first is the application program's name (DOS, within quotation marks); the second is a text string (which could be used for an error message to the user). @APP normally returns the text string if the application has already been attached; if not, it returns an "ERR". Because the second argument is irrelevant, I have used empty quotation marks to satisfy the @function's requirements.

The @APP is "nested" within an @ISERR. The @ISERR translates an ERR returned from the @APP into a "true" (that is, a one). If an ERR is not returned by @APP, then the @ISERR returns a false (zero). Thus, the value returned from @APP is translated into something that an {IF} command can understand.

An {IF} command allows a macro program to perform a specific task based on the result of a TRUE/FALSE test. If the result of the expression within the {IF} command does not have the numeric value of zero, Lotus considers the expression to be TRUE and the macro will execute the instructions in the same cell, immediately after the {IF} command. If the result of the expression within the {IF} command has the numeric value zero, Lotus considers the expression to be FALSE and the macro will continue to the cell below the {IF} command.

Whenever you want to call a second macro from an {IF} test, put the name of the second macro in brackets { }. For example, to call HOOKUP, type {HOOKUP} after the {IF} test.

The "IF" command in this example calls upon the HOOKUP macro if the argument specified in the command is TRUE. By doing this, the DOS application program is attached by the HOOKUP macro. If the argument to the {IF} test is FALSE, then DOS has already been attached and the program just skips down to the next command line.

The next line in the auto-executing macro is equivalent to pressing {SERVICES} (the F9 key), choosing A (for Application), I (for Invoke), typing DOS, pressing return, typing MODE COM1:4800,E,7,2 and pressing return again (~ means Return in the Lotus Command Language). Notice that by using the Invoke command from the macro, MODE.com has been treated as if it were another macro program. After MODE.com has been completed, control returns to your macro program and it continues.

You can create a similar macro program for your instrument application by modifying the example macro for your particular needs. To create the macro, start by using the arrow keys to move the cell-pointer to cell N3. Type a name for the macro. The name can be a single quote-backslash ('\) followed by a letter or a zero (\0 can be used for an auto-executing macro only). Alternatively, the name can be a string of text characters (without the backslash). When choosing a name for a macro, be as descriptive as possible. For this reason, I chose the name AUTOEXEC. To use it, you would type AUTOEXEC into cell N3.

Move the cell-pointer to cell N6. Type HOOKUP into cell N6. This macro will "hook up" the DOS application program if necessary. Move the cell-pointer to cell 03. Type in the macro programs listed (modified to your needs) in cells 03, 04, and cell 06. It is very important that there be at least one blank row between the two macros because a blank row signals the end of a macro to Symphony.

Before you can use the AUTOEXEC and HOOKUP macros, you must first assign them formal names (the names in column N). These names tell Symphony where the macros begin. Assigning a formal name to the macro will "activate" the macro. Name a macro by issuing the / Range Name Label Right (/RNLR) command. When prompted for the range of labels, press ESCape, use the arrow keys to move to the cell containing the AUTOEXEC, type a period (.), and arrow down to highlight both the AUTOEXEC and the HOOKUP cells. Press return.

Symphony needs to know more about the AUTOEXEC macro before it can be used for autoexecution. Specifically, it needs to know that the macro called AUTOEXEC is the one to run when the spreadsheet is retrieved.

Unlike Lotus 1-2-3, you are not required to name an auto-executing macro \0. (You could have called this macro \0, but AUTOEXEC is a much better name.) Symphony only allows \0 to be an auto-executing macro, but does not allow \0 to be executed by typing [ALT] 0. If you were to try this, Symphony would just type an @ sign into the current cell. Naming the macro AUTOEXEC would allow you to run the macro with the USER key (F7) anytime during your session.

To achieve the flexibility that allows a user to specify names for auto-executing macros, Lotus requires you to place the name of the auto-executing macro into the spreadsheet's Settings. Do so by issuing the {Services} Settings Auto-Execute Set command and typing AUTOEXEC. Store the spreadsheet.

GIVING A SECOND NAME TO \0 MACROS

Neither Lotus spreadsheet will allow you to invoke a macro by pressing [ALT] 0. In the example macro that I have given, it is not likely that you will re-run the macro while using the spreadsheet. However, you may want to re-execute some of the future auto-executing macros after you have retrieved the worksheet that contains them.

If you want to manually invoke a macro called \0, you must assign a second name to the macro. In Lotus 1-2-3 or Symphony, the macro name should consist of a backslash followed by a letter. With such a name, you can invoke the macro by pressing [ALT] followed by the letter. In Symphony, you could alternatively assign a \0 macro a descriptive name (such as RERUN). If you assign a descriptive name in Symphony, you can then invoke the macro by pressing the User key (F7), typing in the name, and pressing return.

Because assigning a range name to a cell does not cancel any names that have been previously assigned, assigning another, new name does not cancel its auto-executing properties. You can assign the new name by typing the new name into the cell in which the \0 resided. Then, issue the / Range Name Label Right (/RNLR) command.

A FEW COMMENTS ON COMMENTING

You should add comments to your programs for one VERY important reason: Comments make it easier for someone (including yourself) to read and understand your program. Just try to return to a program after several weeks and attempt to remember what you were trying to accomplish in a certain section of the program! Or try to remember a pitfall that you thought about while you were programming. If you have fully documented a certain section at the time that you wrote it, the program would be at its freshest in your mind and the comments would make the best sense. That way, the comments would be more "readable" and it, in turn, would be easier to correct or modify the program, if necessary.

Commenting your program also helps to clarify your own concept of what the program does.

Commenting is such an essential part of every program that for every line of the program that you write, you should try to write a line of comments.

What is the best way to comment your program? No best way exists because there is so much flexibility in what you can do. Recall that when Lotus reads a macro, it reads straight down a column. If a comment is placed just one column over, it will be ignored. The /*comment*/ format that you see in this book is the standard way to comment in C. Figures 3–6 and 3–7 are examples of how to comment the Lotus 1-2-3 and Symphony auto-executing macros using this convention.

You can use any format that you like, as long as the comments are not in a cell within the program.

Another commenting practice to follow is to write a brief synopsis of the function of the program at the beginning of the program. A good synopsis contains the objectives of the program, instrument(s) controlled, type of testing performed, samples tested, cable requirements, etc.

Because macro execution begins at the cell that corresponds to the macro's name, any text above the cell will be ignored. Just make sure that you leave at least one row between the synopsis and the macro above it. Otherwise, Lotus will think that the comment is part of the previous macro.

```
--------N--------O--------P--------Q--------R--------S--------T--------U-----
1                          /*LOTUS 1-2-3*/
2             /*AUTO-EXECUTING MACRO (USER SETS BAUD RATE, ETC.*/
3              /*USING THE "System" MENU CHOICE AND MODE.COM)*/
4 \0          /S          /*ISSUE THE /S MENU CHOICE TO GO TO SYSTEM;*/
5                         /*USER MUST MANUALY SET MODE AT DOS PROMPT*/
6
7
```

Figure 3–6

```
--------N--------O--------P--------Q--------R--------S--------T--------U-----
 1                           /*LOTUS SYMPHONY*/
 2          /*AUTO-EXECUTING MACRO: TESTS TO SEE IF DOS.APP ATTACHED; */
 3          /*IF NOT, ATTACHES IT.  THEN SETS BAUD RATE, ETC.*/
 4 AUTOEXEC {IF @ISERR(@APP("DOS",""))}{HOOKUP} /*SEE IF DOS.APP ATTACHED*/
 5          {SERVICES}AIDOS~MODE COM1:4800,E,7,2~          /*SET COMM PORT*/
 6
 7 HOOKUP   {SERVICES}AADOS~Q /*ATTACHES DOS.APP*/
 8
 9
10
```

Figure 3-7

WHAT'S IN A NAME?

Another recommendation for making your program more understandable is to use meaningful names for your macros. You will normally use a backslash (\) and a letter for the first macro in your program. This designation will allow you to start a chain of programs by using the [ALT] key and its corresponding letter. However, that macro will usually call upon other macros, and those macros should have descriptive names.

Don't get too fancy with names. Choose names that relate to the processes that you are trying to carry out. Try to use names that remind you of the macro's purpose. And above all, don't use YOUR name as a name for a macro. It's embarrassing to come back to a program once you have become an accomplished macro writer and find your name as a macro. It's even more embarrassing to have a really good macro programmer find it.

Also, observe the following words of caution when you are deciding on a name: Do not use names that look like cell addresses (e.g., N6, AB23); do not use one of the Lotus special keynames, keywords, or macro names (e.g., PUT, QUIT); do not use spaces and symbols within a name; and do not exceed 15 characters. Sometimes you can get away with names that break the rules; but most of the time you can't. Remember, unconventional names can lead to dangerous programming situations.

WHAT'S NEXT?

In the next chapter you will learn how to create macro programs that initialize instruments, prompt them to send out data, and receive data.

4

The Core Program

You are now ready to develop a program that will control an instrument and retrieve its data. The programming that you will learn in this chapter will form a core onto which you can add many capabilities specific to your application. It will also form a framework that you can modify and apply to communications with just about every instrument that you want to integrate. In fact, there may be so many possible additions and changes to this framework that you will not even recognize it in many programs; but the basic missions of each section of the program will be the same.

By the end of this chapter you will see just how easy it is to send prompts to an instrument, receive strings that contain the data that you want, convert portions of the strings into their values, and place these values into the template. After you have completed this chapter, you will be able to interface most scientific equipment to a personal computer using Lotus macro programs.

In addition to teaching you some basic programming skills, another objective of this chapter is to teach you to start small and then expand. When you develop the program for an instrument, you will want to program just enough to get the communications up and running and then thoroughly test the program. Then you can add enhancements to the program. Following this strategy, you will not have to worry about errors in an unrelated section of the program causing problems with the communications. It is also much easier to pinpoint and solve problems if a program is kept to a minimum.

Figure 4-1 shows where this chapter fits into the development process.

Again, the example will use the ACRO-931 Themocouple Input Module.

```
                    STEPS TO A SUCCESSFUL APPLICATION

   Assemble The Instrument
   Test Instrument In Manual Mode
   Determine Wiring Configuration
   Get Cable
   Connect Cable To Computer And Instrument
   Secure Cable
   Determine How To Set Communication Parameters
   Set Module Addresses
   Reset Instrument
   Add PATH Command To AUTOEXEC.BAT File
   Add MODE Command to AUTOEXEC.BAT File
   Design Template And Macro Program
   Start The Spreadsheet
   Type Template Into Spreadsheet
   "Activate" Template
   Test The Template
   Create And "Activate" The Auto-executing Macros
   (Symphony): Specify The Name Of The Auto-executing Macro
   Use The DOS MODE Program To Set Communication Parameters
 ->Type In A "Bare Bones" Program
 ->"Activate" The Test Program
 ->Try To Establish Minimal Communications
 ->Evaluate The Test Program
 ->Expand Test Program To Retrieve Data
   Add Amenities And Menus To The Program
   "Activate/Re-activate" Macro Program
   (Symphony): Create Windows
   Create A Table Of Range Names
   Save The File
   Evaluate The Program
   Increase Communications Speeds
   Final Test The Program
   Save/Backup Several Copies Of The File
   Save Several Copies In Separate Safe Places
```

Figure 4-1

SOME ALTERNATIVES

Basically, two different approaches exist for programming the communications with your RS-232 instruments. The first approach is to use Lotus Measure™ as an interface. The second approach involves an unconventional use of the Lotus 1-2-3 and Symphony file input/output (I/O) commands. Nearly all instrument applications will work comparably with both approaches. However, the second approach will prove less costly because you will not need to purchase additional software.

Chapters 4 through 8 describe, in detail, how to use the file I/O approach to instrument control and data acquisition.

Although it is quite uncommon, the file operations technique does not work for some instrumentation and you will need to use Lotus Measure for those applications. In Chapter 9, I will show you how to identify these instruments and how to modify the macro programs presented in Chapters 4 through 8 to utilize Lotus Measure.

Even if you are planning to use Lotus Measure from the start, work through this and the next few chapters. You will need to learn the concepts presented in each chapter. Programs utilizing Lotus Measure are nearly identical to the ones presented in these chapters and you will get a much better feel for the structure of a program. It will also give you experience in creating macro programs.

THE STRUCTURE OF A PROGRAM

Nearly every program that you write will require the same basic components. The first part is a macro (or set of macros) that is executed at the beginning of the program. Macros of this nature ''initialize'' the spreadsheet, the main program, and your instrument. The purpose of this macro is to combine all of these components into a system that will carry out the tasks specified by the rest of the program.

As you shall see, initialization macro(s) are important because they set all of the starting values within the system to the prescribed conditions needed by the system to operate. The auto-executing macros that you developed in the previous chapter are examples of a common way to initialize communications parameters. Over the next few chapters you will see ways to initialize other factors for both your program and the instrument that your program is controlling.

The next component of a program is the ''main'' program. This component program is the central control for the remainder of the program and will be returned to whenever the tasks in other modules have been completed. View the ''main'' program as the central hub from which subroutines are controlled.

A subroutine is just another macro program that is apart from the main program and can be called upon by the main program. A subroutine usually performs just one very specific task. When a subroutine is called, control is transferred from the main program to the subroutine, the subroutine carries out the task that it was programmed to perform, and then the subroutine returns control back to the main program. For example, some subroutines display custom menus, print, store files, perform statistical calculations, and the like.

One of the most important subroutines is the one that obtains data for you. This subroutine is typically called from a Lotus {FOR} command. I will explain this command in detail below; but for now just keep in mind that the {FOR} command has the ability to call a subroutine repeatedly. Thus, you can get repetitive samplings of your data.

The subroutine that gets your data depends upon your instrument and how it communicates. It usually has four parts:

- The first part sends a string of characters out to the instrument to command it to send back data.
- The second part receives a string of data from the instrument and places it into a cell of the spreadsheet.

- The third part converts the string of data into a value.
- The fourth part transfers the information into one of the cells of your template's matrix.

The last section of a program is known as the "scratch-pad" section. The scratch-pad is a set of cells that temporarily stores data.

View a scratch-pad as a temporary work area or the electronic equivalent of a note-pad. Data will be placed into the scratch-pad from a variety of sources. For example, a {FOR} command will keep track of how many times its loop has been executed in one of the cells of the scratch-pad. (Programs that are executed in a repetitive manner are referred to as "loops".) In this context, the scratch-pad will have cells that hold numbers that are used for counting in programs that repeat a prescribed number of times.

Data that has been input by a user will also be stored in the scratch-pad.

Command strings that are sent out to the instrument are often stored in the scratch-pad area. In fact, it is a very good idea to try to keep all command strings in the scratch-pad if you can. By assembling all of your data and program tables in one location, it makes it easier to view what is happening during program execution. That is, scratch-pads and tables make it much easier to locate and correct errors.

As you can deduce from the description, the scratch-pad is usually in a state of constant flux. Information that is placed into cells is overwritten time after time. The scratch-pad is one of the most active portions of the entire spreadsheet and can be informative to watch.

For you to fully understand the function of a scratch-pad, you must first know the concept of a "buffer". The concept of a buffer is perhaps the most important concept of this entire chapter. A buffer is merely a temporary storage area for data and is an excellent example of one of the primary functions of a scratch-pad. A buffer is usually empty at the beginning of a program. When data comes in from an instrument, it is stored in the buffer. When more data comes in, it overwrites whatever was there.

You will need to use a buffer any time that you get data from an instrument. In this context, the buffer serves as a staging area for the data. You need a staging area because the command that receives the data (the Lotus {READLN} command) needs a specific location to place the data that it has read in. Likewise, the Lotus Measure commands also require specific locations to receive their data. Once the data has been placed in the buffer, you can move it anywhere you want in your spreadsheet template with a {PUT} or a {LET} command.

ON WITH THE EXAMPLE!

Figures 4-2 and 4-3 show programs that illustrate each of the points made in the last section. These programs control the ACRO-900 and receive its data. They are simple programs that are easy to understand. It should be easy for you to modify

```
---------N--------O--------P--------Q--------R--------S--------T--------U-----
1  /*"BARE BONES" PROGRAM: TRIES TO GET PC AND INSTRUMENT TO JUST TALK*/
2
3                          /*LOTUS 1-2-3*/
4            /*AUTO-EXECUTING MACRO (USER SETS BAUD RATE, ETC.*/
5             /*USING THE "System" MENU CHOICE AND MODE.COM)*/
6  \0        /S        /*ISSUE THE /S MENU CHOICE TO GO TO SYSTEM*/
7                      /*USER MUST MANUALLY SET MODE AT DOS PROMPT*/
8
9                              /*LOTUS SYMPHONY*/
10           /*AUTO-EXECUTING MACRO: TESTS TO SEE IF DOS.APP ATTACHED; */
11           /*IF NOT, ATTACHES IT.  THEN SETS BAUD RATE, ETC.*/
12 AUTOEXEC {IF @ISERR(@APP("DOS",""))}{HOOKUP} /*SEE IF DOS.APP ATTACHED*/
13           {SERVICES}AIDOS~MODE COM1:4800,E,7,2~        /*SET COMM PORT*/
14
15 HOOKUP    {SERVICES}AADOS~Q /*ATTACHES DOS.APP*/
16
17                     /*MAIN PROGRAM*/
18 \R        {OPEN "COM1",M}    /*OPEN COM PORT AS DEVICE FILE NAME*/
19           {WRITELN "@CPU"}   /*TRY TO GET VERSION NUMBER OF CPU SOFTWARE*/
20           {READLN BUFFER}    /*TRY TO READ IN THE VERSION NUMBER*/
21           {CLOSE}            /*CLOSE THE PORT*/
22           {CALC}             /*FORCE RECALCULATION OF THE SPREADSHEET*/
23
24 BUFFER                       /*BUFFER FOR INCOMING DATA*/
25
26
```

Figure 4-2

the programs for your particular instrument. In the next chapters, I will give you some other representative examples to further aid you.

WHERE TO START

Start with an instrument connected to your computer. Make sure that both are working correctly. Do not try to integrate an instrument unless you are absolutely certain that it is working correctly. That is, thoroughly check the instrument out in manual mode (if possible) before trying to communicate with it from your macro program. If your instrument is not working perfectly in manual mode, adding communications will only compound its problems and make it difficult to pinpoint the cause of any problem.

Next, create the simplest program that you can. If your instrument allows you to write a string of characters to its display or has a command that returns the instrument's software version number, start there. These modest functions are usually the easiest ways to make sure that your personal computer and the instrument are talking properly. Next, try to get some data . . . any data . . . from your instrument into a cell of your spreadsheet.

Once you are able to prompt the instrument and/or get data back from it reliably, you can expand the control, extend the acquisition to retrieve data that is more useful, and add all of the amenities. You should increase the program in small

```
---------N--------O--------P--------Q--------R--------S--------T--------U-----
1  /*PROGRAM TO GET TEMPERATURES FROM ACRO-931 THERMOCOUPLE INTERFACE*/
2
3                           /*LOTUS 1-2-3*/
4              /*AUTO-EXECUTING MACRO (USER SETS BAUD RATE, ETC.*/
5               /*USING THE "System" MENU CHOICE AND MODE.COM)*/
6  \0          /S        /*ISSUE THE /S MENU CHOICE TO GO TO SYSTEM; */
7                        /*USER MUST MANUALLY SET MODE AT DOS PROMPT*/
8
9                             /*LOTUS SYMPHONY*/
10             /*AUTO-EXECUTING MACRO: TESTS TO SEE IF DOS.APP ATTACHED; */
11             /*IF NOT, ATTACHES IT.  THEN SETS BAUD RATE, ETC.*/
12 AUTOEXEC {IF @ISERR(@APP("DOS","")}{HOOKUP} /*SEE IF DOS.APP ATTACHED*/
13             {SERVICES}AIDOS~MODE COM1:4800,E,7,2~        /*SET COMM PORT*/
14
15 HOOKUP      {SERVICES}AADOS~Q /*ATTACHES DOS.APP*/
16
17                       /*MAIN PROGRAM*/
18 \R          {OPEN "COM1",M}    /*OPEN COM PORT AS DEVICE FILE NAME*/
19             {GOTO}START~       /*MOVE CELL-POINTER TO STARTING POSITION*/
20             {WRITELN "#4"}     /*INITIALIZE ACRO-931 (MODULE 4)*/
21             {FOR ROW,0,4,1,GET_ROW}     /*GET 5 ROWS OF DATA*/
22             {CLOSE}            7*CLOSE THE PORT*/
23             {CALC}             /*FORCE RECALCULATION OF THE SPREADSHEET*/
24
25                       /*LOOP THAT GETS ONE ROW OF DATA*/
26 GET_ROW     {DOWN}                       /*MOVE DOWN TO NEXT ROW*/
27             {FOR PROBE_NUM,0,7,1,GET_TEMP}       |*GET A ROW'S WORTH*|
28                                          |*OF DATA:  NUMBER OF PROBES*|
29                                          |*(COLUMNS) MINUS 1 = 7.    *|
30                                          |*(CNTR STARTS AT 0)        *|
31
32                  /*LOOP TO GET DATA FROM ONE OF THE CHANNELS*/
33                    /*IMPORTANT NOTE TO READER: SEE TEXT IN*/
34                  /*CHAPTER FOR AN EXPLANATION OF THE 3 DOTS*/
35 GET_TEMP {WRITELN @CHOOSE(PROBE_NUM,"TEMP1","TEMP2",...,"TEMP8")}
36             {READLN BUFFER}    /*READ IN A TEMPERATURE VALUE*/
37             {RECALC PARSED}    /*FORCE PARSING OF THE STRING*/
38             {PUT C8..J3000,PROBE_NUM,ROW,PARSED:VALUE}/*PUT DATA IN SHEET*/
39
40             /*SCRATCHPAD OF NAMED RANGES*/
41 ROW                   /*COUNTER FOR FIRST LOOP; ROW OFFSET FOR {PUT}*/
42 PROBE_NUM             |*COUNTER FOR 2ND LOOP; CURRENT PROBE NUMBER; LIST*|
43                       |*OFFSET FOR THE @CHOOSE; COLUMN OFFSET FOR {PUT} *|
44 BUFFER                /*A BUFFER FOR INCOMING TEMPERATURE DATA*/
45 PARSED           ERR /*PARSER FOR BUFFER: @VALUE(@LEFT(BUFFER,7))*/
46
47
48
```

Figure 4–3

steps, writing only one functional unit at a time, testing it, and then proceeding to the next functional unit.

Remember, the larger the program, the harder it will be to find the cause of a problem if one occurs. Also, increasing the size of a program can sometimes affect its timing. By increasing a program's size in steps, you can monitor these timing changes, see how they affect performance, and then optimize the program's speed according to need.

The programs in Figures 4–2 and 4–3 are good examples of this plan of attack. Figure 4–2 contains the bare minimum of what is needed to get a response from an instrument and place it into a buffer. In this example, the "@CPU" is a string that prompts for the version number of the software in the ACRO-900.

Once you have a program like the one in Figure 4-2 up and running (and thoroughly tested), you can expand it slightly by adding a limited number of other fundamental functions. These additions are shown in Figure 4-3.

After you have a program like the one in Figure 4-3 working correctly, you can add features like those shown in Chapter 5.

Although the program in Figure 4-2 is very important in the evolutionary process of creating an application, the program in Figure 4-3 will be the focus of the rest of this chapter. All of the information supplied for Figure 4-3 will apply directly to Figure 4-2.

GETTING ACQUAINTED WITH CREATING MACROS

Even if you are not using the ACRO-900 interface, type the program in Figure 4-3. Use the template spreadsheet that you created in the last chapter. Make certain that the macro is to the right of the last column of cells in the template. Also, make sure that you leave at least one blank row between each macro; Lotus 1-2-3 and Symphony look for these blank rows to determine the end of a macro.

Please make special note of the {PUT} command in Figure 4-3. I had to modify it so that it would fit on the page of this book and have used three dots (. . .) in place of some of the text that should appear in the command. Therefore, you need to add the text back in when you type the command for your program. To add the missing text, type "TEMP3", "TEMP4", "TEMP5", "TEMP6", "TEMP7" into the area where the three dots appear. This text is part of the same cell containing the three dots (i.e., do not type it into an adjacent cell).

It is easy to prepare a section of the spreadsheet for your scratch-pad. Type the names of the variables into the same column as the macro names.

The next task in creating any program is to "activate" the program and scratch-pad. To accomplish this task, assign label names to the macros and the scratch-pad cells. Do so by issuing the /Range Name Label Right (/RNLR) command. When the spreadsheet prompts you for the range of labels, press ESCape, move the cell-pointer to the cell containing the first macro name, type a period (.), use the down arrow key to highlight all of the macro names AND the scratch-pad names, and press return. This defines the cells to the right of the column as having the names specified in the name column. You should now be able to appreciate the convenience of aligning the range names into a single column.

Finally, give cell C8 the name START. Do so by issuing the /Range Name Create (/RNC) command. When the spreadsheet prompts you for the name, type START and press return. When prompted for the range of cells, type C8 and press return.

Now, save the spreadsheet. As we proceed through the following sections, try to see how you might modify the program for *your* instrument.

INITIALIZATION USING AUTO-EXECUTING MACROS

As stated earlier in this chapter, the first part of your program "initializes" the entire system. The easiest and safest way to accomplish initialization is to use auto-executing macros.

The '\0 and AUTOEXEC macros in Figure 4–3 are the auto-executing macros that you created in the last chapter for Lotus 1-2-3 and Symphony, respectively. They perform the critically important task of setting the communications parameters that DOS will use. This initialization step ensures that the parameters in the instrument and personal computer are identical.

As in the last chapter, the Symphony auto-executing macro detects the status of the DOS application program, attaches it (if it has not already been attached), and invokes it to set the communications parameters. The Lotus 1-2-3 auto-executing macro issues the /System command and brings you to a DOS prompt so that you can set the port manually. (Because of this limited usefulness, you may want to omit this macro temporarily.)

You can include both macros in the same worksheet if you like and the two macros will not interfere with each other. That is, if you have Symphony and have specified the name AUTOEXEC as the auto-executing macro in the spreadsheet's Settings, Symphony will ignore the \0; while Lotus 1-2-3 will only recognize \0. (See Chapter 3.)

If you are writing programs that are designed to run within both Lotus 1-2-3 and Symphony spreadsheets, including two different auto-executing macros in the same spreadsheet is a good way to ensure "portability". In other words, the spreadsheet will be independent of whether it is working under Lotus 1-2-3 or Symphony. This flexibility will allow the spreadsheet to be moved back and forth from one Lotus program to the other. Make sure, however, that the auto-executing macro for Lotus 1-2-3 is named '\0 and the auto-executing macro for Symphony has another, different name. Later, I will show you how to detect which spreadsheet is being used and switch automatically between macros that have been customized specifically for the Lotus program being used.

OVERVIEW OF THE "MAIN" PROGRAM

In Figure 4–3, the main program is called \R. It also performs some initialization tasks. Before you can transmit or receive data from an RS-232 communications port (or file) you must first "open" it.

When you open a communications port or file, you are actually telling DOS to set up a table of parameters that specify various attributes of the communications port or file. This table will contain such specifications as the name of the communications port or file, where its DOS buffers are located, the address of the communications port or file, and how the communications port or file can be accessed (read only, write only, or both). Thereafter, you can issue commands in your program that will transfer data to and from the communications port or file and DOS will know exactly how to carry out the specifics of your commands.

Another important initialization task performed by the main program is the one that initializes the ACRO-900. Once initialized, an instrument will operate in the exact fashion that you specify. The initialization performed here merely sends the address of the thermocouple module (#4).

Although this example illustrates an uncommonly simple initialization, it does show the nearly universal method that instruments use to set up their starting values to the ones that you prescribe. More commonly, you will be sending string commands that will move motors to specified starting positions, assign numbers to be used for mathematical manipulations, change operating modes, switch options on and off, designate operating times, select voltages, select scaling factors, etc.

The next task of the main program is to obtain data. As stated previously, {FOR} commands and subroutines are usually used to accomplish this task. The {FOR} command in this macro is an example of this concept.

The main program concludes by closing the port.

The next section takes a closer look at each of these functions and the programming techniques that are used to execute them.

OPENING COMMUNICATIONS PORTS

The {OPEN "COM1",M} command is actually a modification of the Lotus command for opening files. In its usual context, the command looks like the following:

```
{OPEN ''filename'',M}
```

where, "filename" is the name of the file that you want to write to or read from, and the "M" in the command means to use the "modify" mode. In this mode, you can both write to and read from the file (or communications port). Other modes are R, W, and A, representing Read Only, Write-Only, and Append, respectively.

The "COM1" in Figure 4-3 is what is known as a "Device File Name". A Device File Name is a file name that has been reserved by DOS for specific devices that are used to transfer data into and out of the computer. The Device File Name "PRN", for example, is reserved for redirecting normal file output to the printer of your personal computer. Other Device File Names are: LPT1, LPT2, AUX, CON, and COM2. Each of these file names can be used to direct data to printers, the console (display), modems, etc.

You used an example of a Device File Name in Chapter 2 when you issued the "copy con: autoexec.bat" command. In that example, you literally told DOS to copy the contents from a file named con: (i.e., the console/keyboard) to the file named autoexec.bat (i.e., a file on the disk).

The two most important Device File Names that are used in this book are

COM1 and COM2. They can be used as file names to send data to, or receive data from, the RS-232 communications ports.

Lotus 1-2-3 and Symphony use DOS whenever they perform file operations. Therefore, if you call out a "COM1" or a "COM2" in an {OPEN} command, you can use the DOS redirection feature to open one of the RS-232 ports. Thereafter, you can write to, and read from, the port just as if it were a file.

SETTING THE INITIAL CELL-POINTER POSITION

When you set up the example program, you gave cell C8 the name START by issuing a /Range Name Create (/RNC) command. The {GOTO}START~command is equivalent to pressing the GOTO key (F5), typing START, and pressing return (recall that the tilde, ~ ;means "execute a carriage return"). This sequence resets the cell-pointer to the row and column into which the first eight channels of data will be placed. By using this command, you will ensure that when you start the macro, you will be able to view the data as it arrives.

INSTRUMENT INITIALIZATION

Most instruments need to be initialized (configured) before they can give meaningful data. The initialization is performed by sending the instrument string commands.

Usually one command exists for each parameter that needs to be set. Typically, the command starts with a special character (such as a $, @, %, &, *, etc.). A string of characters that specify which parameter is to be set follows the starting character. This string is followed by the value(s) for the parameter. The string is usually terminated with a carriage return (ASCII 13) or a carriage return/line feed (ASCII 13 10) combination.

The sequence of a "header" character, a command mnemonic, a value (or values) and termination by a carriage return/line feed is a VERY common, if not universal, format for command strings that control instruments.

Strings that terminate with a carriage return/line feed combination are transmitted using the {WRITELN string} command. The string specification in this command may be either literal text (within quotation marks), a formula, an @function, or the name of a cell.

Sometimes the "header" character is not one of the usual keyboard characters, but is a non-printing control character (such as a CONTROL-B; which is also called STX, or ASCII 2). When such is the case, the terminating character is also usually a non-printing control character (commonly a CONTROL-C; which is also called ETX, or ASCII 3).

These non-printing control characters fall within the first 26 ASCII character codes. Therefore, you cannot just type the character but will need to use the Lotus @CHAR function. The @CHAR function takes a three-digit ASCII numeric input of the character that you want and returns the character. For example, if you used a 002 in an @CHAR function, the function would become a CONTROL-B (STX).

If you use the @CHAR, you will also need to "add" the character to the string that you are sending out. This process is called "string concatenation". When you string concatenate, you use the ampersand (&). You must also place literal strings within quotation marks. For example, if an instrument required a CONTROL-B header, a "GM" for a command mnemonic, and a value of "458" in its command string, you could use the following formula to assemble the final string:

```
@CHAR(002)&''GM 458''
```

Below is another example to ensure that you know how to use string concatenation to assemble command strings. In this example, your command strings need to start with an STX (ASCII 2) and end with an ETX (ASCII 3). Also assume that you do *not* want to send a carriage return/line feed to the instrument and that the initialization string is in a cell called INITSTRNG.

In this example, you would assemble the string by concatenation and use the {*WRITE*} command to send the string out to the port. In other words, you would NOT use the {WRITELN} if you did not want a carriage return/line feed to be transmitted. Here is an example of a {WRITE} command that illustrates this process:

```
{WRITE @CHAR(002)&INITSTRNG&@CHAR(003)}
```

Take a moment to place string concatenating formulas into some cells of your spreadsheet and make sure that you know how to use them.

If you are using numbers in cells, you will need to convert the numbers to strings before you can concatenate them. Use the @STRING function. For example, if an integer number was in a cell called MODULE, you would use the following formula to concatenate it into the last {WRITE} example:

```
{WRITE @CHAR(002)&INITSTRNG&@STRING(MODULE,0)&@CHAR(003)}
```

Take some time to review your User's Manual to determine which parameters need to be initialized and the string commands that you need to send to the instrument to set them. Sometimes, this information is supplied in a separate, "Programmer's Manual". After you have obtained this information, you can begin to plan the initialization part of your macro program.

INITIALIZING THE ACRO-931

The ACRO-931, like most other instruments, came with a "Programming Guide". In the guide it stated that before a module can be used, it must be initialized by sending the ACRO-900 the address of the module. The command consists of a pound sign (#), followed by a number and a terminating carriage return/line feed.

The {WRITELN"#4"} command in Figure 4-3 sends the string #4 out the COM1 port, followed by a carriage return (ASCII 13) and a line feed (ASCII 10). This command will select the module at address number 4 (the address number, if you recall, that I set it to in Chapter 2). I use type T thermocouples and Celsius data. These are the default settings for the module, so I do not have to send out any more strings.

If, by chance, you are using the ACRO-931 and some other kind of thermocouple, you can look into the ACROSYSTEMS instruction manual for the strings to send out to the module to change its output.

Likewise, your particular instrument will probably have its own series of initialization strings. If so, just create a series of {WRITELN} commands to send each of them out to the port.

If your instrument cannot tolerate the carriage return/line feed, use the {WRITE string} command. If your instrument does not require initialization strings, skip this step.

LOOPING TO PROMPT AN INSTRUMENT FOR DATA

The construction of string commands that prompt for data is usually in the same format as those that are used to initialize an instrument. Therefore, transmitting a string that prompts for data will follow the same rules as just described. However, prompting for data is usually repetitive.

There are two "nested" {FOR} loops in the program shown in Figure 4-3. Nested {FOR} loops are perhaps the most commonly used methods for obtaining a matrix of data. You will probably use them often for your instruments, so we will take a closer look at how they work.

The outer {FOR} loop is in the main program and simply executes the inner loop five times (zero through four). By constructing this combination of commands, you will get five readings from each of the eight channels. The results are placed into sequential rows of the spreadsheet template.

Just before each row of eight results is started, the {DOWN} command moves the cell-pointer down to the next row. This movement allows you to view the results as they arrive. This feature is relatively unimportant in this example, but will become very important when you need to scroll the screen during an experiment that contains more than 20 rows of data.

The format of the outer loop is as follows

```
{FOR ROW,0,4,1,GET_TEMP}
```

where, GET_TEMP is the name of the macro that you want the spreadsheet to execute repeatedly, and the arguments ROW, 0, 4, and 1 represent the scratch-pad cell that will contain the current loop number, the starting value, the stop value, and the size of the step, respectively.

The inner {FOR} loop prompts each of the eight channels of the module. The

{WRITELN @CHOOSE(PROBE_NUM,"TEMP1","TEMP2","TEMP3","TEMP4", "TEMP5","TEMP6","TEMP7","TEMP8"} command in GET_TEMP sends out the appropriate string, followed by a carriage return/line feed combination.

The transmitted string acts as a prompt to the ACRO-931, telling it to make a temperature reading on the appropriate channel, convert it to degrees Celsius, and transmit it back to the personal computer.

The @CHOOSE is the simplest @function in a class of @functions that Lotus calls "lookup functions". The @functions in this class take a table and "look up" one of the pieces of data in that table based on its position. Thus, these @functions require a definition of the table AND the offset(s) to use for pinpointing the piece of data that you want.

Other @functions in this class include @INDEX, @VLOOKUP, and @HLOOKUP. They are all very handy programming tools and you will run into them often in this book. The tables that these three @functions use are named ranges in the spreadsheet.

The @CHOOSE function is a little different. The table that the @CHOOSE uses is a list of options that are contained within the @function.

The @CHOOSE lookup function allows the spreadsheet to select the result from the list contained within the @CHOOSE. The result is based on the position of the piece of data within that list. In this example, PROBE_NUM (the counter for the loop) specifies the position in the list of the item that you want to sent out to the port. The "TEMP1", "TEMP2", . . . is the list.

At the beginning of the {FOR} loop, PROBE_NUM is zero. The first item in the list has an offset of zero. Therefore, the @CHOOSE evaluates to the string "TEMP1" and the {WRITELN} sends TEMP1 out to the port. The next time through the loop, PROBE_NUM is one. The second item on the list ("TEMP2") has an offset of 1. The @CHOOSE evaluates to "TEMP2" and the {WRITELN} sends TEMP2 out to the port. This process continues until the {FOR} loop ends and all eight prompts have been transmitted.

The @CHOOSE is by no means the only way that you can select the string that you want to send out to the port. You can also use @INDEX, @VLOOKUP, @HLOOKUP, or even eight {WRITELN "string"} commands. You can also use lists within named ranges instead of literal strings within quotes. In later chapters, I will show you how to use some of these other functions and also how to use named ranges.

RECEIVING AND PARSING DATA FROM AN INSTRUMENT

When an instrument receives a command to return data, it will perform the tasks that are required to obtain the data and transmit the data back to the personal computer. The command that you use to receive the data is the {READLN} command.

Usually when an instrument transmits data, the data is embedded in a string of ASCII characters that also contains other information. Also included in the string may be characters that report on the status of the instrument, characters that separate the data, spaces, and/or a terminating carriage return/line feed combination. You therefore need to use one (or more) of the Lotus string @functions to retrieve just the data that you want.

These string @functions will cut ("parse") the appropriate characters out of the string that comes into your BUFFER cell and transform characters into a second string. Parsing is a very common process that you will need to perform for just about every instrument application that you create.

After you have formed the second string, you will need to use an @VALUE function to change the string into a number. The use of an @VALUE function is another process that is almost universally required.

The requirement for the use of @VALUE is associated with an earlier discussion in Chapter 2 related to how numbers were transmitted across RS-232 lines. That is, if the number 527 is transmitted, it is transmitted as the ASCII characters 53 50 55, NOT 527. Therefore, your program receives a character 5, a character 2, and a character 7. The string looks a lot like a number, but it is not. In order for you to be able to do mathematical manipulations or graphics on the data, you must first transform the strings into numbers. The @VALUE function does this translation for you.

As you can see, the string @functions and @VALUE functions must work together to first pick out the appropriate characters from a string and then change the characters into a value. In the next section, we will explore some specifics on how to apply them to an instrument's strings.

CUTTING A PIECE OF THE PIE

Lotus 1-2-3 and Symphony have a wealth of "string functions" that help you to retrieve information that you need from a long string of characters. Three of these string functions are called @LEFT, @RIGHT, and @MID.

You can use the @LEFT(string,number_of_characters) function to take the left-most number of characters or the @RIGHT(string, number_of_characters) function to take the right-most number of characters in a string. If the characters that you want are in the middle of a string (or even at the ends for that matter), you can use the @MID(string,starting_position,number_of_characters) function.

Lotus has two other very useful string @functions. The first one is called @CLEAN.

@CLEAN provides a very easy way to remove a carriage return from a string. The @CLEAN function also removes any other control characters and non-printing characters (ASCII 0 to 31) from the string. The @CLEAN function will likewise remove some of the "noise" (nonsensical characters) that can be received at the port when an instrument is switched on. Nesting this function within the @VALUE often produces very reliable results and is often more efficient than using @LEFT, @RIGHT, and @MID functions. A parsing function using @CLEAN would look like the following

```
@VALUE(@CLEAN(BUFFER))
```

Sometimes the safest way to get the value of your string is to combine the @CLEAN with an @MID, @LEFT, or @RIGHT function. This combination would result in a "nested" group of @functions and would look like this

```
@VALUE(@MID(@CLEAN(BUFFER),0,7))
```

This function often gives you a very specific "piece of the pie" because it cuts out control characters and then takes a very specific number of characters for conversion into a number. Thus, if any characters or numbers existed outside of the range that is expected, they would be ignored.

The @VALUE(@CLEAN(BUFFER)) function usually provides an adequate degree of safety and it is much faster. For this reason, the remaining examples will use the function without the @MID to parse the data.

The @TRIM(string) function is an @function that is similar to @CLEAN. @TRIM removes leading and trailing spaces from a string. @TRIM also compresses multiple spaces within the string and converts them into single spaces. If the string that you are dealing with has a variable number of (or a large number of) spaces in it, then you will want to use the @TRIM function to remove the spaces. A function using all four of the parsing tools would look like this

```
@VALUE(@MID(@CLEAN(@TRIM(BUFFER)),0,7))
```

This function would begin by removing all non-printing and control characters. It would then remove all extra spaces. Finally, it would take the first 7 characters of the remaining string and convert them into a number. Fortunately, this degree of nesting is usually not needed for scientific data.

HOW TO DETERMINE A PARSING FORMULA

A couple of methods are at your disposal to determine how to parse a string of data from your instrument. The first one is to look into the instrument's User's Manual for the format of the transmitted string and to count the characters. Using this information, you could create the parsing @function in the PARSE cell and hope that it works. This method usually takes an exorbitant amount of time because of inaccuracies in the information given in most User's Manuals.

The second method is trial and error. This method usually works too, but it is even more time-consuming.

The method that I have developed takes the least amount of time. To use this method, get some data from your instrument into the BUFFER cell of your scratchpad using your macro program. Then use the {EDIT} key (F2) to expose the real contents of the cell. By moving the cursor back and forth through the cell, count the number of spaces and "non-printing" characters before and after the string that you are interested in. "Non-printing" characters will be easy to see because they appear as Chinese letters in the cell.

After viewing a cell's contents, you should be able to decide whether to use the @CLEAN and/or the @TRIM function(s). If you need to use the @CLEAN and/or the @TRIM function(s), prepare a second cell containing the @function(s) before you proceed. This cell should be near the BUFFER cell so that you can observe both cells concurrently. For example, if you need both @CLEAN and @TRIM, then place the following equation in a cell near the BUFFER cell:

```
@CLEAN(@TRIM(BUFFER))
```

If you use one or both of these formulae, issue the /Range Value (/RV) command to transform the formula back into a string. When prompted for the cell to copy the value to, specify a convenient unoccupied cell near the cell with the formula.

Use the {EDIT} key (F2) again and move the cursor back and forth through the appropriate cell. If you applied the @TRIM and/or @CLEAN functions, use the transformed cell. Otherwise, use the BUFFER cell directly. Count the number of characters before and after the string that you are interested in. Remember that the first character in the cell is at offset zero, not one. From these two counts, decide whether @LEFT, @MID, or @RIGHT is the appropriate function to use for parsing and where to "cut" the string for the information that you want. Jot down these beginning and ending offsets for the string.

Next, type a nested @VALUE function into the PARSED cell. The function should include the appropriate @LEFT, @MID, or @RIGHT function nested within it. Be sure to include the @TRIM and/or @CLEAN function, if appropriate. The cell to refer to in the innermost function is BUFFER.

The value that you see in the PARSED cell should be the correct one. If it is not, repeat the process to find the source of the problem and use the {EDIT} key (F2) to edit the function(s) in PARSED until the value is correct.

Be sure to erase all test cells using the / Erase command (Symphony) or the /Range Erase command (Lotus 1-2-3). These cells will have no future use and will slow your spreadsheet.

EXPLANATION OF THE PARSING FUNCTION IN THE EXAMPLE PROGRAM

The program in Figure 4-3 illustrates how all of these concepts fit together.

Each time the ACRO-931 receives a string that prompts for data, it determines the temperature of the appropriate probe and sends a string of data back to the spreadsheet. The {READLN BUFFER} command in the macro waits until the string comes into the port (terminated by a carriage return/line feed), retrieves it, and places it into a cell in the scratch-pad called BUFFER.

The strings that are received from the ACRO-900 are strings of character digits followed by carriage returns (ASCII character 13).

Assuming the equation @VALUE(@LEFT(BUFFER,7)) has been placed into the PARSED cell of the scratch-pad, the @LEFT function returns the first seven characters of the string that come into BUFFER. These seven characters form a new string, the value of which is the temperature of the probe. (The @VALUE function will transform the seven characters into a number.)

If I had tried to use the string directly in the @VALUE function, I would have received ERRors because the @VALUE function can only use digits. For this reason, I used the @LEFT function to stop just before the carriage return terminator. Alternatively, I could have used an @VALUE(@MID(BUFFER,0,7)) to perform the parsing function. This function would have started at position zero of the BUFFER and then would have taken the next seven characters.

Using parsing cells with active @functions presents a problem that you need to contend with. This problem is the subject of the next section.

FORCING RECALCULATIONS

When a string of data comes into a buffer from one of the RS-232 ports, the macro program executes the {READLN} command without recalculating the spreadsheet. Thus, none of the other cells are aware that the value of the buffer has changed.

Not recalculating the spreadsheet is not specific to just the {READLN}. Lotus suspends spreadsheet recalculation while almost all macro commands are being executed. For this reason, you must specifically recalculate all formulae and @functions before they will update.

The {RECALC PARSED} command is needed to force a recalculation of the PARSED cell (since it has @functions in the cell). If this command is not present, the value of PARSED does not change after each subsequent string is received. You can force recalculation of formulae and @functions with any of three different macro commands:

- The {RECALCCOL range} command recalculates entire columns in the specified range, moving from left to right (column by column).
- The {CALC} command recalculates the entire spreadsheet.
- The {RECALC range} command recalculates only the cells in a specified range.

Whenever possible, it is best to use the {RECALC} command because it only recalculates a specific region of the spreadsheet and is much faster than recalculating the entire spreadsheet with a {CALC} command. In this case, the {RECALC} command recalculates the PARSED cell only.

MOVING DATA INTO THE TEMPLATE

Each time a BUFFER has been updated, parsed, and the value determined, your program will need to use a {PUT} command to transfer the data into the appropriate cell of the template. The format of the {PUT} command is as follows:

```
{PUT C8..J3000,PROBE_NUM,ROW,PARSED:VALUE}
```

where, C8..J3000 is the range that defines the block of cells in which the {PUT} could place the result; PROBE_NUM (the small loop counter) specifies the column (left/right) offset within the range; ROW (the large loop counter) specifies the row (up/down) offset within the range; and PARSED is the result that is to be placed into the cell specified by the column and row offsets. As usual, the first row and column of the range have the offset of (zero,zero). In this example, that cell is C8.

Let us examine how the {PUT} command and the two {FOR} loops relate. At the beginning of both loops, both ROW and PROBE_NUM are zero. When the {PUT} command is encountered, it will take the value of PARSED and place it into cell C8 (column-offset=zero, row-offset=zero). Next, PROBE_NUM increments to 1; while ROW stays at zero. When the {PUT} command is encountered, it will take the current value of PARSED and place it into cell D8 (column-offset=one, row-offset=zero). This process will continue down row 8 until the inner loop reaches eight. Then the outer loop will increment to one. The inner loop will then re-initialize to zero. This time when the {PUT} command is encountered, it will take the current value of PARSED and place it into cell C9 (column-offset=one, row-offset=zero). This process will continue until both loops have reached their limits. At this point, you will have a matrix that is 8 columns wide by 5 rows deep.

Notice that I had a value of 3000 for the last row of the range. I specified this value because it does not use any memory while it is in the {PUT} command; but allows for the number of readings to be increased to a very large number if need be. Later, you will learn how to prompt users to enter the number of readings that they want to take. Making the upper row limit of 3000 will provide the flexibility to allow them to specify a number up to 3000.

The {PUT} command has another built-in feature that can provide a measure of safety to your program. Notice that I used column J as the last column for the range of acceptable cells in the {PUT} command. This provides a degree of safety because it does not allow the program to place data in sections of the template or program where it will overwrite formulae and/or macro commands. This feature is an important one that you will want to exploit in your programs.

THE GRAND FINALE

The macro program concludes with a {CLOSE} command and a {CALC} command. Once you open a communications port or a file, it remains open until Lotus 1-2-3 or Symphony encounters a {CLOSE} command. It is very important that you close every communications port or text file that you have opened. This termination will force DOS to read or write any characters that are left over in the buffer that DOS uses as a staging area for sending and receiving characters and it will clear the buffer for future use.

The {CALC} command forces a recalculation of all of the formulae in the spreadsheet and updates, or re-displays, the current data. As you know, when a macro program executes, it does not necessarily recalculate the formulae in a template. Placing a {CALC} at the end of a macro is very important to avoid user confusion.

More importantly, if a user forgets to recalculate a spreadsheet before printing it, the results may not be correct because they may not reflect new values that have been added to the template. To prevent the possibility of catastrophe, get into the habit of placing a {CALC} command at the end of all of your macro programs and, more importantly, before any automated printing is performed by your program.

THE "SCRATCH-PAD"

Earlier in this chapter, we looked briefly at the concept of a scratch-pad. The scratch-pad is a group of cells that temporarily stores data and holds values for loop counters. Your program relies very heavily on the data in your scratch-pad. This relationship makes the scratch-pad an important part of your program and makes it an area where you can greatly enhance the efficiency of your program.

For example, in Figure 4-3 ROW and PROBE_NUM serve three purposes. Their primary function is to serve as loop counters. When a {FOR} loop begins execution, it sets the value of its loop counter to zero. Each time the loop is repeated, the {FOR} command increments the counter by one. Therefore, a {FOR} loop needs a cell in the spreadsheet to hold the current loop value. ROW and PROBE_NUM fulfill this requirement.

Their second function is best illustrated by PROBE_NUM, which serves as the list offset for the @CHOOSE function. As PROBE_NUM increments with the loop, the appropriate string in the @CHOOSE is designated and the string is sent out the communications port.

Their third function is to specify the column and row offset values in the {PUT} command. Thus, as the {FOR} loops increment, they automatically increment the coordinates within the range and the incoming data is placed in the specified cell. This system of {FOR} loops and {PUT} commands is a very handy way to position data and we will use it often throughout the remainder of this book.

The ROW and PROBE_NUM cells thereby make neat programming packages, tying everything together. This kind of package programming is very efficient and decreases the chances of errors.

Another important feature of the scratch-pad is the cell called BUFFER. We discussed the concept of a buffer earlier in this chapter, but it seems appropriate to repeat some of the discussion as it applies to the scratch-pad. If you recall, a buffer is a temporary storage area for data. It is usually empty at the beginning of the program. When data comes in from an instrument, it is stored in the buffer. When more data comes in, it overwrites what was previously there.

You will need to use a buffer any time that you get data from an instrument. In this context, the buffer serves as a staging area for your {READLN} commands. {READLN} needs a specific location to place the data that is read. {READLN} cannot pinpoint data with column- and row-offsets the way that {PUT} can. Once the data has been placed in the buffer, it can be parsed with @functions and moved anywhere in the spreadsheet with a {PUT} command.

The final named cell in the scratch-pad is the PARSED cell. As I described earlier, PARSED is a cell that contains the equation that cuts the first seven characters out of a string that comes in. The @VALUE function transforms them into a number.

The following diagram should help you visualize the flow of data from your instrument to the final cell in your template:

```
             CABLE      DOS  {READLN}        @FUNCTIONS   {PUT}
INSTRUMENT------>PORT------>DOS------>SCRATCH-PAD------>PARSED------>TEMPLATE
                            BUFFER      BUFFER          CELL          CELL
```

NESTING AN @FUNCTION INTO A {PUT} COMMAND

PARSED is a cell that is not actually needed. A better way to have executed the parsing function would have been to modify the {PUT} command to include the @VALUE and @LEFT functions. This would have given the following nested command for the macro shown in Figure 4–3:

```
{PUT C6..J3000,PROBE_NUM,ROW,@VALUE(@LEFT(BUFFER,7)):VALUE}
```

This example shows a more efficient way to parse. This gain in efficiency is because every time a spreadsheet is recalculated, all of the @functions will update. This recalculating takes time. You could just recalculate the specific cell that contains the parsing @function(s) from within the program by issuing a {RECALC} command. Nevertheless, this method of recalculation won't help much because the {RECALC} command takes time too.

If the @function is nested within a macro {PUT} command, the calculation would only be done at the time that the {PUT} command is executed. This change not only proves faster, but it ensures that the correct value is placed into the template cell.

For this reason, the parsing function in all future examples will be performed from within {PUT} commands.

TEST IT!

As with a template, it is better to test macro programs as you create each section rather than waiting until the entire program is complete. If you wait until the end, all of the many subroutines and program lines will make it more difficult to isolate the errors. However, if you test as you build, you can verify that each section works correctly before you add another level of complexity.

Test your macro program under actual experimental conditions. This real-life testing often uncovers problems that cannot be found by just testing the system under contrived conditions.

WHAT'S NEXT?

You now know all of the basics to write a program to run your own instrument. If by chance you have the ACRO-900 interface, you can run the program by retrieving the example spreadsheet and pressing [ALT] and R together. The program is functional and should immediately begin to gather data and place it into the spreadsheet. If you want to abort the macro while it is running, press [CONTROL] [BREAK].

In the next chapter, I will show you how to enhance this basic program. I will show you how to program macros to display custom menus, obtain user input, end the screen flicker that occurs when a macro runs, control and determine time delays, format results, and perform some statistical calculations. These improvements will make your macro programs more flexible, easier to use, and your system will be used more professionally and efficiently.

5

Program Enhancements

In this chapter, I will show you how to polish your basic instrument control program. Most of the improvements will be in the user interface. The user interface is the means by which users interact with a program, communicating specific actions that they want the program to perform.

As a general rule, the easier it is for a user to learn how to operate a program, the less training that you will need to provide. "User friendly" programs—programs that make the user's tasks as easy to discern and accomplish as possible—allow users with little specialized knowledge to use your programs. Programs can be made user friendly by including elements such as menus, prompts to get information from the user, graphics, status messages and help messages. You can also create a system that is able to recover from an error created by a user, the program, or communications with the instrument. This system will make it less frustrating for the users of your program.

This chapter contains instructions on how to create programs that use custom menus, obtain user input, provide status messages, and end the distracting screen flicker that occurs as a macro runs. In subsequent chapters, these features will be expanded upon. This chapter will contain descriptions of how to implement a number of other features as well. One of the features is a way to automatically date-stamp an experiment. This dating will ensure that when a report is printed or filed away, the report will bear the date that the data was actually collected.

Another feature that will be discussed is a system within the program that handles the data once it has been collected. This system includes user friendly ways to store the data in a file and to print it. This chapter will also teach you how to create macro subroutines that can be accessed via your custom menu to complete these two tasks.

In addition to these user friendly features, your program may also need to handle delays between data collection time-points. Many experiments take place over long periods of time and require the data to be taken at specific time intervals. This chapter will teach you how to build time delays into your program to accommodate this requirement.

This chapter will also continue to address the portability issues between the two spreadsheets. As I mentioned in the last chapter, with a few minor modifications, your programs will be able to run from within both Lotus 1-2-3® and Symphony®. Although the Lotus Command Language is the same for both spreadsheets, some program changes are necessary because of the differences in the menu commands that are required to complete certain tasks in Lotus 1-2-3 and Symphony. For example, to erase data from a cell in Lotus 1-2-3, you would issue the / Range Erase (/RE) command. In Symphony, the equivalent command is / Erase (/E). This chapter will show you how to take advantage of the different ways that the two spreadsheets handle auto-executing macros to determine which set of commands to use.

All of the information contained in this chapter will make your macro more flexible, easier to use and your program will be more polished and professional.

Figure 5–1 shows where this chapter fits into the development process.

BRIEF OVERVIEW OF THE ENHANCEMENTS

Figures 5–2 and 5–3 show the program that we have been developing. Figure 5–4 is a view of MENU1 with column widths enlarged.

As you can see, several enhancements have been added. You can apply these enhancements to your program to achieve the goals that I outlined at the beginning of this chapter. I will describe the enhancements briefly in this section and in more detail subsequently.

First is the menu itself. You can create your own custom menus that act just like the menu that appears when you press slash (/). In general a menu (such as the one in Figure 3) is a good means by which a user can switch between different functions of a program.

The next enhancement is the addition of the {LET} commands in the two auto-executing macros of Figure 5–2. These commands specify which Lotus spreadsheet is being used and are part of an extensive system that determines the proper menu commands to issue for the Lotus program being used.

The THERMO program shown in Figure 5–2 has a number of improvements over the corresponding program of Chapter 4. This program was called \R in the last chapter. The \R allowed the program to be started by pressing [ALT] R. This capability is no longer needed because the program can be started from the menu;

```
                STEPS TO A SUCCESSFUL APPLICATION

          Assemble The Instrument
          Test Instrument In Manual Mode
          Determine Wiring Configuration
          Get Cable
          Connect Cable To Computer And Instrument
          Secure Cable
          Determine How To Set Communication Parameters
          Set Module Addresses
          Reset Instrument
          Add PATH Command To AUTOEXEC.BAT File
          Add MODE Command to AUTOEXEC.BAT File
          Design Template And Macro Program
          Start The Spreadsheet
          Type Template Into Spreadsheet
          "Activate" Template
          Test The Template
          Create And "Activate" The Auto-executing Macros
          (Symphony): Specify The Name Of The Auto-executing Macro
          Use The DOS MODE Program To Set Communication Parameters
          Type In A "Bare Bones" Program
          "Activate" The Test Program
          Try To Establish Minimal Communications
          Evaluate The Test Program
          Expand Test Program To Retrieve Data
        ->Add Amenities And Menus To The Program
        ->"Activate/Re-activate" Macro Program
        ->(Symphony): Create Windows
        ->Create A Table Of Range Names
        ->Save The File
        ->Evaluate The Program
        ->Increase Communications Speeds
        ->Final Test The Program
        ->Save/Backup Several Copies Of The File
        ->Save Several Copies In Separate Safe Places
```

Figure 5–1

hence, the program's new, more descriptive name. Using descriptive names for macros is a good way to make programs easier to follow.

The other changes in the THERMO program meet several of the objectives that were set forth previously in this chapter. They are

- Date stamping of the template to provide an accurate record of when the experiment was run.
- Obtaining user input. (In this case, the user will specify the number of data points to collect and the delay between the data points.)
- Providing status reports to the user.
- Eliminating screen flicker.
- Obtaining the precise time that the experiment started. (This information will be used to determine elapsed time for a delay arrangement that pauses execution of the program.)
- Time formatting of the elapsed time column.

```
---------N-------O---------P--------Q--------R--------S--------T--------U--------
 1 /*PROGRAM TO GET TEMPERATURES FROM ACRO-931 THERMOCOUPLE INTERFACE*/
 2
 3                         /*LOTUS 1-2-3*/
 4          /*AUTO-EXECUTING MACRO (USER SETS BAUD RATE, ETC.*/
 5           /*USING THE "System" MENU CHOICE AND MODE.COM)*/
 6 \0       {LET VERSION,0}    /*TELLS PROGRAM IT'S RUNNING UNDER 1-2-3*/
 7          /S        /*ISSUE THE /S MENU CHOICE TO GO TO SYSTEM*/
 8          {MENUBRANCH MENU1}          /*PUT UP USER'S MENU*/
 9
10                          /*LOTUS SYMPHONY*/
11          /*AUTO-EXECUTING MACRO: TESTS TO SEE IF DOS.APP ATTACHED; */
12          /*IF NOT, ATTACHES IT.  THEN SETS BAUD RATE, ETC.*/
13 AUTOEXEC{LET VERSION,1}    /*TELLS PROGRAM IT'S RUNNING UNDER SYMPHONY*/
14          {IF @ISERR(@APP("DOS",""))}{HOOKUP} /*SEE IF DOS.APP ATTACHED*/
15          {SERVICES}AIDOS~MODE COM1:4800,E,7,2~        /*SET COMM PORT*/
16          {MENUBRANCH MENU1}          /*PUT UP USER'S MENU*/
17
18 HOOKUP   {SERVICES}AADOS~Q           /*ATTACHES DOS.APP*/
19
20                /*MACRO FOR THE ACRO-931 THERMOCOUPLE MODULE*/
21 THERMO   {LET DATE,@DATEVALUE(@NOW)}          /*DATE STAMP THE EXPT*/
22          {OPEN "COM1",M}    /*OPEN COM PORT AS DEVICE FILE NAME*/
23          {GETNUMBER "How Many Time Points?",POINTS}
24          {GETNUMBER "How Long Is The Interval Between Measurements?",DELAY}
25          {INDICATE I/O}              /*INDICATE INPUT-OUTPUT SECTION*/
26          {GOTO}START~       /*MOVE CELL POINTER TO STARTING POSITION*/
27          {WINDOWSOFF}{PANELOFF}      /*FREEZE THE DISPLAY*/
28          {WRITELN "#4"}     /*INITIALIZE ACRO-931 (MODULE 4)*/
29          {LET BEGIN,@TIMEVALUE(@NOW)}         /*GET THE STARTING TIME*/
30          {IF DELAY>0}{FOR ROW,0,POINTS,1,BIGLOOP}      |*THESE GIVE MAX *|
31          {IF DELAY=0}{FOR ROW,0,POINTS,1,SMALLLOOP}    |*PERFORMANCE FOR*|
32          {CLOSE}            /*CLOSE THE PORT*/         |*ZERO DELAY     *|
33          {CALC}             /*FORCE RECALCULATION OF THE SPREADSHEET*/
34          {GOTO}START~{LEFT}          /*RESET CELL-POINTER TO START*/
35          {IF VERSION=0}/RFDT3{DOWN POINTS}~  /*TIME FORMAT, 1-2-3*/
36          {IF VERSION=1}/FT3{DOWN POINTS}~    /*TIME FORMAT,SYMPHONY*/
37          {WINDOWSON}{PANELON}        /*THAW THE DISPLAY*/
38          {INDICATE}                           /*FREE INDICATOR BOX*/
39
40                     /*LOOP THAT GETS ONE ROW OF DATA*/
41 BIGLOOP  {INDICATE PAUSE}{PANELON}{PANELOFF} /*INDICATE DELAY SECTION*/
42          {WAIT @NOW+@TIME(0,0,DELAY)}         /*WAIT FOR SPECIFIED TIME*/
43 SMALLLOOP{LET CURRENT,@TIMEVALUE(@NOW)}       /*GET CURRENT TIME*/
44          {LET DELTA,CURRENT-BEGIN}            /*FIND ELAPSED TIME*/
45          {PUT ELAPSCOL,0,ROW,DELTA}           /*PUT ELAPSED TIME IN SHEET*/
46          {PUT PTCOL,0,ROW,ROW}                /*PUT POINT NUMBER IN SHEET*/
47          {WINDOWSON}{WINDOWSOFF}              /*ADD DATA TO SCREEN*/
48          {DOWN}                               /*MOVE DOWN TO NEXT ROW*/
49          {FOR PROBE_NUM,0,7,1,GET_TEMP}       |*GET A ROW'S WORTH OF DATA:*|
50          {INDICATE NEXT}{PANELON}{PANELOFF}   |*NUMBER OF PROBES (COLUMNS)*|
51                                               |*MINUS 1 (CNTR STARTS AT 0)*|
52
53
54
```

Figure 5-2

The BIGLOOP macro produces a delay in a program, using the Lotus {WAIT} command. The final two macros (PRINT and SAVE), print the results and file them away, respectively.

Can you identify each of the program elements from the last chapter? They are still there, but they have become relatively buried in all of the enhancements

```
---------N-------O--------P--------Q--------R--------S--------T--------U--------
 54          /*LOOP TO GET DATA FROM ONE OF THE CHANNELS OF MODULE #4*/
 55 GET_TEMP{WRITELN @INDEX(TMPSTRNGS,0,PROBE_NUM)}/*ASK ACRO-931 FOR A TEMP*/
 56          {READLN BUFFER}    /*READ IN A TEMPERATURE VALUE*/
 57          {PUT RAWDATA,PROBE_NUM,ROW,@VALUE(@CLEAN(BUFFER)):VALUE}
 58                   /*PARSER FOR BUFFER: @VALUE(@CLEAN(BUFFER))*/
 59                   /*RAWDATA RANGE IS C8..J3000*/
 60
 61                   /*MAIN MENU; SEE FIGURE 5-3 FOR DETAILS*/
 62 MENU1    RUN      PRINT    SAVE     COMMENT  QUIT
 63          RUN THE TPRINT THESAVE THE ADD COMMEQUIT THE MACRO
 64          {THERMO} {BRANCH P{BRANCH S{GOTO}COM{QUIT}
 65          {MENUBRANCH MENU1}         {?}~
 66                                     {MENUBRANCH MENU1}
 67
 68          /*PRINT MACRO: SET "REPORT" RANGE; PRINT; RETURN TO MENU1*/
 69 PRINT    {CALC}                     /*FORCE UPDATE OF ALL FORMULAS*/
 70          {WINDOWSOFF}{PANELOFF}     /*FREEZE THE DISPLAY*/
 71          {HOME}            /*GO TO CELL A1;SET "REPORT" RANGE*/
 72          /RNCREPORT~{ESC}{HOME}.{DOWN 7+POINTS}{RIGHT 9}~
 73          {IF VERSION=0}/PPRREPORT~AGPQ                /*LOTUS 1-2-3*/
 74          {IF VERSION=1}{SERVICES}PSSRREPORT~QAGPQ     /*SYMPHONY*/
 75          {WINDOWSON}{PANELON}       /*THAW THE DISPLAY*/
 76          {MENUBRANCH MENU1}         /*RETURN TO USER'S MENU*/
 77
 78 SAVE     {IF VERSION=0}/FS{?}~      /*LOTUS 1-2-3*/
 79          {IF VERSION=1}{S}FS{?}~    /*SYMPHONY*/
 80          {RIGHT}~ /*FORCE OVERWRITE; CANNOT BE ON SAME LINE AS THE FS*/
 81          {HOME}                     /*MOVE CELL-POINTER TO CELL A1*/
 82          {MENUBRANCH MENU1}         /*RETURN TO USER'S MENU*/
 83
 84          /*SCRATCHPAD OF NAMED RANGES*/
 85 VERSION           /* 0=>LOTUS 1-2-3; 1=>SYMPHONY */
 86 ROW               /*COUNTER FOR FIRST LOOP; ROW OFFSET FOR {PUT}*/
 87 PROBE_NUM         |*COUNTER FOR 2ND LOOP; CURRENT PROBE NUMBER; *|
 88                   |*OFFSET FOR @INDEX; COLUMN OFFSET FOR {PUT}  *|
 89 POINTS            /*USER DEFINED NUMBER OF DATA POINTS TO COLLECT*/
 90 BEGIN             /*TIME EXPERIMENT STARTED*/
 91 DELAY             /*USER DEFINED DELAY BETWEEN DATA POINTS*/
 92 CURRENT           /*CURRENT TIME*/
 93 DELTA             /*ACTUAL DELAY, FROM TIME ZERO*/
 94 TMPSTRNGTEMP1     \
 95          TEMP2    |
 96          TEMP3    |
 97          TEMP4    >  /*STRINGS TO INITIATE ACRO-931 INTERFACE*/
 98          TEMP5    |  /*TO GET TEMPERATURES FROM EACH OF THE PROBES*/
 99          TEMP6    |
100          TEMP7    |
101          TEMP8    /
102 BUFFER            /*A BUFFER FOR INCOMING DATA*/
103
104
105
```

Figure 5-3

that have been added. If you are having trouble finding them, review "The Structure of a Program" section in Chapter 4.

This new program is a perfect illustration of why you should program and test in small sections and then expand. If you were to start with a program of this complexity and you had a communications problem, it would be much more difficult to pinpont the cause of the problem and correct it.

Now that you have had an overview of what will be discussed, create the example program and look at the specifics.

WHERE TO START

You can start the process of learning how to use the enhancements by creating the program shown in Figures 5–2 through 5–4. Use the spreadsheet from the last chapter and modify it to include the enhancements.

One quick way to add the enhancements is to insert enough empty rows within the existing macros to add the new command lines. Position the cell-pointer at the line of the program where you want to add rows. In Symphony, issue the / Insert Row (/IR) command. In Lotus 1-2-3, issue the / Worksheet Insert Row (/WIR) command. Next, arrow down the number of rows that you want to add and press return. Then type in the additions or changes.

Adding or deleting rows sometimes invalidates the labels that name macros and scratch-pad cells. Because of this possibility, you should develop a policy of always resetting range names any time that you add or delete rows. As you become more experienced, you will be able to identify the situations that invalidate label names and can cut back on this policy accordingly.

To reset old labels and add new ones, issue the / Range Name Label Right (/RNLR) command. When the spreadsheet prompts you for a range of labels, press ESCape, move the cell-pointer to the cell containing the first macro name, type a period (.), use the down arrow key to highlight all of the macro names AND the "scratch-pad" names, and press return. This sequence will overwrite the old ranges and update them to the new ones. Your next step is to assign names to the special cells and ranges in your program and template.

Begin by creating a range named TMPSTRNGS in the indicated area of the scratch-pad. Move the cell-pointer to the cell containing the TEMP1 string. Issue the / Range Name Create (/RNC) command. When prompted for the name of the range, type TMPSTRNGS. When prompted for a range of cells, type a period (.) and arrow down to the cell containing the TEMP8 string. Press return. TMPSTRNGS is a named range of cells that contains the string commands that will be sent out to the ACRO-900.™

Repeat the procedure described above to create a range named RAWDATA. When prompted for a range of cells, type C8. .J3000 and press return. Repeat

```
--------N--------O-----------------P-----------------Q-----------------R----------------------S-----------T--------U--------
60
61
62 MENU1    RUN               PRINT             SAVE              COMMENT                QUIT
63          RUN THE TEST      PRINT THE RESULTS SAVE THE FILE     ADD COMMENTS/CONDITIONS QUIT THE MACRO
62          {THERMO}          {BRANCH PRINT}    {BRANCH SAVE}     {GOTO}COMMENTS~        {QUIT}
63          {MENUBRANCH MENU1}                                    {?}~
62                                                                {MENUBRANCH MENU1}
63
62
```

Figure 5–4

the procedure three more times to create ranges named REPORT, PTCOL and ELAPSCOL for ranges A1..J7, A8..A3000, and B8..B3000, respectively. RAWDATA, REPORT, PTCOL and ELAPSCOL refer to the range for incoming raw data, the range to be printed, the range for the time point number, and the range for elapsed time, respectively.

Two other cells require names. The first cell is the date cell. Issue the / Range Name Create (/RNC) command. When prompted for a range name, type DATE. When prompted for a range of cells, type D4 and press return. The second cell is the comments cell. Repeat the procedure to name cell D3 COMMENTS.

To ensure that all of the ranges required for your program are created and have the appropriate cells assigned within the ranges, create a table of the named ranges. To make a range table in both Lotus 1-2-3 and Symphony, issue the / Range Name Table command (/RNT). When prompted for the location of the table, press ESCape and move the cell-pointer to one of the cells below the macro and the scratchpad. Press return. The table should look like the one shown in Figure 5-5. (I added the RANGE TABLE heading and moved the bottom half of the list to the next column.)

Creating a table of range names is an invaluable piece of documentation that supplements normal commenting. It is also an irreplaceable tool to ensure that you have correct cell specifications for all ranges.

The table in Figure 5-5 contains some extra names. For example, the new program no longer uses the PARSED cell, and the name of the \ R macro has been changed to THERMO. Therefore, these two ranges are no longer needed. Remove the extra names with the / Range Name Delete (RND) command.

Another step in cleaning up a program is to change the format of cells that require special formatting. In this example, you would need to format the BEGIN, DELAY, CURRENT, and DELTA cells. Formatting will make the cells more read-

```
---------N--------O--------P--------Q----------R--------S--------T--------U-----
105                               RANGE NAME TABLE
106                               ================
107                    AUTOEXEC O13          PRINT     O69
108                    BEGIN    O90          PROBE_NUMO87
109                    BIGLOOP  O41          PTCOL     A8..A3000
110                    BUFFER   O102         RAWDATA   C8..J3000
111                    COMMENTS D3           REPORT    A1..J7
112                    CURRENT  O92          ROW       O86
113                    DATE     D4           SAVE      O78
114                    DELAY    O91          SMALLOOP  O42
115                    DELTA    O93          START     C8
116                    ELAPSCOL B8..B3000    THERMO    O21
117                    GET_TEMP O55          TMPSTRNGSO94..O101
118                    HOOKUP   O18          VERSION   O85
119                    MENU1    O62          \0        O6
120                    POINTS   O89
121
122
123
```

Figure 5-5

able. With Lotus 1-2-3, use the / Range Format Data Time 3 command. In Symphony, use / Format Time 3.

To make the template less cluttered, erase the statistical functions at the bottom of, and to the right of, the data section of the spreadsheet template. In Lotus 1-2-3, use the / Range Erase command; in Symphony, use the / Erase command.

The next section examines the program enhancements in detail.

PROGRAM PORTABILITY

One of the goals of this chapter is to show you how to use your programs for both Lotus 1-2-3 and Symphony. This capability is called "portability" and allows you to share your program with colleagues or upgrade Lotus 1-2-3 to Symphony without having to perform an extensive rewrite.

As stated previously, the two spreadsheets have minor differences in the way that some of the menu commands are issued; but the macro commands and @functions are identical. Therefore, if you can detect which Lotus spreadsheet is using a macro program, you can issue the appropriate menu commands. The {LET} commands in each of the two auto-executing macros of Figure 5–2 help you to do just that.

Recall that Lotus 1-2-3 can only use a macro named '\0 as an auto-executing macro; on the other hand, Symphony is configured to use AUTOEXEC as its auto-executing macro. If a spreadsheet is retrieved by Lotus 1-2-3, the {LET} command in '\0 will execute and place a zero in the VERSION cell. Likewise, if a spreadsheet is retrieved by Symphony, the {LET} command in the AUTOEXEC macro will execute and place a one in the VERSION cell.

This system, then, detects the Lotus program being used with the macro and places the information in a cell. Later, you can use this information and Lotus {IF} commands to decide which macro subroutines to execute or which sets of menu commands to issue. (See Chapter 3, "Creating an Initialization Macro for Symphony" for a discussion of the {IF} command.)

With this system, you may need to redefine a few graphics or print settings and rename the file extension (e.g., from ".wk1" or ".wr1" when you go from Lotus 1-2-3 to Symphony), but you will not have to change the program.

The THERMO program in Figure 5–2 contains an example of how this system acquires information that can be used to issue the appropriate set of menu commands to time format a column of the template. The two {IF} commands near the bottom of the macro determine whether the program is running under Lotus 1-2-3 or Symphony and then issue the appropriate commands. As explained above, the {LET VERSION,0} and {LET VERSION,1} commands in the auto-executing macros will place a 0 or 1 into the cell named VERSION, indicating Lotus 1-2-3 and Symphony respectively. By testing for a zero or a one in the VERSION cell with the {IF} commands in THERMO, the program can issue the appropriate set of commands to format the elapsed time column.

{IF} commands can be used in many ways to achieve portability. The PRINT and SAVE macros in Figure 5-3 are two other examples and are described in detail later in this chapter. Future chapters contain numerous examples of other implementations that you can use in your programs.

STARTING MENUS FROM AUTO-EXECUTING MACROS

A feature that you will want to include in your program is one that allows users to retrieve a spreadsheet and be automatically presented with a menu of all functional options available in a program. A program that automatically takes care of all details for the user without the user having knowledge of them is very user friendly. The following discussion describes how this system works.

After VERSION has been updated and the communications parameters have been set in the auto-executing macros, the {MENUBRANCH MENU1} command displays the menu called MENU1 in the control panel. The custom menu (called MENU1) looks and works just like the ones that Lotus 1-2-3 or Symphony presents to you when you press the / key.

Each menu option has its own capsule description (help line) that is displayed when the menu option is highlighted with the menu-pointer. Each menu option is also tied to a macro routine. When a user selects a menu option, the corresponding macro routine executes.

You can execute a menu in two different ways. The first way is to use the {MENUCALL menu_name} command; the second is to use the {MENUBRANCH menu_name} command. These two commands are similar. They both pause macro execution, go to a specified location, and display the menu at the location.

However, each command treats a menu differently. {MENUCALL} treats the menu like a subroutine. That is, it returns to the main macro when the subroutine instructions in the menu are completed. With {MENUBRANCH}, the macro "jumps" to the menu. Execution terminates in the original macro and total control transfers to the menu. Then, the menu executes. When the last instruction in the menu is encountered, macro execution will terminate (unless another branching command is encountered).

As you can see, a significant difference exists between the two menu commands. This dissimilarity can be important because Lotus limits the number of subroutine levels that you can call within a program. A more detailed explanation follows.

If a macro calls another macro as a subroutine, Lotus is supporting two "levels" of subroutines. Likewise, if a second subroutine calls a third subroutine, Lotus is supporting three levels, and so on. The number of subroutine levels that either Lotus 1-2-3 or Symphony can support (without returning to the first macro level) is 31.

If you exceed the subroutine limit by calling a macro from itself and/or by trying to "nest" too many subroutines, the macro aborts and an error message appears. Because a {MENUCALL} acts just like a subroutine call, the same problem would occur if you tried to repeatedly reuse a menu from a {MENUCALL}.

For the reason outlined above, it is advisable to use the {BRANCH} command whenever using macros and the {MENUBRANCH} command whenever using a menu.

MENUS

Figure 5-4 is an example of a menu. Note that this figure is an expansion of the column width of the program menu that appears in Figure 5-3. Figure 5-4 allows you to view details of the menu more easily. MENU1 is a simple menu. Take a moment to examine this menu so that you will know how to construct a menu for your own application.

All custom menus are structured with the same basic format. The first row of cells defines the options that will appear when the menu is displayed. The text for the first option must be in the cell specified in the {MENUBRANCH} or {MENU-CALL} command. The text for the other options in the menu must be in the same row and in the columns immediately to the right of the column that contains the first option. If you want to be able to select a menu option by pressing the first letter of a prompt, you must designate the first letter of each option as a unique uppercase letter, a digit, or a symbol from the keyboard. Option prompts can be of variable length. Symphony and Lotus 1-2-3 automatically place two empty spaces between prompts. The length of the prompts plus spaces must not exceed 80 characters. If the total exceeds 80 characters, the spreadsheet truncates all entries longer than a specific length. The length depends on the number of prompts used.

Your menu can have a maximum of eight selections. If you try to use more than eight selections, the spreadsheet will ignore them. If you have fewer than eight selections, the cell to the right of the last menu choice must be empty. An empty cell signals the end of a menu list to Lotus.

The row immediately below the options contains the prompt's help message. The menu displays a help message when an option is highlighted. The longest possible help message is 75 characters. The spreadsheet will ignore all additional characters when a message is displayed.

The cells below a help message contain the instructions that the menu choice should execute if a user makes that particular choice. These cells may contain a macro, a subroutine cell, a branch to a macro, a branch to another menu, or a branch to the same menu.

CALLING A SUBROUTINE

The cells below help messages of a menu may contain subroutine calls. A subroutine call from a menu is handled in the same fashion as a subroutine call from a macro program: Control of the program is temporarily shifted to the subroutine, the subroutine carries out the tasks it was programmed to perform, and then control is returned to the menu.

The "Run" menu choice in MENU1 is an example of this concept. This menu choice calls the THERMO macro program as a subroutine. Notice the method that is used to call the subroutine. Whenever you call a subroutine from either a macro or a menu, place the name of the subroutine in { } brackets.

For example, to call THERMO, type the following text into your program:

```
{THERMO}
```

If your macro has a backslash-letter name (e.g., \R), type the following text into your program to call the macro subroutine:

```
{\R}
```

In the example, after the THERMO program has completed its execution, program control returns to the menu. The {MENUBRANCH} command at the end of this menu choice is required to continue with the program. Without it, the program would terminate. When the program executes the {MENUBRANCH}, it branches back to the menu and displays the menu again so that the user can make another selection.

OTHER PROGRAMMING TECHNIQUES FOR MENUS

The other menu choices will show you other methods that you can use to program a menu. The first way is to enter commands starting at line three of the menu. In this case, the commands form a macro of their own. The COMMENTS choice is an example of this method. This macro uses the {GOTO} command to position the cell-pointer at the COMMENTS cell and then prompts the user for input.

The Print menu choice shows you another way to execute a macro from within a menu. Print uses the {BRANCH} command. The {BRANCH} command is similar to the {MENUBRANCH} command, except that it branches to a macro program instead of a menu. Like {MENUBRANCH}, {BRANCH}, permanently transfers control of the program to the cell named in the {BRANCH} command. Once control has been shifted to the new cell, it does not return to the original macro unless another {BRANCH} is issued in the macro to go back to the original macro. A subroutine call is different because the macro will automatically return to the calling program. I prefer {BRANCH}ing as opposed to subroutine calls because it allows more levels of macro nesting, thereby allowing programs to be longer and more efficient.

At the end of the PRINT macro is a {MENUBRANCH} command. This command is needed to return to MENU1 because, unlike the Run menu choice that calls THERMO as a subroutine, the Print menu choice {BRANCH}es to the PRINT macro. If {MENUBRANCH} were omitted, the program would terminate after the macro was completed.

QUITTING A MENU

Always include a Quit menu choice at the end of a menu. This choice allows a user to abort the menu's execution. One of the most important reasons for using a spreadsheet is that it allows a user to explore data, manipulate it, graph it, perform statistics on it, and perform unique operations that a menu does not provide. A Quit menu choice allows a user to utilize these capabilities.

The {QUIT} command that appears under the example menu choice terminates all macro and menu execution and returns the user to the spreadsheet.

OBTAINING INFORMATION FROM THE USER

The THERMO program is the \R macro from the last chapter. It has been upgraded to illustrate programming methods that achieve some of the goals outlined at the beginning of this chapter.

For example, a method has been added to the program to prompt the user for input. To illustrate how to accomplish this, a pair of prompts were added to first ask the user how many time-points to collect and then to ask what interval to use between the time-points.

The command that you use to obtain numbers from a user is the {GETNUMBER} command. For text, use the {GETLABEL} command. The syntax of these two commands is

```
{GETNUMBER prompt,location}
{GETLABEL prompt,location}
```

When either of these two commands is executed in a macro, a prompt is displayed on the second line of the control panel and execution of the macro is paused until input is made. When the user presses a carriage return to the keyboard, the input is placed into the cell specified by "location". In the example program, the number of time points will go into the cell named "POINTS" and the interval will go into the cell named "DELAY".

User input obtained from a {GETNUMBER} or {GETLABEL} command can be used later by your program. For example, in this program the "POINTS" number will set the limit for the main loop counter and the "DELAY" number will set the delay in a {WAIT} command. (A description of how this information is incorporated is given below).

The COMMENTS menu choice in Figure 5-4 illustrates another important programming tool that you can use to get input from a user. It is the {?} command.

The {?} command halts macro execution, allowing the user to type any combination of keystrokes, function keys, or arrow keys (to move the cell-pointer). The macro will continue after the user presses return. This allows the user to input text or numbers directly into cells anywhere in the spreadsheet.

Pressing return has no effect other than to end the {?} command. That is, pressing return with the {?} command will NOT enter data into a cell. To enter user input, you must include a tilde (~) after the {?}. Without a tilde after the {?} command, the data that were input may be lost.

PROVIDING STATUS REPORTS

Another feature to include in your program is one that provides users with information about the status of a program as it executes. This information is especially important if your program takes a relatively long period of time to complete its tasks.

The next section of this chapter will explain how to eliminate the screen flicker caused by the control panel and the spreadsheet as they update each time a macro command is issued. If your program takes a relatively long period of time to complete a task and if you have "frozen" the panel and spreadsheet, it may appear to the user that nothing is happening; the user may even think that something has gone wrong or that the program has stopped without notice. Therefore, in these instances you will want to signal a reassurance to the user that the program is still working correctly.

Alternatively, there may be times when you want to give a short prompt to the user to press a start button, load a sample, etc.

The most convenient way to provide status reports and short prompts is with the {INDICATE} command. The {INDICATE} command will place a short string of characters into the mode indicator box. The mode indicator box is the small square box at the top right corner of the display.

The syntax of this command is

```
{INDICATE string}
```

The string specification is optional. If used, it can be up to five characters long. If omitted, the {INDICATE} command will free the indicator box so that Lotus 1-2-3 or Symphony can regain command of it. When you use the {INDICATE} command with a string, the string remains on the screen, regardless of any subsequent changes that Lotus would normally make. For example, in Figure 5-2, the {INDICATE I/O} , {INDICATE PAUSE}, and {INDICATE NEXT} commands in the macros place status messages in the indicator box.

It is important that the control panel be "on" before the mode indicator is updated. Therefore, if you have issued a {PANELOFF} command to freeze the panel (see below), you must issue a {PANELON} command after the {INDICATE} command to update the mode indicator. You will then have to reissue a {PANELOFF} command to "re-freeze" the panel. Use the {INDICATE string} {PANELON} {PANELOFF} instruction judiciously because turning the panel on and off takes time and will tend to slow your program's execution.

ELIMINATING SCREEN "FLICKER"

To a macro programming novice, watching a program execute can be very exciting. However, after you have worked with a macro program for a short period of time, you may become tired of or annoyed at the "flickering" that takes place as the macro executes. "Flickering" is caused by the panel and the spreadsheet updating during normal macro execution. More importantly, flickering will substantially slow the execution of your program.

You can use a {WINDOWSOFF} {PANELOFF} combination to freeze the display. If you freeze the display, you will eliminate the "flicker".

If you use this combination to freeze the display, then you will need to issue a {WINDOWSON} command for any new data to be displayed and the screen refreshed. If you want to proceed with the macro, follow the {WINDOWSON} command with a {WINDOWSOFF} command to freeze the display again.

GETTING THE DATE AND TIME

As previously mentioned, placing a date-stamp on the template is an invaluable aide to keeping records straight. In some experiments you may also want to keep track of the time of day.

In fact, in many experiments, the time of day can be a very useful piece of data. For example, determining the exact moment when a certain observation was made could be an important piece of data if you were collecting data that determined the rate of change of an experiment per unit of time. One instance of this type of experiment is in the area of kinetic studies. In these studies, a user needs to have an accurate system for determining the exact time that a particular data point was taken.

The two {LET} commands in the THERMO macro in Figure 5-2 illustrate how to get the current date and starting time of an experiment. Getting the time at the beginning of an experiment is the key to determining "true" elapsed times. The syntax of this command is:

```
{LET BEGIN,@TIMEVALUE(@NOW)}
```

This command begins by evaluating the innermost @function, @NOW. The @NOW function returns a decimal serial number that represents the current date and time (as they are known to DOS). The @NOW function requires no arguments. The serial number that it returns is based on the date and time you entered when you started your personal computer. If you have a built-in clock/calendar card, the @NOW uses the date and time from the card.

The @TIMEVALUE function in the {LET} command translates the value that is returned by the @NOW into a *time* serial number. The time serial number represents the time of day as a decimal fraction between midnight (00:00:00) and

11:59:59 p.m. (23:59:59). These fractions begin at 0.0000000000 and end at 0.9999884259.

After the time value has been determined by the example program, the {LET} command places the final value into the cell named "BEGIN".

The {LET} that places the *date* in the template works similarly. This {LET} gets the current date and places it into the cell named "DATE". Thus, the experiment is automatically date-stamped each time the THERMO macro is run.

The reason that you want to use the {LET} command from within a program (instead of just placing the @functions directly into the cells) is simple: Placing a {LET} within a program that generates data "freezes" the correct result in the spreadsheet. This result is more desirable than placing @functions in cells because the @functions update each time the spreadsheet updates. Similarly, a {LET} command that has been placed in an auto-executing macro updates the date and time whenever the spreadsheet is retrieved; which may not necessarily coincide with when the data was actually collected.

DETERMINING ELAPSED TIMES

Another important use for time functions is to determine elapsed times. The SMALLOOP program in Figure 5-2 shows you how to implement a system that determines elapsed times. SMALLOOP uses a {LET} command to obtain the current time. Getting the time directly from the personal computer and from within the loop improves accuracy. If you were just to use the amount of time specified in commands that delay the execution of a program, you would not account for the time that the remainder of the program takes to execute.

The first {LET} command in SMALLOOP places the current time into the cell called CURRENT. The second {LET} command in SMALLOOP calculates elapsed time by subtracting the current time from the beginning time. The subsequent {PUT} command places the difference into the appropriate cell of the template. The loop counter (ROW) corresponds to the row offset for the current time point.

USING THE {WAIT} COMMAND TO GENERATE DELAYS

Another example of programming that relates to time is one that places a delay in your program before a certain task is initiated. In this case, a "timer" is set in the program and when the timer goes off, the program gathers data.

The {WAIT} command performs this function. As we shall see later in this section, the {WAIT} command suspends all processing until a certain time has been reached. It then continues with the next program step.

One use of the {WAIT} command is in kinetic studies. Kinetic studies are commonly performed in enzyme assays. In this situation, you could establish a program which "waits" for an enzyme's lag phase to complete before taking readings. You could then take timed readings (say every 20 seconds) over a specified interval. By knowing the exact time that each of the measurements was taken, you could calculate both elapsed time and the rate of change per unit of time.

Another example of the use of the {WAIT} command is for building delays into your experiments. Some experiments take place over long periods of time and require that data be collected only infrequently. If an experiment is of this nature, then you will need to use {WAIT} to slow down the data collection process.

The {WAIT} command has many other uses. Later in this book, you will learn how to use a {WAIT} command to build a time delay into a program that will "wait" for an instrument to catch up to the program.

Many instruments ignore commands that enter their communications ports while they are in the middle of one of their operations. For example, consider an instrument that automatically withdraws samples from test tubes and takes 20 seconds to do so. If you send an instruction to the instrument during this time period, the instrument would ignore it. However, if you place a 21-second delay into your program, you would be sending the instruction to the instrument at a time when the instrument would be able to "listen" to it.

Similarly, some instruments are robotic. That is, you can control their motors from the personal computer. By using {WAIT} commands between commands that you send to the instrument, you can orchestrate motors into performing a precise sequence with the exact timing that you want.

Some instruments have small buffers to store commands that they receive from their RS-232 ports. These buffers are easily overrun, especially at fast Baud rates. By using a {WAIT} between sending commands to an instrument, you can slow the transmission and thereby ensure that the instrument does not receive garbled messages.

As you can see, {WAIT} is a very powerful, yet simple, command. Take a look now at how the {WAIT} command is used in the example program.

In the example program, {WAIT} is used to build a delay into the reading sequence. This delay will allow the long-term temperature variability to be monitored in the processes that are being tested. If readings were taken every second, the amount of data would be massive. By using a {WAIT} command, the amount of data is reduced and data manipulation is manageable.

If a delay is specified by a user of the example program, the program uses the BIGLOOP subroutine. Once in BIGLOOP, the {WAIT} command executes. This line of the program is one that you may use often, so it is explained in detail below.

The command begins by getting the current time using the @NOW function. It then evaluates the @TIME function and adds its value to the value of @NOW. This new value is a serial number for a future timepoint.

The syntax of the @TIME function is @TIME (hours,minutes,seconds). The @TIME function translates the three arguments into a decimal fraction equivalent to the time of day. Like the @TIMEVALUE function, this decimal fraction is between 0.0000000000 (midnight) and .9999884259 (11:59:59 p.m.).

This function can be used in a rather unorthodox way. It can translate delay time from seconds (or minutes) into a form that can be added to the @NOW's. For example, if you were to add 5 (for 5 seconds) directly to the value from @NOW you would be adding 5 days! The @TIME function, therefore, is essential because it translates the number of seconds that have been conveniently specified by the user into the decimal fractions that the program can add to @NOW's time.

Once the serial number is determined by addition, the {WAIT} portion of the command takes over. The {WAIT} command suspends all processing until the serial number of the personal computer's internal clock reaches the calculated serial number. All processing will be suspended until the time indicated by the serial number argument has been reached.

It is important to realize that the {WAIT} command is performing a "wait until a particular time" and not a "wait for a certain amount of time". From a user's point of view however, these two delays appear the same.

PRINTING A REPORT WITH A MACRO

In every program that you write, you will want to have a means to automatically print a report of the data that has been generated in the experiment. The Print menu choice offers you one way to implement this capability in your programs. It also illustrates another example of how to ensure Lotus 1-2-3/Symphony portability.

The Print menu choice {BRANCH}es to the macro named PRINT. The PRINT macro begins by setting the cell-pointer to cell A1 (home). The REPORT range is then reset to include all of the data points in the template. It is necessary to reset the template because the amounts of data that can be collected will vary (because of the {GETNUMBER} input). The / Range Name Created command is used in the program. The following sequence highlights and selects the entire data range and is based on the number of data points that a user specifies in the {GETLABEL} command:

```
{ESC}{HOME}.{DOWN 7+POINTS}{RIGHT 9}~
```

Alternatively, the following sequence can also highlight the range:

```
{ESC}.{DOWN 5}{END}{DOWN}{END}{RIGHT}~
```

This last method is faster. However, if you are not extremely careful, this method can be dangerous and you may end up with a range that goes to the bottom of the spreadsheet. In other words, the program may print reams of blank sheets of paper.

If you choose the second method, you need to initialize the REPORT range to span at least two cells. This is a requirement of the {ESC} in the sequence: If you do not pre-set a "dummy" range for REPORT, the {ESC} is interpreted to mean that the range prompt should back up to the previous menu. If you recall, this requirement is why the range was set to cells A1..J5 at the beginning of this chapter.

Once a range has been initialized, you will not have to respecify it; the program will do it automatically.

This macro also uses two {IF} tests to determine whether a spreadsheet is running under Lotus 1-2-3 or Symphony and calls the appropriate subroutine to perform the print task. Before printing actually begins (the AGP portion of the command), the print source range is updated to the current REPORT settings (the SSRREPORT~ portion of the command).

SAVING A FILE WITH A MACRO

Another feature to offer in your menu system is a way to save the data in a file. This feature will allow your users to store the data away in a file for future recall. The SAVE macro shown in Figure 5-3 is one of several methods that you can use to save a file. Future chapters will show you some other methods that will allow you to save just a template (or data) portion of the spreadsheet. By saving just the template containing the data, you will conserve disk space. Future chapters also contain information about retrieving stored files.

The SAVE macro has a construction that is similar to the PRINT macro. The SAVE macro contains a set of {IF} commands like the PRINT macro. These {IF} commands test to see which set of commands should be issued to save the file. If Lotus 1-2-3 is being used, the familiar /FS command is issued; if Symphony is being used, the familiar Services FS ({S}FS) command is issued. This example again shows how to provide portability between the two Lotus programs.

This macro also contains the presence of two curious commands, the {RIGHT}~ command and the {HOME} command. You need to include this set of commands in your macro to bypass the prompt that asks a user to confirm an overwrite of an existing file. The {RIGHT}~ command forces the macro to overwrite an existing file.

The requirement for this command will become apparent to you if you recall what happens when you try to overwrite an existing file. In Lotus 1-2-3, a prompt asks (left to right): Cancel or Replace; in Symphony, the prompt is: No or Yes.

If you issue an R~ or Y~ in your macro command and the file did not exist, then you would place the R or Y into the cell currently highlighted by the cell-pointer. This would overwrite the contents of the cell, thus destroying any existing data. You can issue the {RIGHT} command to highlight Replace or Yes in the menu and then issue the tilde (return) command. That way, if a file did not exist, the cell-pointer in the spreadsheet would harmlessly move one cell to the right, a return would be issued, and no data would be destroyed.

Issuing the {RIGHT} command with filenames that do not exist will eventually cause a problem if left unchecked. The cell-pointer will move one cell to the right for each time that you select the Save menu option. Soon, your spreadsheet would "walk" to the right unless you reposition the cell-pointer.

The function of the {HOME} command is to move the cell-pointer to cell A1. You can also use the {LEFT} command in this macro to reposition the cell-pointer. However, if the cell-pointer was already at Column A, Lotus would have beeped because the cell-pointer would not be able to be moved left any further.

It is critically important that you do *not* type the {RIGHT}~ command in the same cell as the File Save commands. Due to a peculiarity in the way that macros run in Lotus 1-2-3 and Symphony, if the {RIGHT}~ command is in the same cell as the File Save commands, the file will not be written to the disk.

If you feel uncomfortable with this feature of overwriting without confirmation, you can replace the {RIGHT}~ command with a second {?}~ command to cause the macro to pause until the user specifies which action to take. In this case, if a file does not exist, then you will have to press return twice before the macro will continue.

PROGRAMMING FOR MAXIMUM PERFORMANCE

The rest of this chapter will be dedicated to programming tips. The first tip relates to timing within a program.

Whenever you build a delay into a program, you should provide a means of bypassing the delay altogether. By doing so, you provide a way for the user to get the data as expeditiously as possible. The two {IF} commands in the THERMO macro illustrate a method that you can use to provide this feature to users that need the minimal possible time between data points on occasion.

The {WAIT} command in BIGLOOP takes a second or two to execute (depending on the speed of your personal computer). Therefore, the minimal time to run through one loop is one to two seconds. By testing whether a user wants the delay or minimal time, you can bypass this interval altogether. If the user wants a delay, the loop starts at the beginning of the BIGLOOP macro. If the user wants maximal performance, execution of the loop begins after the {INDICATE} and {WAIT} commands.

Cutting into the middle of a macro is a very useful virtue of the way that programs are executed in Lotus. If you recall, a macro executes until it encounters a blank row. However, if a program calls a subroutine that starts at the middle of another macro, then the subroutine will begin at that point and will ignore whatever is above it.

This technique can be used to cut down on the amount of program cells that must be entered. If a section of a macro subroutine performs a certain task that is the same as the task that is needed elsewhere in a program, you can create a common task at the bottom of the macro, call a subroutine that starts in the middle of the macro, and begin execution at that point.

MAKING YOUR JOB A LITTLE EASIER

Whenever you prepare a program, make the programming as flexible as possible to allow for future modification. That is, if you need to modify a program in the future, you will want to be able to implement the update with a minimum of changes. As a general rule, the fewer changes that are required to implement an update, the less chance of forgetting to edit something that will affect the program's performance. One very good programming technique can be seen in the {PUT} commands that appear in the SMALLOOP and GET_TEMP macros of Figures 5-2 and 5-3.

If you recall, in the last chapter cell addresses were placed into the {PUT} command for the temperature data. Notice that named ranges are now specified in the {PUT} commands: PTCOL translates into (is a synonym for) A8..A3000 and ELAPSCOL translates into B8..B3000.

You should get into the habit of using named ranges instead of cell addresses for {PUT} and {LET} commands. Using named ranges will make your program easier to read and to follow. RAWDATA is certainly more informative than C8..J3000. Names become very important in large spreadsheets and/or macro programs.

Perhaps more importantly, if you change a template (by moving cells, inserting or deleting rows or columns, etc.) and the {PUT} and {LET} commands of your program contain cell addresses, all the addresses must be edited. However, if you change the cell address specifications of a named range and if named ranges have been specified in macro commands, then you do not have to change a thing. For example {PUT C8..J3000,...} would need to be edited if you moved the range over by just one column, but {PUT RAWDATA,...} would not. If you have a large program with many macro commands, editing is not only a massive amount of work but a prime target for bugs in your program.

The {PUT} command in GET_TEMP is another example of using a range name in a macro command. In this case, RAWDATA replaces C8..J3000.

WAYS TO GET MORE EFFICIENCY FROM {PUT} AND {WRITELN}

The {PUT} command in the GET_TEMP macro illustrates another programming technique to use routinely. In the last chapter we created a cell (PARSED) that parsed a string of characters from the instrument and found a value for the data that we were interested in. The {PUT} command then placed the value of PARSED into the appropriate cell.

The {PUT} command in the example program of this chapter is quite different. Notice that the reference to the PARSED cell was removed and replaced with a formula that calculates the value directly from within the {PUT} command. The formula used is the one developed in the last chapter.

You have already seen how to "nest" @functions to parse strings of characters. This example shows another type of "nesting." In this case, an @function is "nested" inside a macro command. By using this technique, you can eliminate the PARSED cell, the {RECALC} command, and the inefficiency associated with having to locate and recalculate intermediate cells each time you need their values. The direct benefits of nesting are the conservation of some memory and an increase in the speed of the program.

The {WRITELN} command has similar improvements. In the last chapter, this command used an @CHOOSE function to determine the string that was going to be sent to the instrument. This command was rather long and hard to modify or troubleshoot. In fact, if the character command strings in the @CHOOSE were made much longer, the {WRITELN} would exceed the size limitations of the cell.

The improvement uses the named range TMPSTRNGS with an @INDEX function. @INDEX is a very useful @function and runs more efficiently than @CHOOSE. Compared to the alternatives, programs that use @INDEX are quite fast. This @function is also easier to fit on a page than @CHOOSE.

The syntax of @INDEX is

```
@INDEX(range,column_offset,row_offset)
```

This @function pinpoints a cell located within a range based on the specific column and row offsets and returns the value of the cell. Using @INDEX, you can access any cell within a two-dimensional array (matrix) and get its contents. In this example, TMPSTRNGS has only one column. The column and row offsets in an @INDEX function begin at zero,zero. Thus, the column offset in the example @INDEX is set to 0. The row offset is the loop counter, PROBE_NUM.

Using PROBE_NUM as the row offset allows you to increment the offset and thereby change the string sent out to the communications port. For example, when PROBE_NUM is zero, TEMP1 is sent out to the port. When PROBE_NUM is 1, TEMP2 is sent out to the port, and so on.

Using a table to select instrument command strings presents many other benefits. A table makes a program more organized, more "readable", easier to follow, and easier to change. These benefits present a powerful incentive to use @INDEX functions and tables routinely.

CRANK UP THE SPEED

After you have your program running successfully, you can increase the Baud rate in a stepwise fashion until you no longer get reliable communications. Then, return to the last Baud rate that gave acceptable performance. This speed test will optimize the performance characteristics of your application.

If you change the Baud rate of your instrument, it is crucially important that you turn your instrument off momentarily and then turn it back on. Because most instruments check the values of their communications parameters only at startup, the old values will be used unless the instrument is restarted again. Resetting an instrument will ensure that it is using new communications parameters.

Likewise, if you change the Baud rate of your instrument, make certain that you also change the Baud rate in your personal computer; otherwise, you may be misled into concluding that the new Baud rate is not working correctly. It is easy to forget to change both your instrument and personal computer when increasing Baud rates to optimize performance. Be extra careful during this phase of your program development.

TEST IT!

A template and macro program that have not been thoroughly tested under actual experimental conditions should not be used to report results. Important decisions may be used on a report that looks formal, but contains erroneous results. For this reason, you should compare the results of the macro program with your manual method, the instrument's display, etc. You should also compare the results of the calculations in your template with the results that you obtain from a calculator, etc.

ANALYSIS AND DOCUMENTATION—THE EASY WAY

The Cambridge Spreadsheet Analyst™ (available from Turner Hall Div., Symantec Corp., 10201 Torre Ave., Cupertino, CA 95014) is a program that will help you document, test, validate, analyze and troubleshoot your Lotus 1-2-3 and Symphony spreadsheet templates and macro programs. This relatively inexpensive program analyzes spreadsheets for 25 different types of error conditions, such as formulae that refer to blank cells, formulae that reference labels, overlapping named ranges, and circular references in formulae (formulae that are dependent on one another, such dependencies creating endless loops). Thus, the program helps you to avoid costly spreadsheet mistakes.

The program provides you with detailed reports in areas such as

- maps of cells
- direct usage and downstream impact
- global settings
- overlapping named ranges
- unreferenced name ranges
- cells not referenced in any formulae
- ERR sources in cells
- cells with a value of ERR
- formulae with questionable references

- invalid formulae
- formulae with reversed ranges
- potential errors
- named range analysis
- ranges with more than one name
- cross references
- cell contents with formats

In addition to the above, The Cambridge Spreadsheet Analyst can analyze your macro programs and graphically map a tree of the nesting, operation and interaction of all macros in your program so that you can better understand the commands that influence your spreadsheets.

These reports will make nice additions to your documentation package.

BACK UP SEVERAL COPIES

Once you have added all of your enhancements and completed final testing, your spreadsheet will represent a considerable investment in time. You should protect this investment. Save the spreadsheet (without data) on several floppy disks. Also, print out the program for hard copy documentation. Save all of the items in several different, safe places.

WHAT'S NEXT?

This chapter is the conclusion of the discussion of the basic programming necessary to get data out of an instrument that takes prompts for its data. It also concludes the use of the ACRO-931 Thermocouple Interface as an example. The next chapter will show you how to make a few adjustments to the program that we have developed to allow it to run analog-to-digital converters. The program will then be able to handle analog output from instruments that do not have RS-232 adapters and from experiments that result in voltages.

Even if you are not interested in this capability, read the opening sections of the next chapter. Analog-to-digital converters can open a whole new world for you. If, after reading the opening sections, you are still interested only in information on how to automate other categories of instruments, proceed to Chapter 7. Chapters 7 through 10 show you how to apply the programming techniques that you already know to most scientific equipment. They will also teach you many new programming techniques.

6

Adding Your Own RS-232 Interface

This chapter will show you one of the ways that you can obtain data from an instrument that does not have RS-232 capability. This method will require your instrument to have either a chart recorder output, an analog voltage output, or a meter display. If your instrument has a digital display (i.e., can display numbers), Chapter 10 shows some alternative methods that will allow you to get data directly from the digital display. You can also use information contained in this chapter if you are conducting experiments that use sensors which produce voltage changes that you want to monitor.

If any of these scenarios describe your situation, then you will need to digitize voltage signals before your computer can receive the data and place it into one of the Lotus spreadsheets for analysis. An "analog-to-digital" converter performs this function.

THE ASSOCIATION OF ANALOG AND DIGITAL SIGNALS

Where do you find analog signals? In just about every experiment from which you collect data, the data originate as analog signals. To be more specific, in almost every experiment that you conduct, the final results begin as continuously variable voltage levels that are proportional to the events being measured. The voltage levels obtained are referred to as "analog" and the analog signals can take on *any* value between a lower and upper limit in the range of acceptable values.

Contrast this with digital signals, which switch suddenly between two very different voltage levels to give either a binary zero or a binary one. That is, while analog signals have continuous values, digital signals take on discrete values. RS-

232 serial communications is a good example of the transmission of digital signals because each signal is either a binary zero or a binary one.

As a general rule, personal computers only understand digital data. Therefore, analog signals need to be converted into digital signals before they can be displayed in numerical form or before their values can be used by personal computers.

Digitizing data results in decimal numbers that correspond to analog voltage levels. The device that performs this conversion is called an "analog-to-digital" (A/D) converter. After an analog signal is digitized, you can receive the information from the converter and place it into your spreadsheet.

ANALOG INSTRUMENTS AND EXPERIMENTS

Although all instruments collect their data as analog, most modern instruments have built-in A/D converters. These converters transform the analog signals that come from their detectors into digital signals. Once data is "digitized" inside an instrument it is in a form that can be displayed. Most modern instruments also have built-in RS-232 ports that can transmit digitized data as character strings for communications to printers, personal computers, etc.

Other instruments can collect analog data, but do not have RS-232 ports for transmitting digitized data to a personal computer. These instruments were commonly manufactured prior to the personal computer revolution. They include spectrophotometric detectors that are currently used for applications such as monitoring chromatographic columns.

With this category of instruments, an intermediate device that has an A/D converter needs to be placed between the instrument and the personal computer. The function of the A/D converter is to convert the analog signals from the instrument's chart recorder output, voltage output, or meter display into digital signals that can be utilized by the personal computer's data processing unit.

Instruments are not the sole sources of analog data. Many experiments can be performed using sensors to collect the data. These sensors produce analog signals that originate from either a natural electrical source (e.g., biopotentials from an electrocardiograph, voltages from a thermocouple, electromotive force of two redox half-cells, pH, etc.) or from a transducer.

A transducer is a sensor that produces an electrical output that is proportional to the physical or chemical property being measured. The following is a partial listing of the measurements that can be made using various transducers:

Pressure	Weight
Flow Rate	Temperature
Force	Light Intensity
Inductance	Position
Velocity	Acceleration
Strain	Depth
Texture	

Analog data derived from sensors must be converted to digital data using an intermediate A/D converter. As a general rule, any time that you use a personal computer to monitor an experiment that results in an analog signal, you will need to digitize the signal before the personal computer can use the information.

BENEFITS OF A/D CONVERTERS

An A/D converter can be invaluable if you have equipment that cannot be updated to RS-232 capability (due to the unavailability of RS-232 options). An A/D converter can make these instruments more functional by automating the data acquisition process. An A/D converter in your laboratory can be advantageous in many other ways.

For example, RS-232 capability is often sold separately as an option for many instruments. These options are often more expensive than purchasing a stand-alone A/D converter from one of several manufacturers and connecting the converter to the instrument's analog output. The economy of stand-alone converters becomes even more pronounced when you consider that a single A/D converter interface can usually be used for several different instruments. For example, the ACROSYSTEMS® ACRO-900™ input system can monitor from 1 to 8 instruments, either individually or simultaneously.

On the other hand, some newer instruments can also benefit from "external" A/D converters. For instance, some instruments have RS-232 boards that must be installed whenever data is transmitted to a personal computer. When an RS-232 board is installed, it often disables manual control of the instrument. The Perkin-Elmer LS-2B Filter Fluorimeter (described in Chapter 7) is an example of this shortfall. Using the analog or chart recorder output allows you to switch between manual and computer data acquisition without having to internally dismantle an instrument to perform the conversion.

Using an A/D converter will also allow you to gather data simultaneously. That is, when the RS-232 control disables the manual mode of an instrument, the instrument normally loses its ability to use its internal printer as well. With an A/D converter you can get a print-out of the data from the instrument while you input data into the personal computer. This means that you can transmit data to your personal computer without worrying about losing the data if a power outage occurs during the experiment.

Furthermore, by taking raw data from an analog or chart recorder output you can often improve the resolution and/or sensitivity of an instrument. Many instruments have only 12-bit converters for their data. There are many stand-alone converters on the market that give 17-bit (or more) resolution. A 17-bit converter permits resolving voltages as small as 1.7 microvolts on a 100 millivolt scale. That translates to a resolution of 0.00075%, or 7.5 parts per million. However, the utility of this increased resolution depends on the contribution of noise in the final signal. Thus, although you may increase resolution, the resolution may be in the noise region of the signal and you will gain nothing. For this reason, you should contact the manufacturer of your instrument and carefully consider whether the additional cost of a 17-bit converter will benefit you.

Using a stand-alone converter also corrects the problem of RS-232 deception. RS-232 is a standard for a physical connection and has no bearing on the program that must reside within an instrument to interface with the commands that you send to it. Having "RS-232 capability" can mean anything from having a connector to having full control over an instrument. Often, it is easier just to get data from an analog output than to depend on your instrument's RS-232 interface.

Getting data when you want it is another major advantage of stand-alone A/D interfaces. The interface can collect and save data independently of your personal computer. When you are ready for the data, you can have your program prompt the interface to send the data that it has collected back to the personal computer. Thus, you can collect data at any time without tying up your personal computer. Very few instruments have this capability. Adding this capability to your laboratory will markedly improve its throughput.

The remainder of this chapter further investigates A/D converters. It will teach you how to choose the correct A/D converter for your laboratory and how to control the converter using a program like the one developed in Chapters 3 through 5.

BASIC DATA ACQUISITION CONCEPTS

A *D*ata *A*cquisition *S*ystem (DAS) is a device that allows you to feed data from the real world into your personal computer. It takes analog signals (either continuous voltages or currents) and converts them into a form that your personal computer can understand. This form is digital. These devices are also commonly referred to as analog-to-digital (A/D) interfaces.

Measurements are coordinated by a controller, which is usually computerized. This part of the system is called the *C*entral *P*rocessing *U*nit (CPU).

Two types of Data Acquisition Systems exist: "plug-in" and "stand-alone." A plug-in system plugs directly into your personal computer. The plug-in card usually has an external terminal board that attaches it to the wires that are attached to the voltage source you are trying to measure. To use plug-in A/D interfaces with Lotus spreadsheets, you need to purchase Lotus Measure™ because plug-in boards do not utilize the RS-232 port. Chapter 9 contains instructions on how to use Lotus Measure with plug-in A/D interfaces. Because the basic theories and programming techniques of plug-in boards and stand-alone systems are similar, using Lotus Measure will require that you have a firm understanding of the concepts of this chapter before you proceed to Chapter 9.

Stand-alone systems are independent of your personal computer. They "stand-alone" and are attached to the personal computer via an RS-232 connection. They can be interfaced using the concepts presented in Chapters 2 through 5 of this book.

Because they are attached via RS-232, they are independent of the personal computer that drives them—you can use an Apple II just as easily as an IBM PC. They can also be placed in almost any environment (like limited access containment rooms). Enclosures are available to provide a barrier for the protection of the environment and/or the system.

Stand-alone systems offer many advantages when compared to plug-in systems. They do not take up a slot in your personal computer and they are expandable. Expandability means that you can start small and build the system as your requirements grow. As you build and expand, your cost per channel decreases.

Stand-alone interfaces are available in many forms—from single channel units designed for a specific type of input, to modular systems that can directly accept input from thermocouples, RTDs (*R*esistance *T*emperature *D*etectors), strain, position and other sensors. Usually, the best choice for a DAS is one that is modular because it allows you to select and purchase the modules that meet your current needs only, and yet are expandable to meet your future needs.

HOW TO CHOOSE AN A/D CONVERTER

Consider a number of factors when selecting a DAS. Begin by determining your requirements. Ask yourself whether you need only analog inputs, whether you also need outputs, and how many inputs and outputs you need. Also consider whether you will ever need to add more input channels, output channels, or other types of channels (e.g., digital channels; see Chapter 10). Price may be an important factor, but you should look to the future. Although different systems have differing initial costs, some are more cost effective in the long run. If you anticipate future needs and requirements, ensure that the system that you select either has them now, or can be easily upgraded to add them later.

To minimize initial expense, one A/D converter is normally used for a number of input channels unless very high measurement speeds are required. Simple units have only one or two input channels and cannot be expanded. With more sophisticated systems, you can expand up to 48 or more channels, including different input types. These systems may be more cost effective for you in the long run.

As previously discussed, plug-in boards and some stand-alone systems need to be in constant communication with a personal computer to operate. However, some stand-alone systems can also collect data independently for a specific period of time. This difference in capability can be profound. You must decide if you are willing to dedicate a personal computer to long-term data collection or if you want to perform data acquisition without a personal computer and then transfer the information into the personal computer later. More enhanced systems are more costly initially, but may be worth the difference for the increased productivity.

If you decide to run your experiments independently of the computer ("in the background''), then you should also consider the amount of on-board random access memory (RAM) that you need. Most units have a standard of two Kilobytes. The standard is sufficient for small experiments. However, if you are using the interface for *H*igh *P*erformance *L*iquid *C*hromatograpahy (HPLC), spectrophotometric scanning, running very long experiments with thousands of data points, screening chromatographic columns, etc., then you should consider expanding the amount of memory. Optional expansion sizes are usually about 32 Kilobytes, which should be sufficient for nearly all experiments. This option will ensure that you do not run out of memory. Running out of memory can cause severe problems. On some interfaces, the interface will lock up; on other interfaces, when the end of memory is reached, the interface will wrap around back to the beginning and start overwriting the original data points.

Brochure specifications for Data Acquisition Systems can be confusing. The A/D converter that a manufacturer uses in an interface will determine the accuracy, resolution, speed and electrical noise tolerance of the system. In general, you will want to purchase a converter with the highest resolution and accuracy that you can afford. If you are measuring a process that changes very rapidly, you should also shop for speed. However, you may be forced to decide between speed and the tolerance of the system to electrical noise (see below).

The accuracy of a converter is usually specified as a percentage of full scale. The resolution is usually specified as a number of bits. The number of bits refer to the number of levels into which an analog signal can be digitized. This number is actually a power of 2. Thus, a 17-bit converter can resolve 2 to the seventeenth power (131,072) levels within the voltage range. A 16-bit converter is half of that, 65,536 levels. This number of levels corresponds to a resolution of 0.0015%, or 15 parts per million.

This digitizing error limits the accuracy (and therefore, the sensitivity and resolution) of a conversion. This limitation can be important in a process that has both very large and very small signals. For example, trace analysis has a very large signal for the primary component and a very small signal (e.g., 0.0001%) for the trace component. This situation is very common for HPLC analyses. For this reason, 20-bit A/D converters are used for this type of application.

The full-scale input voltage range of an A/D converter should match the signals being measured as closely as possible. For example, if the output of your instrument is 0 to 100 millivolts, you would not choose a converter that has a 0 to 20 volt input range; you would choose a converter that spans a 0 to 100 millivolt range. The closer the two full-scale ranges correspond, the better the resolution of the system. Some A/D converters have programmable voltage ranges that can be changed as different input channels are selected. The software switchable ranges that are commonly available are ± 200 mV, ± 2 V, and ± 20 V. Ranges of ± 100 mV, ± 1 V, and ± 10 V ranges are also popular.

The speed of an A/D converter is specified as either the time required for a single conversion or as a frequency. Frequency is the number of conversions per second. However, the overall speed performance of a system is often limited by the software that runs it and/or by the memory where the data is stored. Thus, the speed specification for the software in the DAS is usually the most important specification for you to consider.

The maximum number of conversions per second for each channel of an A/D converter system is the conversion rate divided by the number of channels. If an experiment's input signal changes with time, the A/D conversion rate for that channel should be at least 5 times the highest frequency found in the input signal. When an input signal changes during a conversion, the accuracy of the conversion may be compromised if the conversion is too slow. A "sample and hold" may be required to maintain accuracy.

The "sample and hold" capability takes a "snapshot" of the input signal at a precise point in time and holds the signal until the conversion is complete. A "sample and hold" is usually required only for signals that change very quickly. The most important specification for "sample and hold" is the uncertainty in the sample time. This uncertainty can significantly affect the accuracy of a measurement. Therefore, "sample and hold" accuracy should be at least as high as the accuracy of the A/D converter.

The final criterion of a converter is concerned with electrical noise rejection. To intelligently evaluate this criterion, you need more information about A/D converters. The two most commonly used converters in A/D interfaces are "integrating" and "successive approximation".

Integrating A/D converters have excellent noise immunity, but it is achieved at the expense of speed. Typically, integrating converters operate at only 20 to 200 samples per second. The input voltage is measured by charging a capacitor over a fixed interval of time. The final charge on the capacitor that occurs during this fixed interval is directly proportional to the input voltage. The time required to discharge the capacitor is also proportional to the input voltage. The value reported, therefore, is actually a count of the number of clock pulses required for the capacitor to discharge.

Successive approximation A/D converters use a digital-to-analog (D/A) converter and a voltage comparator to determine voltage. At the beginning of a conversion, the most significant bit of the D/A converter is turned on. The voltage comparator compares this voltage with the input voltage. Depending on whether the D/A voltage is greater or less than the input voltage, the bit is turned off or remains on. Then, the next bit is turned on and the comparison is repeated. This process of turning bits on and off continues until the D/A value matches the input voltage exactly.

The successive approximation A/D converter is quite susceptible to noise, but is much faster than an integrating converter. Therefore, if you acquire data rapidly (100 to 200,000 conversions per second), use an A/D interface based on a successive approximation converter.

Several A/D converters are commercially available. The following is a discussion of some of the stand-alone systems. Chapter 9 has a discussion of plug-in boards.

THE ACRO-900™ SYSTEM FROM ACROSYSTEMS®

The ACRO-900™ is available from ACROSYSTEMS®, Inc., 66 Cherry Hill Dr., P.O. Box 487, Beverly, Mass. 01915. This system is the same as the OM-900 available from Omega Engineering, Inc., P.O. Box 2669, Stamford, CT 06906.

These two Data Acquisition Systems represent top-of-the-line data interfaces. They are the same systems that I have been describing in the last few chapters. Each system has the capability to work independently of a personal computer. You can order 32K of RAM storage to support this capability.

An ACRO-900 or OM-900 system includes a power source, a microprocessor module, and one or more analog input module(s) to meet your specific data acquisition needs. Modules make these systems both expandable and flexible.

Both companies offer several input, output, and I/O modules. Some of the A/D converter modules that ACROSYSTEMS and Omega offer are

Model	911	912	914	916
# Analog Inputs	4	8	4	8
# Analog Output	4	0	2	0
A/D Converter Resolution	12-bits	17	12	12
D/A Converter Resolution	12-bits	—	12	—
Conversion Time	50μs	50ms	50μs	50μs
Input Ranges: ±	200mV	200mV	200mV	200mV
	2V	2V	2V	2V
	20V	20V	20V	20V
Input Accuracy: ±	0.1%	0.03%	0.1%	0.1%

In addition to the above, Models 911, 914 and 916 also have eight digital inputs and eight digital outputs. Model 916 does not have analog output. Model 912 is a high resolution, 17-bit converter that has analog input only.

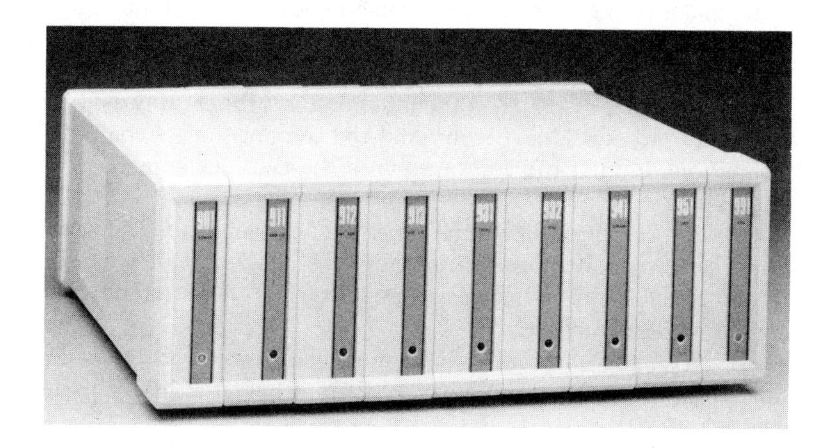

As a reference, the 50 microsecond successive approximation conversions that Models 911, 914, and 916 produce will allow data rates of up to 20,000 samples per second. The 50 millisecond dual slope integration conversions of the Model 912 will only allow data rates of up to 20 samples per second, but it has a very high resolution and good noise rejection.

THE ACRO-400™

This system is a small step down from the Model 900. It is available from ACRO-SYSTEMS, Inc. This system does not feature the modularity or expandability of the Model 900, but it offers 16 analog inputs, one analog output, and 32 lines of digital I/O.

The A/D converter is the same 17-bit integrator that is used in the ACRO-912 and resolves voltages as small as 1.7 microvolt on the ± 100 millivolt scale. The system's price is about half that of the Model 900.

The following table shows some of the other specifications of this interface:

Model	ACRO-400
# Analog Inputs	16
# Analog Outputs	1
A/D Converter Resolution	13- to 17-bits
D/A Converter Resolution	12-bits
Conversion Time	100 millisecs @17 bits; <10ms @13 bits
Input Ranges: ±	100 mV
	1V
	10V
Input Accuracy: ±	0.03% of input +0.01% range + 10μV

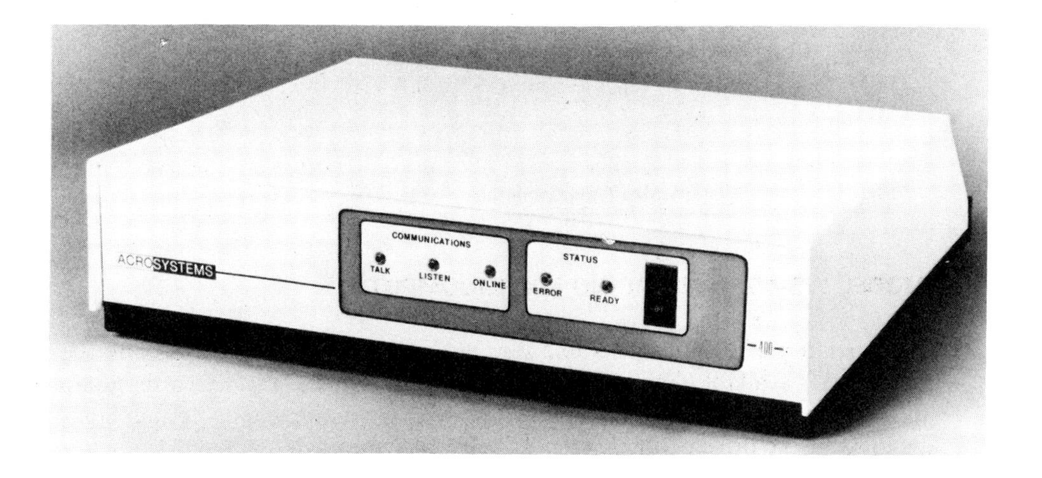

Input ranges and A/D conversions are software selectable. In fact, the commands are nearly identical to those of the ACRO-900. Very few modifications would need to be made to the examples given in this book before the programs could be used with the ACRO-400.

The ACRO-400 can be used as a free-standing data acquisition and control system, capable of monitoring and controlling a complete process or test procedure without personal computer intervention. The system is ideal for controlling fermentors, reactors, and performing biomedical experiments. It is also a good system for chromatographic or spectrophotometric analyses.

THE NELSON ANALYTICAL SERIES 900

The Nelson Analytical PC Integrator is available from Nelson Analytical, 10040 Bubb Rd., Cupertino, CA, 95014. It is a premium quality converter that results in 20-bit resolution. This resolution is appropriate for HPLC and Gas Chromatography.

Like other quality stand-alone converters, this instrument can run in "background" mode (independent of the personal computer). The string commands sent to the PC Integrator are similar to those sent to the ACRO-900 and OM-900. (See Chapter 9.) They consist of a single character with a numeric argument.

The Nelson integrator is limited to one chromatograph and two detectors. It is not expandable.

Below are some additional specifications for this series of converters:

Model	950	960	970
# Analog Inputs	2	2	2
# Analog Outputs	0	0	0
# Relays	1	1	1
# Digital Outputs	7	7	7
A/D Converter Resolution	20 bits	20 bits	20 bits
D/A Converter Resolution	N/A	N/A	N/A
Conversion Time	100/sec	100/sec	100/sec
Input Ranges: ±	0.1v	0.1v	0.1v
	1.0v	1.0v	1.0v
	2.0v	2.0v	2.0v
	10.0v	10.0v	10.0v
Memory:Max # data points	7,500	30,000	120,000

THE MODEL WB-40 INTERFACING MICROCOMPUTER

This interface is a considerable step down from both Model 900s and the ACRO-400. It is available from Omega Engineering, Inc. The WB-40 is a stand-alone interface that has eight analog inputs. The system is not expandable, but costs about a third of the price of the Model ACRO-900 system.

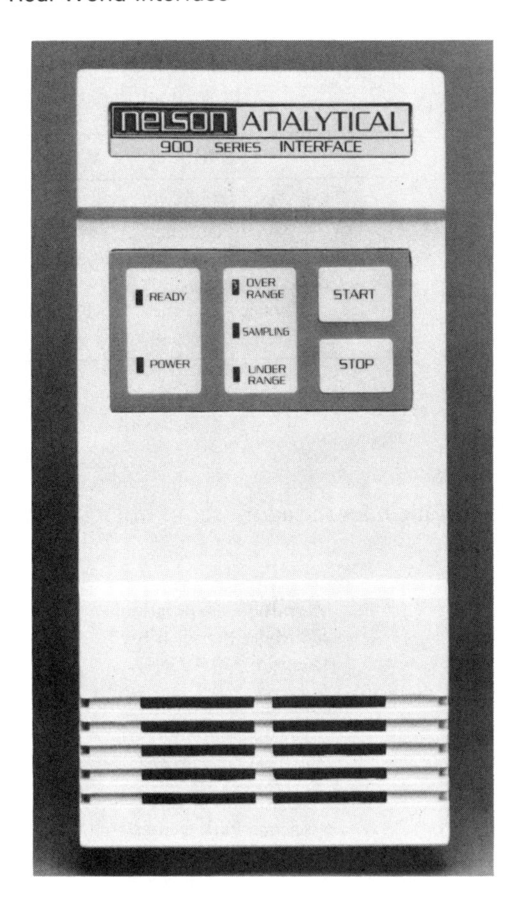

This system has an input range of 0 to ± 4.999 volts with a resolution of 250 microvolts. The following table includes additional specifications of this interface:

Model	WB-40
# Analog Inputs	8
# Analog Outputs	0
A/D Converter Resolution	250 μV; "20,000 part resolution"
D/A Converter Resolution	N/A
Input Ranges: ±	4.999

THE MODEL WB-31 REAL WORLD INTERFACE

This interface is a "bare bones" model from Omega Engineering, Inc. It costs very little and is quite easy to use. It has two analog inputs that have a range of 0 to ± 4.999 volts. The system is not expandable, but it can run independently of a personal computer if you purchase the optional WB-31B package.

The following table includes additional specifications of this interface:

Model	WB-31
# Analog Inputs	2
# Analog Outputs	0
A/D Converter Resolution	0.25 millivolts
D/A Converter Resolution	N/A
Conversion Time	133 milliseconds
Input Ranges: ±	4.999

OTHERS TO TRY

The following list includes the addresses of other manufacturers of Data Acquisition Systems:

MetraByte Corporation
440 Myles Standish Blvd.
Taunton, MA 02780

National Instruments
12109 Technology Blvd.
Austin, Texas 78727 6204

CyberResearch, Inc.
5 Science Park Center
P.O. Box 9565
New Haven, CT 06526

Contec Microelectronics USA, Inc.
3000 Scott Blvd., Suite 211
Santa Clara, CA 95054

Thornton Associates, Inc.
1432 Main St.
Waltham, MA 02154

GW Instruments
P.O. Box 2154
264 Msgr. O'Brien Hwy. 8
Cambridge, MA 02141

Personal Computing Tools, Inc.
101 Church St., Unit 12
Los Gatos, CA 95030

Strawberry Tree Computers
150 North Wolfe Road
Sunnyvale, CA 94086

Many other companies manufacture both plug-in and stand-alone data acquisition interfaces. "Scientific Computing And Automation" is a publication that contains advertisements and articles on both interfaces and scientific equipment. It is published six times per year by Gordon Publications, 13 Emery Avenue, Randolph, NJ 07869-1380. It is a good means of keeping abreast of new developments in this field.

SETTING UP THE INTERFACE

When you receive your stand-alone interface, you will need to set it up. The remainder of this chapter describes the ACRO-912 A/D converter in detail. Chapter 2 had a description of how to set up the ACROSYSTEMS ACRO-900 interface. For other interfaces, consult accompanying User's Manuals and then follow the steps given in Chapter 2 for setting the communications parameters and module addresses and for connecting the interface to your personal computer.

Connecting the interface to an instrument or sensor is simple. Use a two lead cable. If you are interfacing to several voltage sources, use separate cables: Do not just get a cable with more wires. Use cables that are shielded and as short as possible. Following these recommendations becomes critically important when you work with low voltages, because "cross talk" (interference) between the wire that is carrying your signal and adjacent wires (or the outside world) can cause spurious results.

If you are using a sensor, you can usually connect the sensor directly to the interface. Consult your User's Manual to determine the output signals of the sensor's wires and then follow the instructions below.

To monitor an instrument, look for an accessory connection panel on the side or rear of the instrument. The panel is the most convenient place to obtain measurement signals. The output will be labeled "Recorder Signal", "DC out", or the like. This output is usually a strip with screws on it or a connector jack.

Whether monitoring a sensor or chart recorder output, the procedure for connecting the instrument to the converter is the same. One of the connectors will be labeled with a minus sign, "Common", "Negative", "Ground", "Low", or "Reference". This output needs to be connected to the minus terminal on the interface. Connect one end of the black wire of the cable to the instrument's (or sensor's) common terminal and the other end to the minus terminal on the interface. Likewise, connect the terminal labeled with a plus, a "Hi", or a voltage to the plus terminal on the interface.

Many instruments give a choice of full-scale voltage ranges. For example, many instruments provide a 1-volt full-scale and a 100 millivolt full-scale output. If given a choice, use the largest voltage. This choice will normally give the greatest signal-to-noise ratio, and the best results. However, you should confirm your choice by examining the results after you have written your program.

Some instruments have an additional output marked INT. This output provides an external tap into an instrument's INTegration signals. This output often provides useful information. To connect this output to the interface, connect the "Low" output of the instrument to the minus terminal of one of the other channels of the interface. Then, connect the INT output of the instrument to the plus terminal on the corresponding channel.

Ensure that all screws are tightened for good connections. Also, examine the wires to make sure that they are not touching each other and that they did not break when you tightened the connectors.

TAPPING INTO AN INSTRUMENT'S METER

If you do not have a voltage output on your instrument, you will need to take a signal from the meter that displays results. If you are timid about opening an instrument, contact an electronics professional, someone with experience in electronics, your instrumentation maintenance shop, a consultant, or the service department of the instrument's manufacturer.

If your instrument's display has a meter with a needle that deflects along a scale to indicate the value of a measured quantity by the position of the needle on the scale, your instrument's display is an "analog meter".

Analog meters usually measure the *current* produced by the electronic components of your instrument in response to the process being monitored. Your A/D converter measures *voltages*. Therefore, you must convert the meter's current signals into voltage signals before you can digitize them.

Current is converted by measuring the voltage drop across a resistance. These

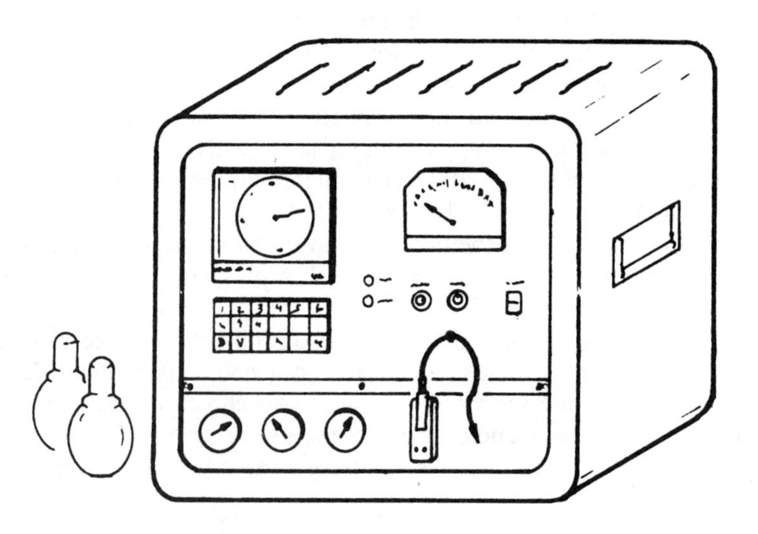

voltages are proportional to the current supplied to the meter (voltage = current * resistance).

You can obtain input from a meter with your A/D converter via two configurations. These configurations are shown in Figures 6–1 and 6–2. The configuration that you use will depend on complex relationships between the meter and the circuit board that supplies the current. Due to the variability that can occur in these relationships, the best way to determine which configuration is appropriate for your instrument is by experimentation.

The first configuration treats the meter itself as a resistor and measures the voltage drop across it. To implement this configuration, unplug the instrument, open the instrument and locate the connectors at the rear of the meter. These connectors are usually threaded bolts with nuts attached. One of the connectors will be labeled with a plus sign. Connect this output to the plus terminal on the interface. Attach one end of a wire in your cable to this connector and the other end to the plus terminal on the interface. Likewise, attach the other connector to the minus terminal on the interface.

Test the instrument's response by writing a program like the one illustrated in Figure 6–15. You might have to modify the {WRITELN} commands so that they use string commands that are appropriate for your A/D converter. You may also need to modify the parsing method in the {PUT} command to correctly handle the format of the data string that your converter returns.

When you monitor the voltage of this arrangement, use an input range of 0 to 100 millivolts (or 0 to 200 millivolts) for your A/D interface. Next, adjust the meter's Zero and Full Scale readings using the protocol in your instrument's User's Manual.

The program in Figure 6–15 monitors the voltage of a meter over a period of time. When prompted for the number of data points, type eleven (11). (By entering 11, you can take readings in 10% increments from Zero to Full Scale.) The time delay that you specify depends on how fast you can adjust the meter reading with the Zero Adjust knob. Specify a delay long enough to allow you to make a 10%

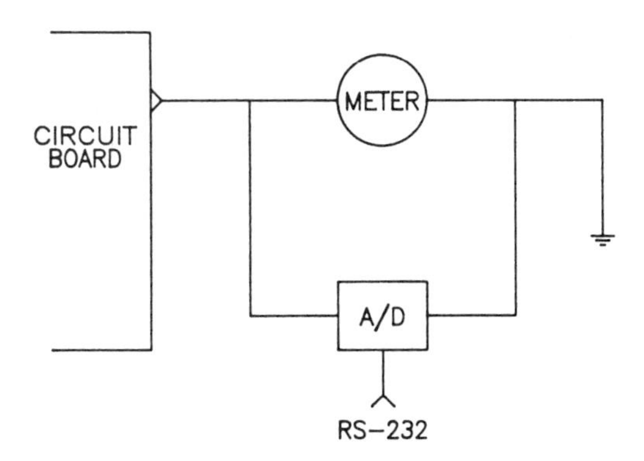

Figure 6–1

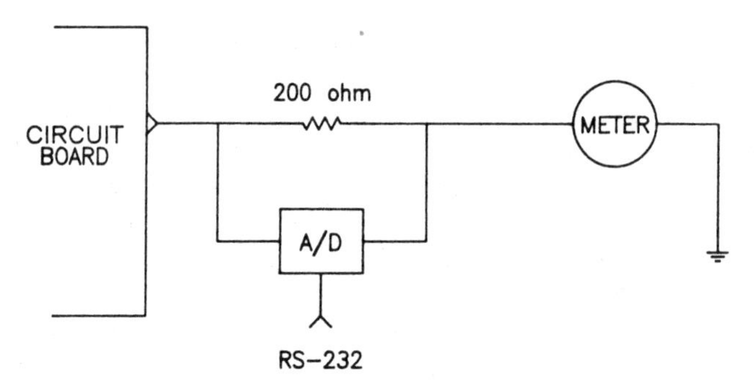

Figure 6–2

increase in the meter's reading and for the meter to reach a steady reading. These procedures usually total about 20 seconds.

Now, run the program to collect some data. After each meter reading has been placed into your spreadsheet template, use the Zero Adjust knob to increase the meter reading by 10%. Continue this data collection until you have 10% increments from Zero to Full Scale.

Your next task is to place a precision resistor between the meter and the circuit board that drives the meter. If this resistor is about 200 ohms and the input impedance of your A/D converter is greater than about 1 megohm, you can connect the converter across the resistor and obtain voltages.

Figure 6–2 illustrates this configuration.

To implement this configuration, unplug the instrument, open the instrument and locate the connectors at the rear of the meter. Remove (or cut) the wire between the meter and the circuit board that drives the meter. Insert a 200 ohm precision resistor (one that has greater than 1% accuracy). If you solder the connections, make sure that you protect the meter and circuit board from excessive heat. Next, place a wire on each side of the resistor. Finally, connect the wire that is between the circuit board and the resistor to the plus terminal on the interface; connect the wire that is between the resistor and the meter to the interface.

Modify the output range (ABSDATA) for your test program so that it is one column to the right of the data from the first configuration. (Use the / Range Name Create command to re-define the range.) Next, reset the Zero and Full Scale readings. Run the program to collect some data using the same protocol given previously.

Graph the data. Figures 6–3 through 6–5 show some common responses. These figures were obtained from a filter photometer, a pH meter, and a dissolved oxygen meter, respectively. In choosing the best configuration, you want to achieve the best combination of linearity and sensitivity.

Figure 6–3 shows a case where both configurations are linear and have the same sensitivity. You could therefore use either configuration, but connecting across the meter is simpler.

Figure 6–4 shows a situation where both configurations are linear, but connect-

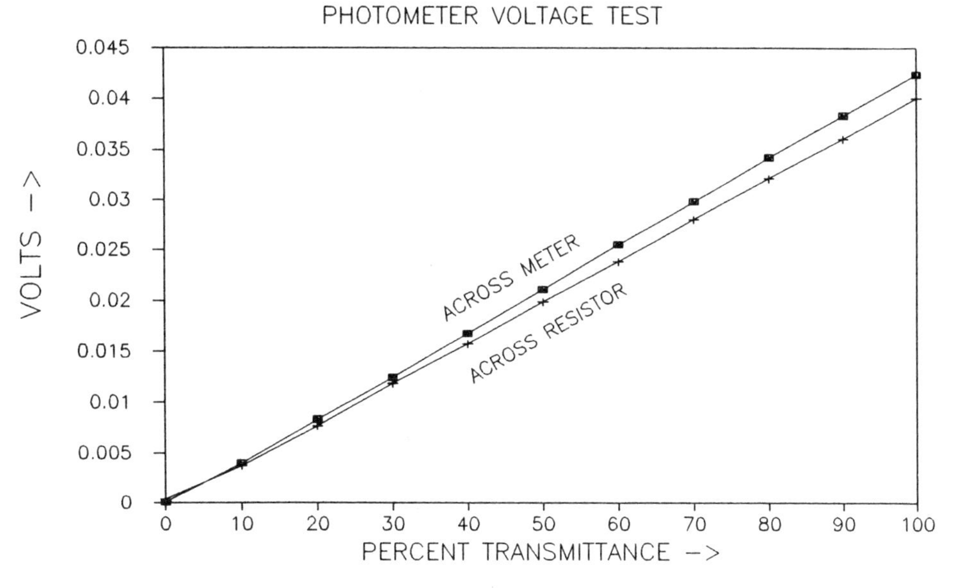

Figure 6-3

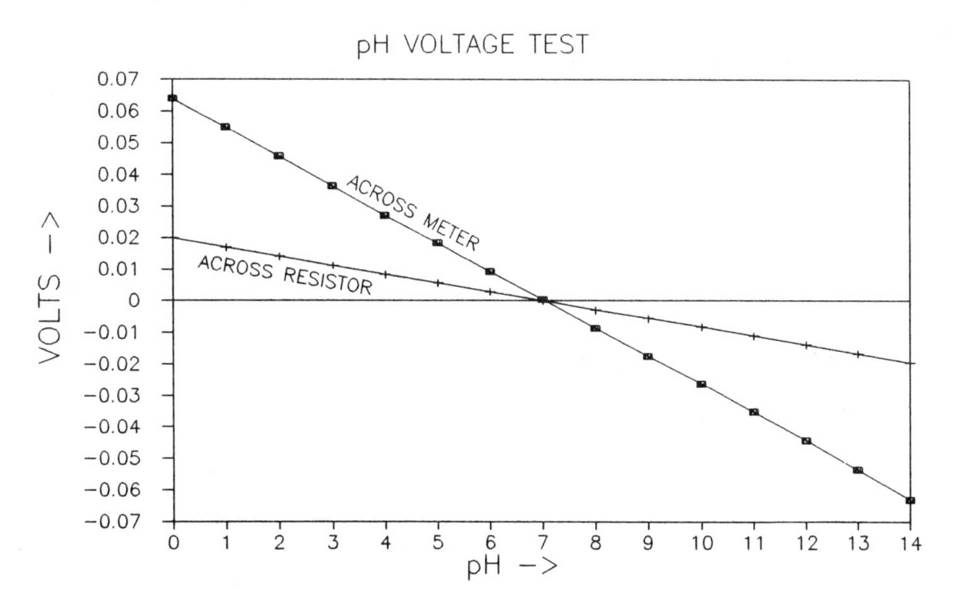

Figure 6-4

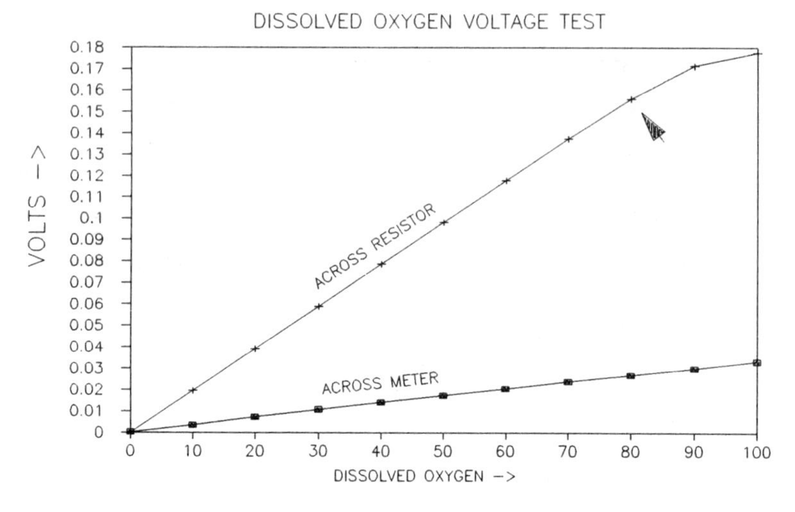

Figure 6–5

ing across the meter gives much better sensitivity. If your instrument's response is similar to this case, connect across the meter.

The most common case is one in which the 200 ohm resistor gives greater sensitivity. Figure 6–5 illustrates this case. However, note the loss of linearity above the 80% mark on the X-axis. If your meter shows this non-linearity, try a 100 ohm precision resistor and repeat the experiment. This change usually rectifies the problem. If not, or if you have trouble setting the Full Scale deflection on the meter, connect across the meter.

One final note: The voltages that you obtain from an interface can be converted back into "meter readings" by using the equation from Ohm's law. The equation is

$$\text{meter reading} = \text{amperes} = \text{voltage} / 200$$

This conversion can be conveniently performed by the {PUT} command that places data into your template.

BRIEF DESCRIPTION OF AN A/D EXAMPLE EXPERIMENT

When the DNA in a sample "melts" or reanneals the sample demonstrates a large change in circular dichroism (CD). Thus, if you monitor the temperature of the sample versus its circular dichroism, you can determine its Tm (melting temperature). The Jobin Yvon Circular Dichrograph is an instrument used for this purpose.

This instrument does not have a built-in RS-232 interface. It also does not have the capacity to determine temperatures of a sample. It is, therefore a prime candidate for automation using an analog-to-digital interface.

Even if you do not own this instrument, it provides a representative programming example for instruments that need to be monitored by an A/D converter.

To get CD measurement from the instrument, you must take the analog output directly from the instrument's analog board. Temperature is obtained by inserting a coated thermocouple directly into the sample cell. The voltage lines and the thermocouple lines are then attached to the appropriate ACROSYSTEMS modules (using the concepts presented earlier in this chapter).

OBTAINING A/D DATA

You can use the basic method described in this section to get data from any stand-alone A/D converter listed in this chapter. If you have another brand or model of stand-alone interface, chances are very high that it will use the same general method.

Basically, to start an experiment, you send a set of string commands to initialize certain parameters in the converter (like module address, voltage range, resolution, conversion rate, etc.). Then, to collect data, send prompts to the A/D converter to signal the converter to return data. Consult your instrument's User's Manual or Programmer's Manual for the appropriate strings to use.

Figures 6–6 through 6–10 show a template and macro programs that can be used to obtain the data for the example DNA melt experiment. From this data, a plot of the voltage versus temperature can be made and the Tm determined.

As you can see, the program is a modification of the program that was used for the thermocouple input. That is, getting A/D data from a stand-alone interface is very similar to the method that was used to get thermocouple data from the ACRO-931. Thermocouples are analog devices; the ACRO-931 Thermocouple Interface is just an A/D converter.

Following from this information, you can apply all you have learned over the last few chapters to stand-alone A/D converters. The only difference is that you will need to change the string prompts to those appropriate to your converter.

Take a look at the changes that have been made to the program in Chapter 5. The only macros that have changed are TEMP_ABS and GET_DATA.

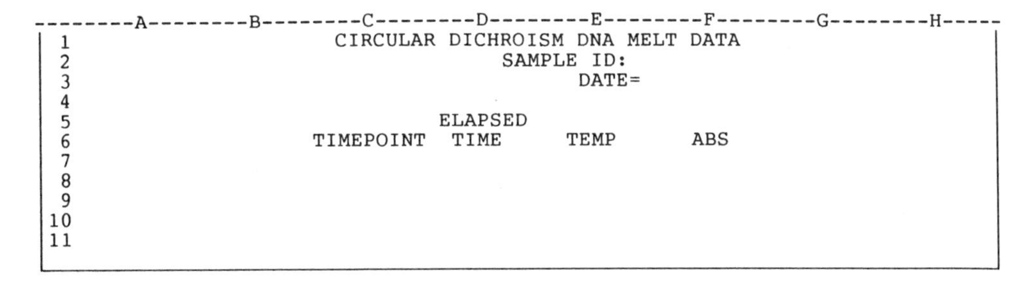
```
--------A--------B--------C--------D--------E--------F--------G--------H-----
 1                   CIRCULAR DICHROISM DNA MELT DATA
 2                                SAMPLE ID:
 3                                       DATE=
 4
 5                             ELAPSED
 6                 TIMEPOINT  TIME      TEMP       ABS
 7
 8
 9
10
11
```

Figure 6–6

```
--------N--------O--------P--------Q--------R--------S--------T--------U--------
1             /*PROGRAM TO GET TEMPERATURES AND ABSORBANCES FROM*/
2            /*JOBIN-YVON CIRCULAR DICHROGRAPH USING THE ACRO-900*/
3
4                 /*AUTO-EXECUTING MACRO FOR LOTUS 1-2-3*/
5  \0        {LET VERSION,0}    /*TELLS PROGRAM IT'S RUNNING UNDER 1-2-3*/
6            /S        /*ISSUE THE /S MENU CHOICE TO GO TO SYSTEM*/
7            {MENUBRANCH MENU1}          /*PUT UP USER'S MENU*/
8
9                              /*LOTUS SYMPHONY*/
10           /*AUTO-EXECUTING MACRO: TESTS TO SEE IF DOS.APP ATTACHED; */
11           /*IF NOT, ATTACHES IT.  THEN SETS BAUD RATE, ETC.*/
12 AUTOEXEC  {LET VERSION,1}   /*TELLS PROGRAM IT'S RUNNING UNDER SYMPHONY*/
13           {IF @ISERR(@APP("DOS",""))}{HOOKUP} /*SEE IF DOS.APP ATTACHED*/
14           {SERVICES}AIDOS~MODE COM1:4800,E,7,2~         /*SET COMM PORT*/
15           {MENUBRANCH MENU1}          /*PUT UP USER'S MENU*/
16
17 HOOKUP    {SERVICES}AADOS~Q           /*ATTACHES DOS.APP*/
18
19           /*MACRO FOR ACRO-931 THERMOCOUPLE AND ACRO-912 A/D CONVERTER*/
20 TEMP_ABS  {LET DATE,@DATEVALUE(@NOW)}          /*DATE STAMP THE EXPT*/
21           {OPEN "COM1",M}    /*OPEN COM PORT AS DEVICE FILE NAME*/
22           {WRITELN "#5;VIR1=-5"}       /*SET FULL SCALE TO 2V*/
23           {GETNUMBER "How Many Time Points?",POINTS}
24           {GETNUMBER "How Long Is The Interval Between Measurements?",DELAY}
26           {INDICATE I/O}{PANELOFF}    /*FREEZE PANEL*/
27           {GOTO}START~        /*MOVE CELL-POINTER TO STARTING POSITION*/
28           {WINDOWSOFF}        /*FREEZE THE DISPLAY*/
29           {LET BEGIN,@TIMEVALUE(@NOW):VALUE}  /*GET THE STARTING TIME*/
30           {IF DELAY>0}{FOR ROW,0,POINTS,1,BIGLOOP}   |*THESE GIVE MAXIMAL*|
31           {IF DELAY=0}{FOR ROW,0,POINTS,1,SMALLOOP}  |*PERFORMANCE FOR   *|
32           {CLOSE}             /*CLOSE THE PORT*/     |*ZERO DELAY        *|
33           {CALC}              /*FORCE RECALCULATION OF THE SPREADSHEET*/
34           {GOTO}START~{LEFT}          /*RESET CELL-POINTER TO START*/
35           {IF VERSION=0}/RFDT3{DOWN POINTS}~  /*TIME FORMAT, 1-2-3*/
36           {IF VERSION=1}/FT3{DOWN POINTS}~    /*TIME FORMAT,SYMPHONY*/
37           {WINDOWSON}{PANELON}        /*THAW THE DISPLAY*/
38           {INDICATE DONE}    /*CHANGE INDICATOR TO "DONE"*/
39
40
```

Figure 6–7

Most A/D converters contain a selection of different built-in full-scale voltage ranges, resolutions, and conversion rates. Each of these parameters is specified by a string command that you send to the converter. These parameters allow you to customize the conversion from your computer without changing switches or jumpers on the interface. Another useful feature is that most stand-alone interfaces include to allow you to set each channel independently.

```
--------N--------O-----------------P-----------------Q-----------------R---------------------S-------------T--------U-------
40
41
42
43 MENU1   MELT                 PRINT              FILE              SAMPLE                QUIT
44         DETERMINE MELT TEMPPRINT THE RESULTS    SAVE THE FILE     SPECIFY THE SAMPLE    QUIT THE MACRO
45         {TEMP_ABS}           {BRANCH PRINT}     {BRANCH SAVE}     {GOTO}SAMPLE~         {QUIT}
46         {MENUBRANCH MENU1}                                        {?}~
47                                                                   {MENUBRANCH MENU1}
48
49
```

Figure 6–8

```
-------N--------O--------P--------Q--------R--------S--------T--------U--------
49
50                     /*LOOP THAT GETS ONE DATA TIME POINT*/
51 BIGLOOP   {INDICATE PAUSE}{PANELON}{PANELOFF} /*SHOW DELAY PHASE*/
52           {WAIT @NOW+@TIME(0,0,DELAY)}        /*WAIT FOR SPECIFIED TIME*/
53 SMALLOOP  {LET CURRENT,@TIMEVALUE(@NOW)}      /*GET CURRENT TIME*/
54           {LET DELTA,CURRENT-BEGIN}           /*FIND ELAPSED TIME*/
55           {PUT ELAPSCOL,0,ROW,DELTA}          /*PUT ELAPSED TIME IN SHEET*/
56           {PUT PTCOL,0,ROW,ROW}               /*PUT POINT NUMBER IN SHEET*/
57           {WINDOWSON}{WINDOWSOFF}             /*ADD DATA TO SCREEN*/
58           {DOWN}                              /*MOVE DOWN TO NEXT ROW*/
59           {GET_DATA}                          |*GET A ROW'S WORTH OF DATA*|
60           {INDICATE NEXT}{PANELON}{PANELOFF}  |*(BOTH TEMP AND ANALOG)   *|
61
62
63                 /*MACRO WHICH GETS ONE TEMP AND ONE ABSORBANCE*/
64 GET_DATA  {WRITELN "#4;TEMP1"}        /*PROMPT ACRO-931 FOR A TEMP*/
65           {READLN BUFFER}    /*READ IN A TEMPERATURE VALUE*/
66           {PUT TEMPDATA,0,ROW,@VALUE(@CLEAN(BUFFER)):VALUE}
67           {WRITELN "#5;VIN1"}             /*PROMPT ACRO-912 FOR A VOLTAGE*/
68           {READLN BUFFER}    /*READ IN A VOLTAGE VALUE*/
69           {PUT ABSDATA,0,ROW,@ABS(@VALUE(@CLEAN(BUFFER))):VALUE}
70                     /*TEMPDATA RANGE IS C7..C3000*/
71                     /*ABSDATA RANGE IS D7..D3000*/
72
73           /*PRINT MACRO: SET "REPORT" RANGE; PRINT; RETURN TO MENU1*/
74 PRINT     {CALC}                    /*FORCE RECALCULATION OF THE TEMPLATE*/
75           {WINDOWSOFF}{PANELOFF}      /*FREEZE THE DISPLAY*/
76           {HOME}            /*GO TO CELL A1; SET "REPORT" RANGE*/
77           /RNCREPORT~{ESC}{HOME}.{DOWN 6+POINTS}{RIGHT 7}~
78           {IF VERSION=0}/PPRREPORT~AGPQ                  /*LOTUS 1-2-3*/
79           {IF VERSION=1}{SERVICES}PSSRREPORT~QAGPQ       /*SYMPHONY*/
80           {WINDOWSON}{PANELON}         /*THAW THE DISPLAY*/
81           {MENUBRANCH MENU1}           /*RETURN TO USER'S MENU*/
82
83 SAVE      {IF VERSION=0}/FS{?}~        /*LOTUS 1-2-3*/
84           {IF VERSION=1}{S}FS{?}~      /*SYMPHONY*/
85           {RIGHT}~ /*FORCE OVERWRITE; CANNOT BE ON SAME LINE AS THE FS*/
86           {HOME}                       /*MOVE CELL-POINTER TO CELL A1*/
87           {MENUBRANCH MENU1}           /*RETURN TO USER'S MENU*/
88
89           /*SCRATCHPAD OF NAMED RANGES*/
90 VERSION             /* 0=>LOTUS 1-2-3; 1=>SYMPHONY */
91 ROW                 /*COUNTER FOR LOOP; ROW OFFSET FOR {PUT}*/
92 POINTS              /*NUMBER OF DATA POINTS TO COLLECT*/
93 BEGIN               /*TIME EXPERIMENT STARTED*/
94 DELAY               /*USER DEFINED DELAY BETWEEN DATA POINTS*/
95 CURRENT             /*CURRENT TIME*/
96 DELTA               /*ACTUAL DELAY, FROM TIME ZERO*/
97 BUFFER              /*A BUFFER FOR INCOMING DATA*/
98
99
100
```

Figure 6-9

For example, each of the ACRO-912 channels can be independently set to ± 20 volts, ± 2 volts, and ± 0.2 volt full-scale. Two conversion rates are available on the ACRO-912. They are 100 ms and 50 ms. At 100 ms, the resolution is 17-bit and at 50 ms, the resolution is 16-bit.

Each of these parameters are selected by sending the module a specific command. The "VIR1 = −5" command that is sent in TEMP_ABS is an example. The command literally means, "set the full-scale range of channel one to 2.0 volts". To

```
---------N--------O--------P--------Q----------R--------S--------T--------U-----
102                            RANGE NAME TABLE
103                            ================
104                  ABSDATA   F7..F3000  PRINT     O74
105                  AUTOEXEC  O12        PTCOL     C7..C3000
106                  BEGIN     O93        REPORT    A1..G9
107                  BIGLOOP   O51        RESULTS   C1..F6
108                  BUFFER    O97        ROW       O91
109                  CURRENT   O95        SAMPLE    F2
110                  DATE      F3         SAVE      O83
111                  DELAY     O94        SMALLOOP  O53
112                  DELTA     O96        START     C7
113                  ELAPSCOL  D7..D3000  TEMPDATA  E7..E3000
114                  GET_DATA  O64        TEMP_ABS  O20
115                  HOOKUP    O17        VERSION   O90
116                  MENU1     O43        \0        O5
117                  POINTS    O92
118
119
120
```

Figure 6–10

use a 17-bit resolution with a 100 ms conversion rate for channel one, the {WRITELN} command could be amended as follows

```
{WRITELN ''#5,VIR1=-5;RATE1=1''}
```

Because the example experiment needs to alternate between temperature and analog readings, the two {WRITELN}s in GET_DATA send the appropriate module addresses to the interface. You can also use an A/D converter. If such is the case, you can move the initializing command for your interface into the TEMP_ABS macro, just below the {OPEN} command.

The VIN1 portion of the command string in GET_DATA makes a request to the module to send the current voltage of channel one. It is equivalent to the TEMP1 command used for the ACRO-931 Thermocouple Interface.

As a general rule, Thermocouple Interfaces use special software to convert the digitized analog signal into temperature. A/D converters often do not contain conversion formulae. This omission keeps them flexible. Therefore, to use voltage data, you may have to make a mathematical conversion. This conversion may be complex (depending on the A/D converter).

Chapter 9 has a thorough discussion of some of the more common schemes used by A/D converters and how to determine which scheme is appropriate for your converter. In the ACRO-912 case, transformation is unnecessary. However, if your A/D converter needs a conversion formula, the formula can be placed in the {PUT} command. For example, if a transformation was to multiply by 1000, then the command would look like the following:

```
{PUT ABSDATA,0,ROW,1000*@ABS(@VALUE(@CLEAN(BUFFER))):VALUE}
```

Scientific data often vary as the logarithm, natural log, exponential, sine, cosine, etc., of the voltage. You can handle these cases by adding the appropriate Lotus mathematical @function(s) into the {PUT} command that parses the string. The following is a list of common @functions:

@LOG(number)	Log of number, base 10
@LN(number)	Log of number, base e
@EXP(number)	The number e raised to the number power
@SIN(number)	Sine of number
@COS(number)	Cosine of number
@TAN(number)	Tangent of number
@ASIN(number)	Arc sine of number
@ACOS(number)	Arc cosine of number
@ATAN(number)	Two-quadrant arc tangent of number
@SQRT(number)	Positive square root of number

For example, to transform a value using the @LOG function use the following {PUT} command:

```
{PUT ABSDATA,0,ROW,@LOG(@ABS(@VALUE(@CLEAN(BUFFER)))):VALUE}
```

Take note of the @ABS function in the {PUT} command. This @function returns the absolute value. Taking the absolute value of voltages is good in many situations because voltage may vary from zero to negative numbers in some experiments and instruments. Negative voltages often have no bearing on measurements. Therefore, the @ABS will ensure that your calculations always work with positive numbers.

RUNNING IN THE BACKGROUND

As explained above, most stand-alone interfaces can run in the background. That is, they can gather data independently of the computer. This feature is especially useful for long experiments that proceed without personal computer intervention. For example, HPLC experiments often take hours and do not require contact with the personal computer.

Running in background mode also allows you to collect data at a rapid rate; in fact data are collected several orders of magnitude faster than prompting for them one data point at a time. The communications process between an instrument and a personal computer is very slow. However, in background mode data collection can be made at the speed of the CPU within the interface. So, if you run experiments that require rapid data collection, consider running in background mode.

To designate a stand-alone interface to run in background mode, send it the appropriate set of string commands. Then, when you want data, just send the interface the appropriate set of strings to prompt the A/D converter to return the stored data.

Figures 6-11 through 6-14 show how the previous DNA melt experiment would have been run in background mode. This example illustrates how to set up

```
---------N--------O--------P--------Q--------R--------S--------T--------U------
 1                   /*BACKGROUND DNA MELT PROGRAM*/
 2
 3                /*AUTO-EXECUTING MACRO FOR LOTUS 1-2-3*/
 4 \0         {LET VERSION,0}    /*TELLS PROGRAM IT'S RUNNING UNDER 1-2-3*/
 5            /S        /*ISSUE THE /S MENU CHOICE TO GO TO SYSTEM*/
 6            {MENUBRANCH BACK_MU}          /*PUT UP USER'S MENU*/
 7
 8                    /* AUTO-EXECUTING MACRO FOR LOTUS SYMPHONY*/
 9            /*AUTO-EXECUTING MACRO: TESTS TO SEE IF DOS.APP ATTACHED; */
10            /*IF NOT, ATTACHES IT.  THEN SETS BAUD RATE, ETC.*/
11 AUTOEXEC  {LET VERSION,1}    /*TELLS PROGRAM IT'S RUNNING UNDER SYMPHONY*/
12            {IF @ISERR(@APP("DOS","")}{HOOKUP} /*SEE IF DOS.APP ATTACHED*/
13            {SERVICES}AIDOS~MODE COM1:4800,E,7,2~          /*SET COMM PORT*/
14            {MENUBRANCH BACK_MU}          /*PUT UP USER'S MENU*/
15
16 HOOKUP     {SERVICES}AADOS~Q             /*ATTACHES DOS.APP*/
17
18                /*MACRO FOR BACKGROUND DATA ACQUISITION */
19 GOBACK     {LET DATE,@DATEVALUE(@NOW)}          /*DATE STAMP THE EXPT*/
20            {OPEN "COM1",M}    /*OPEN COM PORT AS DEVICE FILE NAME*/
21            {GETNUMBER "How Many Time Points?",POINTS}
22            {GETNUMBER "What Is The Interval Between Measurements?",LENGTH}
23            {IF LENGTH=0}{LET LENGTH,0.2}          /*SET MINIMUM WAIT*/
24            {PANELOFF}                    /*FREEZE PANEL*/
26            {GOTO}START~        /*MOVE CELL-POINTER TO STARTING POSITION*/
27            {LET COMMAND,ABS_HDR}         /*MAKE COMND STRNG FOR A->D MODULE*/
28            {LET COMMAND,COMMAND&@STRING(POINTS+1,0)&","}
29            {LET COMMAND,COMMAND&@STRING(LENGTH*50,0)}     /*50 =>1 SECOND*/
30            {LET COMMAND,COMMAND&EOL}                      /*PER CHANNEL*/
31            {WRITELN COMMAND}            /*SEND COMMAND TO VOLTAGE MODULE*/
32            {LET COMMAND,TMP_HDR}         /*MAKE COMND STRING FOR TEMP MODULE*/
33            {LET COMMAND,COMMAND&@STRING(POINTS+1,0)&","}|*NOTE: THIS CALC*|
34            {LET COMMAND,COMMAND&@STRING(LENGTH*50,0)}   |*TAKES ENOUGH    *|
35            {LET COMMAND,COMMAND&EOL}                    |*TIME TO ALLOW   *|
36            {WRITELN COMMAND}                            |*OM-900 TO       *|
37            {CLOSE}                                      |*COMPLETE PRIOR  *|
38            {WINDOWSON}{PANELON}                         |*COMMAND         *|
39            {HOME}              /*REPOSITION CELL-POINTER TO CELL A1*/
40
41
42            /*RETRIEVES THE BACKGROUND DATA*/
43 GETBACK    {GETNUMBER "How Many Data Points?",POINTS}
44            {IF VERSION=0}/REDATA~        /*1-2-3: ERASE THE DATA SECTION*/
45            {IF VERSION=1}/EDATA~         /*SYMPH: ERASE THE DATA SECTION*/
46            {HOME}{DOWN 7}      /*RESET CELL-POINTER*/
47            {OPEN "COM1",M}     /*OPEN THE PORT*/
48            {WRITELN "#4"}      /*SWITCH TO TEMPERATURE MODULE*/
49            {FOR INC1,0,POINTS-1,1,TEMPLOOP}      /*GET BACKGROUND TEMPS*/
50            {HOME}{DOWN 7}      /*RESET CELL-POINTER*/
51            {WRITELN "#5"}      /*SWITCH TO A->D MODULE*/
52            {FOR INC1,0,POINTS-1,1,ABSLOOP}       /*GET BACKGROUND VOLTAGES*/
53            {CLOSE}             /*CLOSE THE PORT*/
54            {CALC}              /*FORCE RECALCULATION OF THE SPREADSHEET*/
55
56
```

Figure 6-11

```
---------N--------O------------------------P----------------------Q------------R----------------S--------------T--------U-----
58
59 BACK_MU  BEGIN                     RETRIEVE                FILE          PRINT             SAMPLE          QUIT
60          START BACKGROUND OPERATION RETRIEVE BACKGROUND DATA SAVE THE FILE PRINT THE RESULTS ENTER SAMPLE ID QUIT THE MACRO
61          {GOBACK}                  {GETBACK}               {BRANCH SAVE} {BRANCH PRINT}    {GOTO}SAMPLE~   {QUIT}
62          {MENUBRANCH BACK_MU}      {MENUBRANCH BACK_MU}                                    {?}~
63                                                                                            {MENUBRANCH BACK_MU}
64
```

Figure 6-12

```
--------N--------O--------P--------Q--------R--------S--------T--------U----
66                      /*LOOP TO RETRIEVE AND PLACE TEMPERATURE DATA*/
67 TEMPLOOP {WRITELN "BUF1"}  /*PROMPT FOR A TEMPERATURE*/
68          {READLN BUFFER}   /*READ IN THE TEMPERATURE*/
69          {PUT DATA,0,INC1,INC1}     /*PUT DATA POINT NUMBER*/
70          {PUT DATA,2,INC1,@VALUE(@CLEAN(BUFFER)):VALUE}
71          {CALC}            /*DISPLAY THE NEW POINT*/
72          {DOWN}            /*MOVE CELL-POINTER DOWN TO SCROLL SCREEN*/
73
74                      /*MACRO WHICH RETRIEVES AND PLACES ABSORBANCE DATA*/
75 ABSLOOP  {WRITELN "BUF1"}  /*PROMPT FOR A VOLTAGE*/
76          {READLN BUFFER}   /*READ IN THE VOLTAGE*/
77          {PUT DATA,3,INC1,@ABS(@VALUE(@CLEAN(BUFFER)))*100:VALUE}
78          {CALC}            /*DISPLAY THE NEW POINT*/
79          {DOWN}            /*MOVE CURSOR DOWN TO SCROLL SCREEN*/
80
81          /*PRINT MACRO: SET "REPORT" RANGE; PRINT; RETURN TO MENU1*/
82 PRINT    {CALC}                   /*FORCE RECALCULATION OF TEMPLATE*/
83          {WINDOWSOFF}{PANELOFF}   /*FREEZE THE DISPLAY*/
84          {HOME}            /*GO TO CELL A1; SET "REPORT" RANGE*/
85          /RNCREPORT~       /*REDEFINE REPORT RANGE*/
86          {ESC}{HOME}.{RIGHT}{DOWN 6}{END}{DOWN}{END}{RIGHT 4}~
87          {IF VERSION=0}/PPRREPORT~AGPQ                  /*LOTUS 1-2-3*/
88          {IF VERSION=1}{SERVICES}PSSRREPORT~QAGPQ       /*SYMPHONY*/
89          {WINDOWSON}{PANELON}     /*THAW THE DISPLAY*/
90          {MENUBRANCH BACK_MU}     /*RETURN TO USER'S MENU*/
91
92          /*SAVE THE FILE*/
93 SAVE     {IF VERSION=0}/FS{?}~    /*LOTUS 1-2-3*/
94          {IF VERSION=1}{S}FS{?}~  /*SYMPHONY*/
95          {RIGHT}~ /*FORCE OVERWRITE; CANNOT BE ON SAME LINE AS THE FS*/
96          {HOME}                   /*MOVE CELL-POINTER TO CELL A1*/
97          {MENUBRANCH BACK_MU}     /*RETURN TO USER'S MENU*/
99
100 \M      {MENUBRANCH BACK_MU}   /*A WAY TO START WITHOUT AUTOEXEC*/
101
102
103         /*SCRATCHPAD OF NAMED RANGES*/
104 VERSION             /* 0=>LOTUS 1-2-3; 1=>SYMPHONY */
105 INC1                /*COUNTER FOR FIRST LOOP; ROW OFFSET FOR {PUT}*/
106 POINTS              /*NUMBER OF DATA POINTS TO COLLECT*/
107 LENGTH              /*USER DEFINED INTERVAL BETWEEN DATA COLLECTION*/
108 BUFFER              /*A BUFFER FOR INCOMING DATA*/
109
110         /*SET OF PROMPTS FROM WHICH TO ASSEMBLE COMMANDS FOR OM-900*/
111 TMP_HDR #4,RESET;BUF1=CLR:DIM800:ON;@REPEAT           /*TEMP HEADER*/
112 ABS_HDR #5,RESET;RATE=1;BUF1=CLR:DIM800:ON;@REPEAT    /*VOLTAGE HDR*/
113 EOL     ;CONV1=1;@UNTIL;BUF1=OFF                /* THE END OF THE LINE*/
114 COMMAND
115
116
```

Figure 6-13

```
---------N--------O--------P--------Q----------R--------S--------T--------U-----
117                         RANGE NAME TABLE
118                         ================
119                ABSLOOP  O75          LENGTH   O107
120                ABS_HDR  O112         POINTS   O106
121                AUTOEXEC O11          PRINT    O82
122                BACK_MU  O59          REPORT   A1..F14
123                BUFFER   O108         RESULTS  A1..F18
124                COMMAND  O114         SAMPLE   D3
125                DATA     C7..F3000    SAVE     O93
126                DATE     D4           START    C8
127                EOL      O113         TEMPLOOP O67
128                GETBACK  O43          TMP_HDR  O111
129                GOBACK   O19          VERSION  O104
130                HOOKUP   O16          \O       O4
131                INC1     O105         \M       O100
132
133
134
```

Figure 6–14

a program for background monitoring of experiments. The pertinent macros are GOBACK and GETBACK. They "go" to background mode and "get" data from the background mode.

The GOBACK macro begins by prompting the user for the number of data points and the interval between data points. A delay may be desirable as an inclusion because data is collected very quickly in background mode. In fact, if you do not slow the A/D converter, you will collect data at the speed of 0.05 seconds per data point when a conversion rate of 50 ms is used. This speed translates to 20 data points per second! Other A/D converters are even faster (up to 200,000 samples per second). Therefore, unless you need this speed, you should collect the data points at a rate that is about 5 times as fast as the change that occurs in the signal.

Command strings are assembled from the user's responses to these two prompts. This assembly allows for customized performance based on user input. The process that is used is called "string concatenation". String concatenation links characters of two or more strings into a single chain. Lotus uses the "&" sign to carry out this process. String concatenation is similar to addition. However, the second string is added onto *the end of* the first one and so on. To better understand this process, take a look at what happens as a string is concatenated for the A/D module.

The final string will be in the cell named COMMAND. The string is assembled by a series of {LET} commands in the GOBACK macro of Figure 6–11.

COMMAND is started by moving the absorbance header (ABS_HDR) string into the cell. At this point, the COMMAND cell looks like the following

```
#5,RESET;RATE=1;BUF1=CLR:DIM800:ON;@REPEAT
```

Next, the number of data points input by the user is converted into a string. Remember a difference exists between a number and a character string. For the same reason that you cannot add an A to a 2.7, you cannot concatenate a number onto a string without first converting the number into characters.

The @STRING function in the {LET} command of GO_BACK performs the conversion. The zero in the @STRING function indicates that the string made from POINTS should have zero decimal places. A comma is also added at this point. The COMMAND string now looks like the following (assuming that the user entered 100 data points at the appropriate prompt):

```
#5,RESET;RATE=1;BUF1=CLR:DIM800:ON;@REPEAT101,
```

Next, the delay portion of the string is added. Again @STRING is used to obtain a string from the user defined delay. Most stand-alone interfaces measure time in millisecond (0.001 second) increments. The standard unit of time for the ACRO-900 is 10 milliseconds. Therefore, your program must multiply the number of seconds for the user specified delay by 100. This multiplication literally gives the number of 10 millisecond intervals to count off for each second ($100 \times 0.01 = 1$ second). Thus, if a user wanted twenty second delays, 2000 (20×100) intervals would have to be specified in the COMMAND string.

Most stand-alone interfaces use this convention and it results in better time resolution. With this method, you can actually program your interface to take readings fractions of a second apart; however, your A/D converter must be able to convert voltages that fast (e.g., the ACRO-900 takes 50 to 100 milliseconds per conversion).

Now, consider the following peculiarity that you need to be aware of and look for in your interface. If two ACRO-900 modules are working in background mode, the interval that is counted off is double that of a single module. That is, the 10 millisecond interval becomes a 20 millisecond interval if two modules are used. For this reason, you would multiply by 50 (instead of 100) to obtain the number of intervals to specify.

It is advisable to use a stop watch to monitor the length of time it takes to complete various combinations of times and numbers of data points. By being accurate, you can ascertain precisely how an interval is defined under the conditions that a module(s) is being used.

After this round of concatenation, COMMAND looks like the following (assuming the user specified a 20 second delay at the appropriate prompt):

```
#5,RESET;RATE=1;BUF1=CLR:DIM800:ON;@REPEAT101,1000
```

Finally, the end of the line (EOL) is concatenated onto the COMMAND string. So, the string that is sent to the OM-912 A/D converter for this example looks like the following:

```
#5,RESET;RATE=1;BUF1=CLR:DIM800:ON;@REPEAT101,1000;CONV1=1;
@UNTIL;BUF1=OFF
```

Take a moment to examine this string because it contains all of the elements that your interface is likely to require.

Commands that are sent to modules of stand-alone interfaces usually start by specifying the module address. In this example, "#5" specifies the address of the OM-912 A/D module. This specification tells the ACRO-900 command module to transmit the remainder of the command to the module at address 5.

It is always a good idea to reset a module to its power-up status before working with it. A reset helps decrease the chances of unpredictable behavior. In this example the RESET command clears the module of all old data and resets all of its parameters to their default states. That is, it resets the module to the state that it would be in if you had just turned it on.

Another parameter that usually needs to be set is the rate and resolution for the A/D conversion. In this example, the RATE command sets the conversion rate to 100 millisecond (17-bit/high resolution mode).

Next, you typically specify the amount of memory that will store the data. In the ACRO-912, the BUF command designates the memory amount. In the example, the buffer is cleared, space is allocated for 800 data points, and the buffer is activated with the ON command. (The ON command does not begin actual data collection.)

Your next step is to specify how data is to be collected. The ACRO-900 system uses a method similar to the Lotus {FOR} loop for these purposes. The equivalent ACRO-900 command is the "@REPEAT...@UNTIL" combination. This combination of commands executes the number of times specified by the digits just after the @REPEAT command and before the comma. A specification for the number of 20 millisecond intervals to count off before repeating the loop follows the comma.

All commands that are to be repeated are found between @REPEAT and @UNTIL. In the example, CONV1=1 appears. This designation directs channel one of the ACRO-912 to convert a single analog voltage in background mode and store the results in the ACRO-912's buffer.

Thus, the @REPEAT command literally says "convert a voltage on channel one, one hundred times, with 20 seconds between each conversion, and store the results in the background buffer".

It is important to turn an A/D converter's background buffer off after all data points from an experiment have been placed in the buffer. This procedure disables the buffer and prevents its data from being overwritten. In effect, turning the background buffer off protects the data until you can retrieve it. This procedure also performs one other very important function. Buffers keep track of the memory address of the piece of data currently being worked on. The method that is used to store this information is called a "buffer pointer" (because it points to a memory location). Turning the buffer off in an interface moves the buffer pointer to the first piece of data collected, thereby allowing you to retrieve data in the order that it was collected. In the example, BUF1=OFF accomplishes this task.

The assembled command is transmitted to the ACRO-912 in the customary fashion. A similar command is assembled and sent to the ACRO-931 Thermocouple

Module. However, sending the ACRO-931 command illustrates an important timing consideration that you need to be aware of. After the command for the ACRO-912 is sent, it takes the ACRO-900 a short period of time to process the command. If the ACRO-900 receives another command while it is processing the current one, it will lock up. This response is typical for many instruments. In the example program, assembling the ACRO-931 command takes just enough time to allow the ACRO-900 to process the ACRO-912 command. If the time interval was not sufficient, a {WAIT} command would have been necessary to produce a delay that was long enough to allow the ACRO-900 to complete the processing of the previous command.

RETRIEVING DATA FROM BACKGROUND MODE

You usually retrieve data from the background buffer in the same fashion that you do in normal (''interactive'') mode. You send a string command to the interface and prompt it for the data, one data point at a time. The GETBACK macro illustrates how this process is accomplished for the ACRO-912 and ACRO-931 example. A closer look at the TEMPLOOP and ABSLOOP macros that are used by GETBACK reveals that they are almost identical to their ''interactive'' equivalents in Figure 6-9. However, BUF1 commands prompt for data instead of TEMP1 and VIN1 commands.

The BUF1 command prompts the module to return a single data point from the buffer area to the personal computer. After the data point is returned, the buffer pointer is automatically incremented by one so that the next prompt will return the next piece of data in the buffer.

TAKING THE SNAPSHOT

This section discusses how your programs should select the data that is to be placed in your template. Screening a chromatographic column, collecting data from an HPLC, scanning a spectrum, determining melting point curves, monitoring the flow rate through a pipe, measuring accelerations, monitoring enzyme kinetics, and monitoring a fermentor are all continuous processes. For a continuous process, you can give a program the number of time points to collect and the interval between the time points and signal the program to start. The programs shown in Figures 6–7 to 6–9 and Figure 6–15 are examples of this form of data collection.

However, obtaining a single representative value for a sample after an instrument has reached a steady state for the sample is different and is dependent on the type of instrument and your expected results.

Some instruments, like the Perkin-Elmer LS-2B Fluorimeter (described in detail in the next chapter), have a ''sample and hold'' output. This type of output is similar to the ''sample and hold'' that was described for A/D converters. It takes

```
--------V--------W--------X--------Y--------Z--------AA-------AB-------AC-------
 1   /* TEST PROGRAM TO MONITOR VOLTAGE OR CHART RECORDER OUTPUTS: */
 2        /*TAKES READINGS AND DETERMINES THE PERIODIC SEQUENCE OF*/
 3                  /*EVENTS DURING THE SAMPLING PROCESS*/
 4 VOLT_TEST{LET DATE,@DATEVALUE(@NOW)}           /*DATE STAMP THE EXPERIMENT*/
 5         {OPEN "COM1",M}    /*OPEN COM PORT AS DEVICE FILE NAME*/
 6         {GETNUMBER "How Many Time Points?",POINTS}
 7         {GETNUMBER "How Long Is The Interval Between Measurements?",DELAY}
 8         {INDICATE I/O}{PANELOFF}     /*FREEZE PANEL*/
 9         {WRITELN "#5;VIR=-5"}        /*MODULE 5: FULL SCALE = 2 VOLTS*/
10         {GOTO}START~        /*MOVE CELL-POINTER TO STARTING POSITION*/
11         {WINDOWSOFF}{PANELOFF}       /*FREEZE THE DISPLAY*/
12         {LET BEGIN,@TIMEVALUE(@NOW):VALUE}   /*GET THE STARTING TIME*/
13         {IF DELAY>0}{FOR ROW,0,POINTS,1,BIGLOOP}   |*THESE GIVE MAXIMAL*|
14         {IF DELAY=0}{FOR ROW,0,POINTS,1,SMALLLOOP} |*PERFORMANCE FOR   *|
15         {CLOSE}             /*CLOSE THE PORT*/     |*ZERO DELAY        *|
16         {CALC}              /*FORCE RECALCULATION OF THE SPREADSHEET*/
17         {GOTO}ELAPSCOL~              /*RESET CELL-POINTER TO START*/
18         {IF VERSION=0}/RFDT3.{DOWN POINTS}~ /*TIME FORMAT, 1-2-3*/
19         {IF VERSION=1}/FT3.{DOWN POINTS}~   /*TIME FORMAT, SYMPHONY*/
20         {WINDOWSON}{PANELON}         /*THAW THE DISPLAY*/
21         {INDICATE DONE}     /*CHANGE INDICATOR TO "DONE"*/
22
23                  /*MACRO THAT GETS ONE DATA TIME POINT*/
24 BIGLOOP {INDICATE PAUSE}{PANELON}{PANELOFF} /*SHOW DELAY PHASE*/
25         {WAIT @NOW+@TIME(0,0,DELAY)}        /*WAIT FOR SPECIFIED TIME*/
26 SMALLLOOP {LET CURRENT,@TIMEVALUE(@NOW)}    /*GET CURRENT TIME*/
27         {LET DELTA,CURRENT-BEGIN}           /*FIND ELAPSED TIME*/
28         {PUT ELAPSCOL,0,ROW,DELTA}          /*PUT ELAPSED TIME IN SHEET*/
29         {PUT PTCOL,0,ROW,ROW}               /*PUT POINT NUMBER IN SHEET*/
30         {WINDOWSON}{WINDOWSOFF}             /*ADD DATA TO SCREEN*/
31         {DOWN}                              /*MOVE DOWN TO NEXT ROW*/
32         {GET_VOLT}                          /*GET A VOLTAGE READING*/
33         {INDICATE NEXT}{PANELON}{PANELOFF}
34
35                  /*MACRO WHICH GETS ONE VOLTAGE*/
36 GET_VOLT {WRITELN "VIN1"}            /*PROMPT ACRO-912 FOR A VOLTAGE*/
37         {READLN BUFFER}              /*READ IN A VOLTAGE VALUE*/
38         {PUT ABSDATA,0,ROW,@ABS(@VALUE(@CLEAN(BUFFER))):VALUE}
39                  /*ABSDATA RANGE IS C7..C3000*/
40
41 \M      {VOLT_TEST}         /*CALL UP VOLTAGE TEST AS A SUBROUTINE*/
42
43         /*SCRATCHPAD OF NAMED RANGES*/
44 VERSION          /* 0=>LOTUS 1-2-3; 1=>SYMPHONY */
45 ROW              /*COUNTER FOR LOOP; ROW OFFSET FOR {PUT}*/
46 POINTS           /*NUMBER OF DATA POINTS TO COLLECT*/
47 BEGIN            /*TIME EXPERIMENT STARTED*/
48 DELAY            /*USER DEFINED DELAY BETWEEN DATA POINTS*/
49 CURRENT          /*CURRENT TIME*/
50 DELTA            /*ACTUAL DELAY, FROM TIME ZERO*/
51 BUFFER           /*A BUFFER FOR INCOMING DATA*/
52
53
```

Figure 6–15

a snapshot of the fluorescence at some pre-defined time after the sampling is completed and sends a steady voltage to the recorder output. After the voltage representing fluorescence units is held for a short period of time, the recorder output returns to a zero voltage (or, at most, a small positive voltage). With this type of experiment, a separation between useful data and the "idle" mode is clear enough to implement a "trigger" to start data collection.

The "sample and hold" feature makes it easy to predict when to collect the data. Unfortunately, most other instruments do not have this "sample and hold" feature. They continually send data and you must characterize the pattern that exists in the output voltage at the point in time when the valid sample reading is being presented. This characterization allows you to develop a program that utilizes the pattern to take a snapshot reading for the data that is to be placed in your template.

You can test your instrument to see how it reacts by writing a program like the one shown in Figure 6–15. You may have to modify the {WRITELN} commands so that they use the string commands appropriate for your A/D converter. You may also need to modify the parsing method in the {PUT} command to correctly handle the format of the data string that your converter returns. The program in Figure 6–15 monitors the analog output of an instrument over a period of time. When prompted for the number of data points, a user types in a number large enough to take readings for three or four sample cycles. The time delay that the user specifies depends on the speed of the process being evaluated. The user must specify a delay that is short enough to allow the program to take at least three or four voltage measurements during the steady state phase.

Next, run the example program shown in Figure 6–15 and your instrument simultaneously to collect some data.

Look at the data to determine if any correlation between the periodic behavior of the voltage output and the data that you want to use exists. Figure 6–16 is an

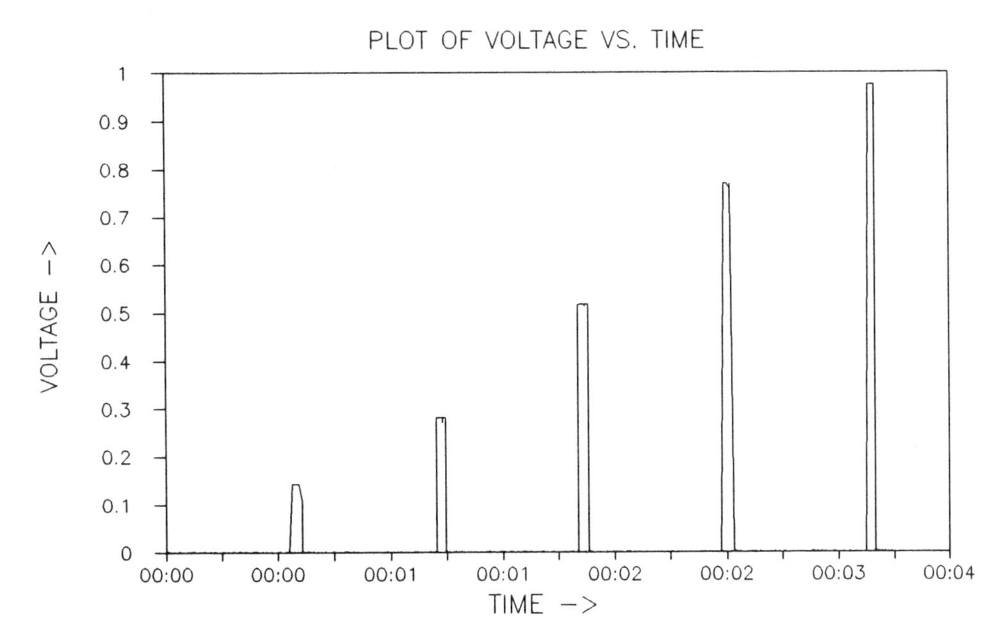

Figure 6–16

illustration of what you may find. As you can see, using Lotus graphics to plot the raw data can be a helpful aid to reviewing it.

Typically the program that you write for your instrument will have three to five different sections. Each of these sections will correspond to a sequence of events that takes place when repetitive samplings are made. To get a feel for what each section of the program will need to accomplish, think about how analog output usually reacts during a model experimental session.

At the beginning of an experiment, a user usually sets an instrument to give a zero signal using some sort of blanking procedure. This procedure sets the display to 0 and the output voltage to a very low value. The display and output voltage remains at these low settings until the user starts the test on the first sample.

When the first sample is introduced, the display and output voltage moves through a series of intermediate values until they reach a steady state. This steady state contains the data to be collected.

Next, the user would remove the first sample from the instrument; the display returns to 0 and the voltage output returns to a low value again.

All subsequent samples are taken in a similar fashion.

Each section of your program needs to correspond to a division in this process. Each section must test for the completion of a particular division of the cycle before continuing on to the next section.

Compare the Perkin Elmer LS-2B program example in Figure 6–17 with the plot in Figure 6–16 to identify each program section. The macro that gets data is called VOLT_TRIP.

The VOLT_TRIP macro begins by performing some housekeeping tasks. It prompts the user to enter the number of samples to be assayed and issues the appropriate erase command to clear out any old data. It then initializes the sample (ROW) counter to zero.

Note that the program follows the general scheme of incrementing the ROW counter and then issuing a {BRANCH GET_VOLT} command at the bottom of the macro to loop back to GET_VOLT. Another means of looping to collect data is a {FOR ROW,1,SAMPLES,1,GET_VOLT} command. However, following the scheme of the example program is required in this situation. Data are received continually from an instrument. Each of the following program sections perform what amounts to an idle phase wherein the section waits until a specific goal is met before progressing to the next section. To accomplish this scheme, {BRANCH} commands return to the beginning of the section and read in another value to be tested. Subroutine calls cannot be used because of the nesting limits imposed by Lotus. That is, if a delay prevented a section from achieving its goal and the subroutine would have had to have been called repeatedly, a nesting error would have occurred and the macro would have been aborted. {BRANCH} commands do not have this nesting limitation. On the other hand, they also do not have a "memory" for where they were branched from; so you cannot "backtrack". If the {FOR} command were used in the program, it would have called GET_VOLT

```
--------V--------W--------X--------Y--------Z--------AA-------AB-------AC-------
 1          /*PROGRAM TO MONITOR VOLTAGE OR CHART RECORDER OUTPUTS TO*/
 2          /*IDENTIFY STEADY STATE READINGS AND TAKE A DATA POINT. IE,*/
 3          /*VOLTAGE MONITORED AND A VOLTAGE OF GREATER THAN 0.001*/
 4          /*"TRIPS" THE DATA COLLECTION PROCESS.*/
 5 \0       {LET VERSION,0}    /*RUNNING UNDER 1-2-3*/
 6          {VOLT_TRIP}        /*START THE PROGRAM*/
 7
 8 AUTOEXEC {LET VERSION,1}    /*RUNNING UNDER SYMPHONY*/
 9          {IF @ISERR(@APP("DOS",""))}{HOOKUP}
10          {S}AIDOS~                  /*INVOKE DOS APPLICATION*/
11          MODE COM1:4800,E,7,2~      /*SET COMMUNICATION PARAMETERS*/
12          {VOLT_TRIP}        /*START THE PROGRAM*/
13
14 HOOKUP   {S}AADOS~          /*ATTACH DOS.APP*/
15
16 VOLT_TRIP{GETNUMBER "How many Samples?  ",SAMPLES}
17          {IF VERSION=0}/REDATA~     |*ERASE DATA SECTION   *|
18          {IF VERSION=1}/EDATA~      |*0=>1-2-3; 1=>SYMPHONY*|
19          {OPEN "COM1",M}            /*OPEN COMMUNICATIONS PORT*/
20          {WRITELN "#5;VIR1=-5"}     /*ACTIVATE ACRO-912 A/D CONVERTER*/
21          {LET ROW,0}                /*INITIALIZE ROW COUNTER*/
22 GET_VOLT {LET ROW,ROW+1}            /*INCREMENT ROW COUNTER*/
23          {WINDOWSOFF}{PANELOFF}     /*FREEZE THE DISPLAY*/
24 WAIT_LO  {IF ROW>SAMPLES}{QUIT}     /*LAST SAMPLE?*/
25          {WRITELN "VIN1"}           /*PROMPT FOR A VOLTAGE*/
26          {READLN VOLT_BUF}          /*READ IN A VOLTAGE*/
27          {LET VOLT_VAL,@VALUE(@CLEAN(VOLT_BUF))}  |*WAIT UNTIL VOLTAGE  *|
28          {IF VOLT_VAL>0.001}{BRANCH WAIT_LO}      |*RETURNS TO LOW VALUE*|
29 WAIT_HI  {WRITELN "VIN1"}           /*PROMPT FOR A VOLTAGE*/
30          {READLN VOLT_BUF}          /*READ IN A VOLTAGE*/
31          {LET VOLT_VAL,@VALUE(@CLEAN(VOLT_BUF))}
32          {IF VOLT_VAL<0.001}{BRANCH WAIT_HI} /*WAIT UNTIL VOLTAGE CLIMBS*/
33          {WAIT @NOW+@TIME(0,0,1)}   /*WAIT 1 SEC FOR EQUILIBRATION*/
34          {WRITELN "VIN1"}           /*PROMPT FOR A VOLTAGE*/
35          {READLN VOLT_BUF}          /*READ IN THE DATA VALUE*/
36          {PUT DATA,0,ROW,@VALUE(@CLEAN(VOLT_BUF))} /*ADD DATA TO TEMPLATE*/
37          {WINDOWSON}{WINDOWSOFF}    /*REFRESH THE DISPLAY*/
38          {BRANCH GET_VOLT}          |*LOOP BACK TO GET_VOLT*|
39                                     |*FOR NEXT SAMPLE      *|
40 \M       {VOLT_TRIP}        /*CALL VOLTAGE TRIP AS A SUBROUTINE*/
41          {CALC}             /*FORCE SPREADSHEET UPDATE*/
42          {CLOSE}            /*CLOSE THE PORT*/
43
44          /*SCRATCH PAD*/
45 VOLT_BUF                    /*BUFFER FOR INCOMING VOLTAGES*/
46 VOLT_VAL                    /*CELL FOR PARSING VOLTAGE STRING*/
47 SAMPLES                     /*NUMBER OF SAMPLES TO BE RUN*/
48 ROW                         /*CURRENT SAMPLE NUMBER*/
49 VERSION                     /*0 => 1-2-3; 1 => SYMPHONY*/
50
51
```

Figure 6–17

as a subroutine; but, only one data point would have been obtained because the program would have had no way of returning to {FOR}.

The GET_VOLT cell contains a {LET} command that increments the counter. At the bottom of the macro, an {IF} tests to see if the appropriate number of samples has been analyzed. If the numbers do not match, the program {BRANCH}es back to the GET_VOLT cell.

The sections that are crucial to getting a steady state value for the template are the WAIT_LO section, the WAIT_HI section, and a {WAIT} command. Each of these sections waits until a certain, specific section goal is achieved before allowing the program to advance to the next section. For example, the WAIT_LO section waits until the output voltage drops below 0.001 volts. This "program trip" voltage is set to a value slightly above the zero background (but well below potential sample values). This limit provides a "barrier zone" and also accounts for small drifts in the instrument's zero. Thus, the "barrier zone" ensures that a program does not function uncontrollably. Use the test program in Figure 6–15 to determine the value of the voltage to use as a "barrier zone" for your particular instrument and A/D converter combination.

The preliminary function of the WAIT_LO section is to ensure that the instrument has been set to zero. Thereafter, the function of this section is to detect when a sample has been removed from an instrument. When removal is detected, the voltage drops to a very low value. Therefore, a signal below 0.001 volts provides a barrier that can be used as a separation between subsequent samples.

The test in the WAIT_HI section detects a signal above 0.001 volt. In this example, after the LS-2B has determined the fluorescence of the sample, it sends a voltage to the chart recorder output that is proportional to the fluorescence. This signal is always greater than 0.001 volt. Therefore, if the test detects a value greater than 0.001 volt, a reading can be made for the template. If the test shows a value less than 0.001 volt, the chart recorder output is monitored again for another voltage to test.

When the 0.001 barrier is exceeded, the {WAIT} command allows time for the signal to reach a steady state. Once the steady state has been achieved, the program reads in another voltage, parses it, and places it into a cell of the template.

If your instrument presents its data in this fashion, use a program like the one presented in Figure 6–17. However, you may have to make some minor string command, voltage barrier, timing, parsing and/or sequence modifications to customize the program to your particular instrument and A/D converter combination.

OTHER WAYS TO TAKE A SNAPSHOT

Waiting for a specific interval of time after an output voltage exceeds a certain value works well for most instruments that perform in a well-defined, periodic fashion. However, your instrument may require other methods.

For example, your program may need to include one section that continually takes readings until the voltage exceeds a certain value and a second section that begins to take sequential readings that determines when the steady state has been achieved. In this case, the voltage value at the end of each pass through the second section is saved and compared to the next value read in to see if the change is smaller than the acceptable limit. When such a comparison is proven acceptable, a reading is taken for the template.

Figure 6–18 is an example of this method. After a voltage of 0.001 is exceeded,

```
--------V--------W--------X--------Y--------Z--------AA-------AB-------AC-------
 1 /*PROGRAM TO MONITOR CHART RECORDER OUTPUTS TO IDENTIFY STEADY-STATE*/
 2 /*READINGS AND TAKE A DATA POINT.  IE, VOLTAGE MONITORED; STEADY STATE*/
 3 /*DEFINED AS A CHANGE OF LESS THAN (0.1 * VOLT_VAL) COMPARED TO PREVIOUS*/
 4 /*VOLTAGE.*/
 5 VOLT_TRIP{GETNUMBER "How many Samples?  ",SAMPLES}
 6          {IF VERSION=0}/REDATA~      |*ERASE DATA SECTION    *|
 7          {IF VERSION=1}/EDATA~       |*0=>1-2-3; 1=>SYMPHONY*|
 8          {OPEN "COM1",M}            /*OPEN COMMUNICATIONS PORT*/
 9          {WRITELN "#5;VIR1=-5"}     /*ACTIVATE ACRO-912 A/D CONVERTER*/
10          {LET ROW,0}                /*INITIALIZE ROW COUNTER*/
11 GET_VOLT {LET ROW,ROW+1}            /*INCREMENT ROW COUNTER*/
12          {WINDOWSOFF}               /*FREEZE THE DISPLAY*/
13 WAIT_LO  {IF ROW>SAMPLES}{QUIT}     /*LAST SAMPLE?*/
14          {WRITELN "VIN1"}           /*PROMPT FOR A VOLTAGE*/
15          {READLN VOLT_BUF}          /*READ IN A VOLTAGE*/
16          {LET VOLT_VAL,@VALUE(@CLEAN(VOLT_BUF))}  |*WAIT UNTIL VOLTAGE  *|
17          {IF VOLT_VAL>0.001}{BRANCH WAIT_LO}      |*RETURNS TO LOW VALUE*|
18 WAIT_HI  {WRITELN "VIN1"}           /*PROMPT FOR A VOLTAGE*/
19          {READLN VOLT_BUF}          /*READ IN A VOLTAGE*/
20          {LET VOLT_VAL,@VALUE(@CLEAN(VOLT_BUF))}
21          {IF VOLT_VAL<0.001}{BRANCH WAIT_HI} /*WAIT UNTIL VOLTAGE CLIMBS*/
22 WAIT_STDY{LET REF,VOLT_VAL}         /*ARCHIVE THE PREVIOUS READING*/
23          {WAIT @NOW+@TIME(0,0,1)}   /*WAIT FOR 1 SEC*/
24          {WRITELN "VIN1"}           /*PROMPT FOR A VOLTAGE*/
25          {READLN VOLT_BUF}          /*READ IN THE DATA VALUE*/
26          {LET VOLT_VAL,@VALUE(@CLEAN(VOLT_BUF))}
27          {IF @ABS(REF-VOLT_VAL)>(0.1*VOLT_VAL)}{BRANCH WAIT_STDY}
28          {PUT DATA,0,ROW,VOLT_VAL}  /*ADD DATA TO TEMPLATE*/
29          {WINDOWSON}{WINDOWSOFF}    /*REFRESH THE DISPLAY*/
30          {BRANCH GET_VOLT}          |*LOOP BACK TO GET_VOLT*|
31                                     |*FOR NEXT SAMPLE      *|
32
33 \M       {VOLT_TRIP}          /*CALL VOLTAGE TRIP AS A SUBROUTINE*/
34          {CALC}               /*FORCE SPREADSHEET UPDATE*/
35          {CLOSE}              /*CLOSE THE PORT*/
36                   /*SCRATCH PAD*/
37 VOLT_BUF                      /*BUFFER FOR INCOMING VOLTAGES*/
38 VOLT_VAL                      /*CELL FOR PARSING VOLTAGE STRING*/
39 SAMPLES                       /*NUMBER OF SAMPLES TO BE RUN*/
40 ROW                           /*CURRENT SAMPLE NUMBER*/
41 VERSION                       /*0 => 1-2-3; 1 => SYMPHONY*/
42 REF                           /*PREVIOUS VOLTAGE VALUE*/
43
44
```

Figure 6-18

the present voltage is transferred to REF cell. Another voltage is then read in and compared to REF. The comparison that is made is the one in the {IF} command near the bottom of the macro. The following is the test that is made

```
{IF @ABS(REF-VOLT_VAL)>(0.1*VOLT_VAL)}
```

The @ABS function takes the absolute value of the difference between the archived REF cell and the current voltage value (VOLT_VAL). The absolute value is required because REF may be smaller than VOLT_VAL and the difference would be a negative number. The absolute value of the difference is then compared to a percentage of the current voltage value. A percentage is used because the voltage samplings can vary extensively in value; a percentage scales the test to a level that is appropriate for the present voltage value.

If the voltage is still changing by more than the percentage, VOLT_VAL is moved into the REF cell and a new VOLT_VAL is read in. This process continues until the voltage change is less than the acceptable limit. When the limit is reached, the value of VOLT_VAL is placed into the template.

You will need to determine the percentage that gives reliable results in your application. Use the data from the test program in Figure 6–15 to determine the percentage to use initially.

The acceptable voltage change limit that you place on this test will depend on several factors. For example, it will depend on the magnitude of the change that occurs in the readings when an instrument has reached the steady state for a sample. It will also depend on the combination of how quickly the voltage changes as it rises up to the steady state and the speed of your A/D converter. That is, if the change is slow and the converter is fast, a steady state may be prematurely detected.

However, the most important factor to consider is the combination of how quickly the voltage changes *just before* it reaches the steady state and the speed of your A/D converter. Voltage changes tend to be very rapid at the start of a sampling, but will slow as the voltage approaches the steady state. Therefore, you need to pay particular attention to changes near the steady state when you determine the limit to use for detecting the steady state.

Finally, the limit depends heavily on your requirements for acceptable accuracy. If your requirements are tight, you will want to make the limits very narrow. However, setting the requirements too tight may mean that you will miss some of the data.

You should confirm a percentage's appropriateness under actual experimental conditions with several samples at both low and high levels. Based on this test, you should fine tune the percentage for maximum reliability.

One variation of this method is to save sequential readings and test them to see when the voltage starts back downward. That is, when REF > VOLT_VAL. When this state is reached, you can take the value of the REF cell (or another of the previous readings before the start in the drop of the signal) and place that value into the template. This method is illustrated in Figure 6–19. The {IF} test at the bottom of the macro keeps branching back to WAIT_STDY until a downward trend is detected:

```
{IF REF<VOLT_VAL}
```

Alternatively, you could get very sophisticated by saving several of the previous readings and using the maximum or average of those readings as the voltage begins to turn downward. This method is illustrated in Figure 6–20. In this case, you could use the Lotus @MAX or @AVG functions to determine the appropriate value to use. The range within the @function should include all of the cells contain-

```
--------V--------W--------X--------Y--------Z--------AA-------AB-------AC-------
 1 /*PROGRAM TO MONITOR CHART RECORDER OUTPUTS TO IDENTIFY STEADY-STATE*/
 2 /*READINGS AND TAKE A DATA POINT.  IE, VOLTAGE MONITORED; WHEN*/
 3 /*DOWNWARD TREND IS DETECTED, SECOND-TO-LAST VOLTAGE IS PLACED*/
 4 /*INTO TEMPLATE*/
 5 VOLT_TRIP{GETNUMBER "How many Samples?  ",SAMPLES}
 6           {IF VERSION=0}/REDATA~      |*ERASE DATA SECTION    *|
 7           {IF VERSION=1}/EDATA~       |*0=>1-2-3; 1=>SYMPHONY*|
 8           {OPEN "COM1",M}             /*OPEN COMMUNICATIONS PORT*/
 9           {WRITELN "#5;VIR1=-5"}      /*ACTIVATE ACRO-912 A/D CONVERTER*/
10           {LET ROW,0}                 /*INITIALIZE ROW COUNTER*/
11 GET_VOLT  {LET ROW,ROW+1}             /*INCREMENT ROW COUNTER*/
12           {WINDOWSOFF}                /*FREEZE THE DISPLAY*/
13 WAIT_LO   {IF ROW>SAMPLES}{QUIT}      /*LAST SAMPLE?*/
14           {WRITELN "VIN1"}            /*PROMPT FOR A VOLTAGE*/
15           {READLN VOLT_BUF}           /*READ IN A VOLTAGE*/
16           {LET VOLT_VAL,@VALUE(@CLEAN(VOLT_BUF))}  |*WAIT UNTIL VOLTAGE  *|
17           {IF VOLT_VAL>0.001}{BRANCH WAIT_LO}      |*RETURNS TO LOW VALUE*|
18 WAIT_HI   {WRITELN "VIN1"}            /*PROMPT FOR A VOLTAGE*/
19           {READLN VOLT_BUF}           /*READ IN A VOLTAGE*/
20           {LET VOLT_VAL,@VALUE(@CLEAN(VOLT_BUF))}
21           {IF VOLT_VAL<0.001}{BRANCH WAIT_HI} /*WAIT UNTIL VOLTAGE CLIMBS*/
22 WAIT_STDY{LET REF,VOLT_VAL}           /*ARCHIVE THE PREVIOUS READING*/
23           {WAIT @NOW+@TIME(0,0,1)}    /*WAIT FOR 1 SEC*/
24           {WRITELN "VIN1"}            /*PROMPT FOR A VOLTAGE*/
25           {READLN VOLT_BUF}           /*READ IN THE DATA VALUE*/
26           {LET VOLT_VAL,@VALUE(@CLEAN(VOLT_BUF))}
27           {IF REF<VOLT_VAL}{BRANCH WAIT_STDY} /*DETECT DOWNWARD TREND*/
28           {PUT DATA,0,ROW,REF}        /*ADD DATA FROM REF CELL TO TEMPLATE*/
29           {WINDOWSON}{WINDOWSOFF}     /*REFRESH THE DISPLAY*/
30           {BRANCH GET_VOLT}           |*LOOP BACK TO GET_VOLT*|
31                                       |*FOR NEXT SAMPLE      *|
32
33 \M        {VOLT_TRIP}         /*CALL VOLTAGE TRIP AS A SUBROUTINE*/
34           {CALC}              /*FORCE SPREADSHEET UPDATE*/
35           {CLOSE}             /*CLOSE THE PORT*/
36                  /*SCRATCH PAD*/
37 VOLT_BUF                      /*BUFFER FOR INCOMING VOLTAGES*/
38 VOLT_VAL                      /*CELL FOR PARSING VOLTAGE STRING*/
39 SAMPLES                       /*NUMBER OF SAMPLES TO BE RUN*/
40 ROW                           /*CURRENT SAMPLE NUMBER*/
41 VERSION                       /*0 => 1-2-3; 1 => SYMPHONY*/
42 REF                           /*PREVIOUS VOLTAGE VALUE*/
43
44
```

Figure 6–19

ing the archived data, separated by commas. For example, if you saved three previous data points (called REF1, REF2, REF3) you could use one of the following functions to find the maximum or average value:

```
@MAX(REF1,REF2,REF3,VOLT_VAL)
@AVG(REF1,REF2,REF3,VOLT_VAL)
```

The following is a sequence of commands from Figure 6–20. The sequence shows you how to move data upward into an archive system when a new value of VOLT_VAL is read in and is not less than the previous reference value (indicating a downward trend):

```
--------V--------W--------X--------Y--------Z--------AA-------AB-------AC-------
 1 /*PROGRAM TO MONITOR CHART RECORDER OUTPUTS TO IDENTIFY STEADY-STATE*/
 2 /*READINGS AND TAKE A DATA POINT.  IE, VOLTAGE MONITORED; WHEN DOWNWARD*/
 3 /*DOWNWARD TREND IS DETECTED, THE LARGEST OF THE LAST 4 READINGS*/
 4 /*IS PLACED INTO THE TEMPLATE*/
 5 VOLT_TRIP{GETNUMBER "How many Samples?  ",SAMPLES}
 6          {IF VERSION=0}/REDATA~      |*ERASE DATA SECTION   *|
 7          {IF VERSION=1}/EDATA~       |*0=>1-2-3; 1=>SYMPHONY*|
 8          {OPEN "COM1",M}            /*OPEN COMMUNICATIONS PORT*/
 9          {WRITELN "#5;VIR1=-5"}     /*ACTIVATE ACRO-912 A/D CONVERTER*/
10          {LET ROW,0}                /*INITIALIZE ROW COUNTER*/
11 GET_VOLT {LET ROW,ROW+1}            /*INCREMENT ROW COUNTER*/
12          {LET REF1,0}               |*                            *|
13          {LET REF2,0}               |*INITIALIZE DATA ARCHIVE SYSTEM*|
14          {LET REF3,0}               |*                            *|
15          {WINDOWSOFF}               /*FREEZE THE DISPLAY*/
16 WAIT_LO  {IF ROW>SAMPLES}{QUIT}             /*LAST SAMPLE?*/
17          {WRITELN "VIN1"}           /*PROMPT FOR A VOLTAGE*/
18          {READLN VOLT_BUF}          /*READ IN A VOLTAGE*/
19          {LET VOLT_VAL,@VALUE(@CLEAN(VOLT_BUF))}  |*WAIT UNTIL VOLTAGE  *|
20          {IF VOLT_VAL>0.001}{BRANCH WAIT_LO}      |*RETURNS TO LOW VALUE*|
21 WAIT_HI  {WRITELN "VIN1"}           /*PROMPT FOR A VOLTAGE*/
22          {READLN VOLT_BUF}          /*READ IN A VOLTAGE*/
23          {LET VOLT_VAL,@VALUE(@CLEAN(VOLT_BUF))}
24          {IF VOLT_VAL<0.001}{BRANCH WAIT_HI} /*WAIT UNTIL VOLTAGE CLIMBS*/
25 WAIT_STDY{LET REF1,REF2}            |*SHUFFLE PREVIOUS READINGS*|
26          {LET REF2,REF3}            |*UP INTO THE ARCHIVE      *|
27          {LET REF3,VOLT_VAL}        |*SYSTEM                   *|
28          {WAIT @NOW+@TIME(0,0,1)}   /*WAIT FOR 1 SEC*/
29          {WRITELN "VIN1"}           /*PROMPT FOR A VOLTAGE*/
30          {READLN VOLT_BUF}          /*READ IN THE DATA VALUE*/
31          {LET VOLT_VAL,@VALUE(@CLEAN(VOLT_BUF))}
32          {IF REF3<VOLT_VAL}{BRANCH WAIT_STDY}/*DETECT DOWNWARD TREND*/
33          {PUT DATA,0,ROW,@MAX(REF1,REF2,REF3,VOLT_VAL)}
34          {WINDOWSON}{WINDOWSOFF}    /*REFRESH THE DISPLAY*/
35          {BRANCH GET_VOLT}          /*LOOP BACK TO GET_VOLT*|
36                                     |*FOR NEXT SAMPLE      *|
37 \M       {VOLT_TRIP}        /*CALL VOLTAGE TRIP AS A SUBROUTINE*/
38          {CALC}             /*FORCE SPREADSHEET UPDATE*/
39          {CLOSE}            /*CLOSE THE PORT*/
40
41                   /*SCRATCH PAD*/
42 VOLT_BUF                    /*BUFFER FOR INCOMING VOLTAGES*/
43 VOLT_VAL                    /*CELL FOR PARSING VOLTAGE STRING*/
44 SAMPLES                     /*NUMBER OF SAMPLES TO BE RUN*/
45 ROW                         /*CURRENT SAMPLE NUMBER*/
46 VERSION                     /*0 => 1-2-3; 1 => SYMPHONY*/
47 REF1                        /*PREVIOUS VOLTAGE VALUES*/
48 REF2
49 REF3
50
```

Figure 6-20

```
{LET REF1,REF2}
{LET REF2,REF3}
{LET REF3,VOLT_VAL}
```

One other method can be used to detect when to take a reading. Many instruments have event markers to indicate the beginning of a process on the chart recorder. If your instrument has this feature, you can take advantage of it to trigger data collection. Use an {IF} test to look for a voltage spike. When a spike is detected, the program must wait a specified period of time before taking a reading.

LOOKING FOR INTANGIBLES

As illustrated, some very sophisticated techniques are available to pick the data point to use for your template. By looking at how data is received in your test program, you should be able to decide which method to use. You may even need to develop your own method.

Whichever method you use, it is important that you test your program and "fine tune" it under actual data collection conditions for your instrument. Test the system with a series of samplings that have both high and low values in the range that you need to detect. These samples will help you to "fine tune" the numbers in the {WAIT} commands and {IF} tests. By testing, you can ensure that an "intangible" does not affect your data collection.

For example, in the programs described in this chapter, the limits for the {IF} commands were 0.001 volt. For other instruments and A/D converters, you may find experimentally that limits of 0.02 volt get more reliable testing. That is, if you use 0.001 for other combinations of A/D converters, full scale voltage ranges, and instruments, you may find that one of the section tests trips prematurely and places extraneous data in your template. This type of problem can only be revealed under actual experimental conditions.

Another important test element is timing. Twelve-bit A/D converters and converters running in background mode tend to be much faster than the 17-bit converter of the ACRO-912. Furthermore, the other category of A/D converters (plug-in boards) do not use the slow RS-232 port. Any of these three situations can make data acquisition faster. When data acquisition is faster, tests for steady states and increasing/decreasing signals can be unreliable (because small differences caused by the fast sampling rate can be affected by noise and resolution inaccuracies).

If you experience unreliability in your data acquisition "triggers", you may want to slow the acquisition rate by placing {WAIT} commands just prior to each data acquisition command in your program. These {WAIT} commands will slow the reading process and will help ensure that rejection limits will not be prematurely exceeded. To use {WAIT} commands, examine each section of your program to ascertain where pauses are needed.

RS-232 AND PLUG-IN BOARD SNAPSHOTS

You can also use the methods presented in this chapter to determine when to take readings in instruments that have RS-232 capabilities. For example, the snapshot methods presented in this chapter can be invaluable in instruments that continually send out RS-232 data without being prompted.

Chapter 9 will show you how to apply the concepts of this chapter to the process of weighing samples on an analytical balance. It will also show you some advanced techniques for taking snapshots with plug-in A/D converters.

MONITORING AND CONTROLLING PROCESSES

Although not covered specifically in this book, you can use stand-alone interfaces to monitor processes and control them based on collected data. If your application fits into this category, you need a module that also has analog or digital output (depending on your application's requirements).

In this instance, you would develop a program (similar to one of the programs presented in this chapter) to monitor the process. You would use the Lotus {IF} command to test whether a certain condition was met. When the condition was met, you would activate an output from the appropriate module using string commands similar to the ones that you would send to obtain data. This output would activate a motor, piston, valve, pump, solenoid, etc., so that a correction could be made on the process.

In such a situation, you would get a copy of the actual history of the process in your spreadsheet, as well as being able to control the science.

Some examples of continuous processes that can be controlled in the manner described above are

Fermentors
Mammalian Cell Culture
Chemical Reactors
Fraction Collectors
Titrations
pH buffer preparation
Automatic bottle filling
Temperature control
Flow rates
Level control

WHAT'S NEXT?

The next chapter will show you how to control more than just the data collection process. For instruments that have motors that can be controlled from a personal computer, you can govern the entire data collection process—from setup of the physical parameters in the instrument to automatically drawing samples into the instrument, taking readings, and getting the data from the instrument into the personal computer. And, the entire process can be controlled from Lotus.

7

Total Automation: Controlling Instruments That Have Motors

In the last few chapters, you learned about the general steps that you would need to perform to turn a data collection device, a personal computer and a Lotus spreadsheet program into an integrated system that gets data and places it directly into a spreadsheet. This chapter will show you how to interface instruments that have a much higher level of automation. That is, you will learn how to control instruments that have motors, solenoids, valves, pistons, pumps, etc.

Another goal of this chapter is to review the steps that you use when you integrate any instrument with a personal computer and a Lotus spreadsheet. Previously, separate chapters focused on each separate task. (See Figure 7–1 for an outline that reviews the process.)

This chapter will bring all the integration tasks together to give you a better feel for the actual flow of events.

Also, by repeating the method of hooking up components of a system and developing a program, you should find it easier to apply the principles to your own instruments.

This chapter also contains more programming techniques that will give your user more flexibility.

BRIEF DESCRIPTION OF THE EXAMPLE INSTRUMENT

The Perkin-Elmer LS-2B Filter Fluorimeter was referred to in the last couple of chapters and it was used as an example. The LS-2B is a good example for this chapter because it illustrates how to completely control an instrument from a

```
                    STEPS TO A SUCCESSFUL APPLICATION

->Assemble The Instrument
->Test Instrument In Manual Mode
->Determine Wiring Configuration
->Get Cable
->Connect Cable To Computer And Instrument
->Secure Cable
->Determine How To Set Communication Parameters
->Set Module Addresses
->Reset Instrument
->Add PATH Command To AUTOEXEC.BAT File
->Add MODE Command to AUTOEXEC.BAT File
->Design Template And Macro Program
->Start The Spreadsheet
->Type Template Into Spreadsheet
->"Activate" Template
->Test The Template
->Create And "Activate" The Auto-executing Macros
->(Symphony): Specify The Name Of The Auto-executing Macro
->Use The DOS MODE Program To Set Communication Parameters
->Type In A "Bare Bones" Program
->"Activate" The Test Program
->Try To Establish Minimal Communications
->Evaluate The Test Program
->Expand Test Program To Retrieve Data
->Add Amenities And Menus To The Program
->"Activate/Re-activate" Macro Program
->(Symphony): Create Windows
->Create A Table Of Range Names
->Save The File
->Evaluate The Program
->Increase Communications Speeds
->Final Test The Program
->Save/Backup Several Copies Of The File
->Save Several Copies In Separate Safe Places
```

Figure 7-1

spreadsheet. In this capacity it is a superb first example for beginners because it not only illustrates the methods used to get data out of an instrument, but it also shows how the same methods can be used to set up the basic parameters that are needed for the instrument to carry out a specific assay.

The LS-2B requires that you send commands that set the wavelength, filter response, photomultiplier voltage, sample time, delay time, purge time, and calculation methods. These commands must be received prior to sampling. Thereafter, the instrument requires that you send commands before each sampling and the assay. That is, the LS-2B must receive commands before it activates the sampler, aspirates, triggers integration, and returns results for the spreadsheet.

This scenario is typical of nearly all highly automated instruments under personal computer control.

However, the example instrument may not be ideal. This is because the LS-2B is not a ubiquitous instrument. The following is a brief overview of LS-2B and its use. Keep in mind how you might modify the example to fit your needs.

The LS-2B is available from Perkin Elmer, Corp., 761 Main Ave, Norwalk,

CT, 06859-0124. The LS-2B measures the luminescence of a sample at a selected wavelength. The instrument can also scan emission wavelengths. A narrow band of light energy is used to excite a sample. Energy emitted from the sample passes through a filter to a photomultiplier (PM) tube. The output of the photomultiplier tube is digitized and then used by the instrument's microprocessor to produce a measurement value (in "fluorescence units").

The fluorimeter can measure both fluorescence and phosphorescence. Phosphorescence is distinguished from fluorescence by the length of emission time after the excitation source is removed. Generally, fluorescence occurs within 10E-9 to 10E-7 seconds of excitation, while phosphorescence has a longer lifetime varying from 10E-6 to several seconds depending on the structure of the molecule. The xenon source of the LS-2B is pulsed, thereby allowing both modes to be available.

The electronics (and therefore functions) of the LS-2B are controlled by a microprocessor. The excitation wavelength is selected using a narrow band pass interference filter. A motorized, continuously variable, interference filter is used for the emission wavelength. This filter provides scanning capability for determination of emission wavelengths.

The instrument has two optional accessories: a 40 position Autochanger and an RS-232 interface board. The Autochanger has a carousel with positions for 40 cups. The sample platform is rotated to bring each sample cup in turn into alignment with a sampling canula. The sampling canula is immersed in the sample and an aliquot of the sample is drawn into a flowcell via a peristaltic pump. The canula is withdrawn at the end of the purge period. The Autochanger thereby allows automatic sampling. Combining the basic instrument and the Autochanger with the RS-232 interface board will allow you to control an entire assay from an external personal computer.

Fluorescence spectroscopy is a valuable tool in almost every type of laboratory, including biochemistry, molecular biology, enzymology, immunology, pharmacology, environmental analysis, clinical pathology, and food analysis. The example presented in this chapter is one in which a fluorescent dye binds to DNA. When the dye binds, a fluorescence appears at 458 nanometers. A standard curve is generated using a set of known standards and a linear regression is performed on the curve. Unknown samples are assayed in replicate and their concentrations are determined from the regression. This is a convenient way to determine the amount of DNA in a sample. More importantly, you can automate the entire process . . . from configuring the instrument's parameters; to drawing the sample into the cuvette; to the final results . . . using variations of the methods that you have already learned.

INSTALLATION

Most instruments that have motorized automation are installed by a manufacturer's service technician. Having a service representative install the equipment ensures that the equipment is working correctly before you start programming. Do not try to interface an instrument unless you are absolutely certain that it is working correctly.

Personal computer control will only compound problems and make it difficult to pinpoint the cause of problems.

Perhaps just as importantly, take the opportunity during installation to inquire about the cable and setting RS-232 communications parameters. (If your instrument has already been installed, call the manufacturer.) Your inquiry will confirm if the information in the User's Manual is correct.

Many manufacturers change the designs of the RS-232 interfaces. These changes usually reflect differences in alternative supplier's costs for the semiconductors that are used on the interface board. For this reason, different interface versions can often produce differences in the methods you use to set communications parameters. More importantly, different versions may require different cables. These differences are a common source of error in User's Manuals because the manuals are not always revised in parallel with interface updates. Fortunately, service representatives usually have up-to-date information.

MAKING THE CONNECTION

Next, decide on the cable to use. Determining the "pin-out" of the connector for the LS-2B is easy: Perkin-Elmer supplies the cable. If you are a novice at interfacing, you may also want to purchase the cable from the manufacturer. However, this purchase can be quite costly. You should be able to determine a cable configuration easily after reading Appendix A.

Once you have a cable, just plug one of the ends of the cable into the personal computer, the other end into the LS-2B and secure it with screws. Sometimes (but very rarely) the two ends of a cable are wired differently; so pay particular attention to any markings on the cable.

The LS-2B transmits data on pin 3. Therefore it is a Data Communications Equipment (DCE) device; the cable that Perkin-Elmer supplies is straight-through. Straight-through cables are inexpensive and can be purchased at almost any computer shop.

The manual that accompanies the LS-2B lists the instrument as a Data Terminal Equipment (DTE). The manual therefore is incorrect. This occurrence is common. Manufacturers often incorrectly list instrument cabling requirements. If you have problems, do not disregard the cable as the problem, even if you are following the recommendations in the User's Manual.

SETTING THE COMMUNICATIONS PARAMETERS

Most RS-232 interfaces allow the Baud rate, character bits, stop bits, and parity to be set. For interfaces that do not have DIP switches, setting these parameters is easy. They are "hard wired" to specific settings. (The LS-2B is an example of this category.) You can then used the MODE command in a program to set the communications parameters of your personal computer to these values.

For interfaces with DIP switches, your next step in interfacing your instrument would be to set the DIP switches or jumpers to the communications values that you want to use in your program. You may also have a service representative do it for you.

The LS-2B RS-232 interface board has one major caveat that some other instruments have. This caveat is not common, but can be confusing unless you are aware of its presence. If an RS-232 interface board is installed and active in an LS-2B, you cannot use the instrument manually. To return to manual operation, you would need to turn the instrument off, open the instrument's cover and unplug the single lead "+5v" connector from the P5V pin on the circuit board. Other instruments may require that the ribbon cable connecting the interface board with the main processing board be disconnected or that a switch be moved and the instrument reset.

DESCRIPTION OF AN ASSAY TEMPLATE

Your next goal is to design and create a template to receive the data from your instrument. To review the guidelines for designing templates, refer to Chapter 3.

Figure 7–2 shows the template created for the DNA application. As you can see, it is slightly more elaborate than the previous examples. It is also more useful and more typical of what a "real life" template would look like.

The example template is divided into four sections. These sections are typical of those found in many templates that you will design for applications that are used for assaying samples.

The Configuration Section

The first section of an assay template is usually a configuration section. You initialize an instrument by sending it string commands. Each time an instrument receives a string command, it performs the initialization task specified by the string. After the instrument completes the task, the instrument generally responds back to the personal computer by returning a string that contains an error code. Sometimes the string also contains some other printing and/or non-printing characters.

For example, the LS-2B returns a four-digit error code: If an input is valid and an action is successful, the error code returned is 0000.

The method of receiving a command and returning a status string is common to most instruments and can be useful to you as a programmer. It is not only an invaluable debugging (error identification) tool, but it will allow you to show a user the status of an instrument. This display will allow the user to know that the instrument has been set up correctly and that no errors have occurred.

To create a more elaborate program, you could test a status string, flash a warning message and/or abort a program if it is not the same as the one expected from successful completion of a task. This "safety valve" will make sure that users do not try to start their experiments with improperly initialized instruments. An

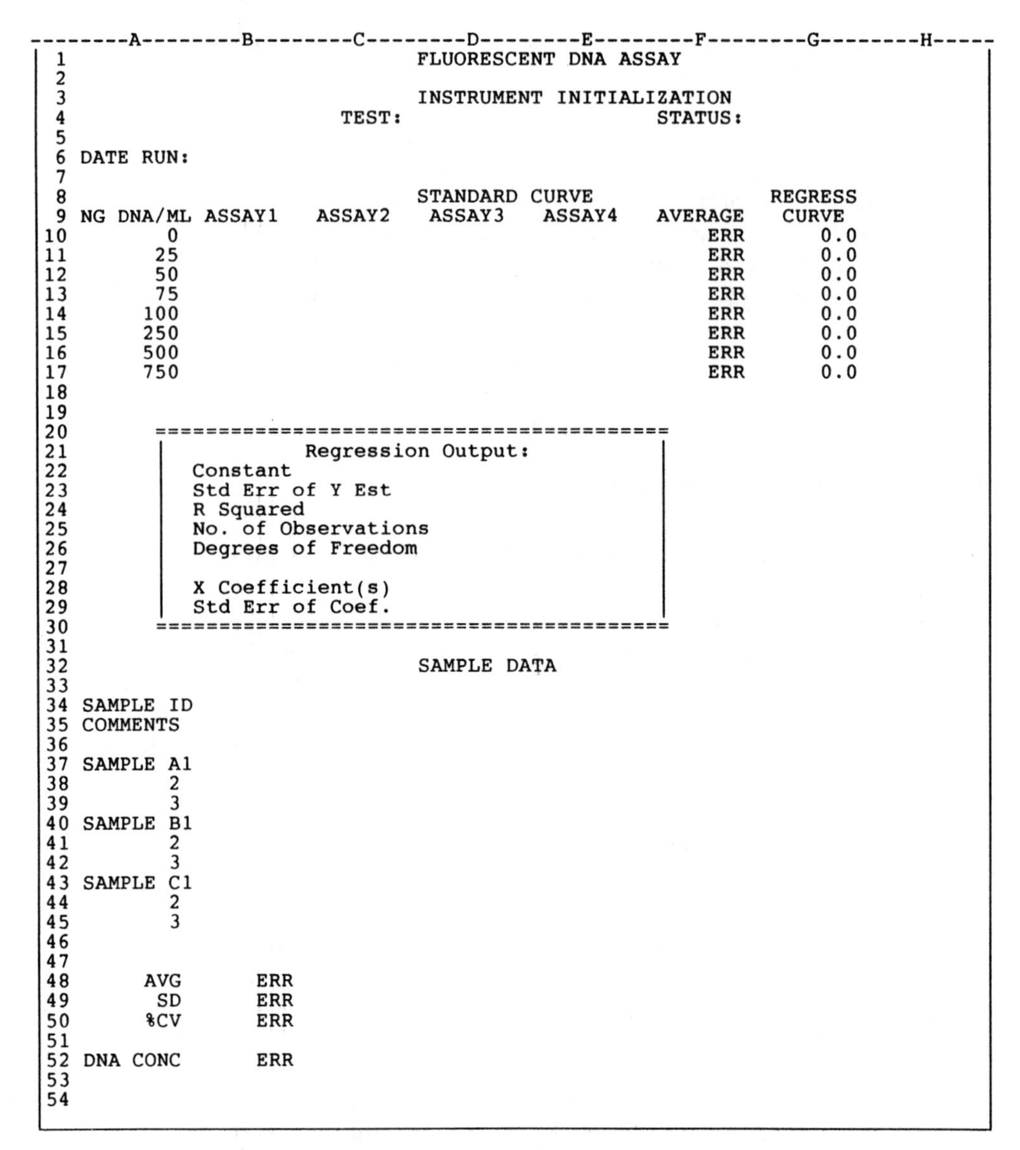

```
--------A--------B--------C--------D--------E--------F--------G--------H-----
 1                                    FLUORESCENT DNA ASSAY
 2
 3                                    INSTRUMENT INITIALIZATION
 4                              TEST:                      STATUS:
 5
 6 DATE RUN:
 7
 8                                    STANDARD CURVE                REGRESS
 9 NG DNA/ML ASSAY1    ASSAY2    ASSAY3    ASSAY4    AVERAGE    CURVE
10         0                                              ERR        0.0
11        25                                              ERR        0.0
12        50                                              ERR        0.0
13        75                                              ERR        0.0
14       100                                              ERR        0.0
15       250                                              ERR        0.0
16       500                                              ERR        0.0
17       750                                              ERR        0.0
18
19
20       =========================================
21       |           Regression Output:          |
22       |  Constant                             |
23       |  Std Err of Y Est                     |
24       |  R Squared                            |
25       |  No. of Observations                  |
26       |  Degrees of Freedom                   |
27       |                                       |
28       |  X Coefficient(s)                     |
29       |  Std Err of Coef.                     |
30       =========================================
31
32                                    SAMPLE DATA
33
34 SAMPLE ID
35 COMMENTS
36
37 SAMPLE A1
38         2
39         3
40 SAMPLE B1
41         2
42         3
43 SAMPLE C1
44         2
45         3
46
47
48      AVG        ERR
49       SD        ERR
50      %CV        ERR
51
52 DNA CONC        ERR
53
54
```

Figure 7-2

example of this test is shown in the INITLOOP macro in Figure 7-5. The mechanism of the test will be explained later in this chapter.

The Standard Curve Section

The next section of an assay template is a Standard Curve section. This part of your template should be set up to receive the standardization data from your instrument.

In this example, the template is set up to receive data from the LS-2B when the standard curve is being run with known amounts of DNA. It allows for quadruplicate determinations; to obtain more replicates you can add more columns. If you want fewer replicates, use the template as is. No calculations will be adversely affected by fewer data points.

The column to the right of the raw data section (entitled "AVERAGE") contains active formulae that gives the averages of the raw data points.

The last column (entitled "REGRESS CURVE") gives the concentration values that the known amounts would have given if they were calculated from the regression curve. The last column provides a good visual check for how well the regression curve "fits" the data. That is, this column will allow you to compare the values predicted from a regression curve to actual values. The closer that actual values agree with the predicted values, the more accurate your regression analysis is, and the more accurate that the predictions for your samples will be.

The Regression Section

The next section of an assay template is the Regression section. Both Lotus spreadsheets have the ability to perform regression analyses on data. This example shows only simple linear regression. The Lotus regression application program however is very powerful and will support almost any type of regression that you would want to perform. For example, Lotus can do exponential, polynomial, and log-logit regression analyses.

The Regression section sets up an area to receive data from the Lotus regression programs. However, when the numbers for the regression parameters are returned from the regression program, the program places text into the spreadsheet along with the numbers. The text appearing in the box of Figure 7-2 originated from the regression program.

If you were to enter your own text in this area, it would be overwritten. You can get around this problem by building a box the way that you want, picking an area of the spreadsheet outside the viewing area, directing the program to place the regression numbers into that area, and placing formulae into the cells in the template's box that refer to the corresponding cell in the regression area.

For example, if your regression area had its "constant" cell set to H5, you would place +H5 in the cell of your box that corresponds to this piece of data.

The definitions of the information returned are listed below. The form of the line that is returned in the example is: y = "X Coefficient" * X + "Constant" (i.e., $y = mx + b$). Each of the lines of text in the box have the following meaning:

- **Constant:** The same as the bias; that is, the y-intercept.
- **Std Err of Y Est:** The "Standard Error of Y Estimate" and a measure of the certainty with which the independent values in the sample can be used to predict the dependent values. In general, the larger the value of the Std Err of Y

Est compared to the predicted y values, the less certain that you can be about the prediction.

- **R Squared:** The "coefficient of determination" or the "square of the correlation coefficient". This statistic tells you what percentage of the variation in the dependent (y) variable is "explained" by the variation in the independent variable. If the relationship between the two variables is strong, the R Squared value will be close to 1. If the relationship between the variables is weak or non-existent, the value will trend toward zero. For example, an R Squared value of 0.721833 means that about 72% of the variation in y is explained by the change in x.
- **No. of Observations:** The number of observations. It is simply the total number of values in the Y range.
- **Degrees of Freedom:** For small populations, the degrees of freedom is computed by subtracting TWO from the number of observations.
- **X Coefficient(s):** The slope of the line. If quadratic or polynomial regressions are performed, the coefficients for each power of X are given.
- **Std Err of Coef:** More commonly known as the "Standard Error of X". It is an estimate of the standard deviation of the distribution of the X Coefficient. Generally, the larger the Standard Error of Coefficient in relation to the X Coefficient, the less certain that you can be about the prediction. With this test, the larger the sample, the smaller the Standard Error of Coefficient.

The last column of the Standard Curve section (the one entitled "REGRESS CURVE") relates to the Regression section. Later, when you enter the template for this example, you will place a Lotus equivalent to the formula $x = (y-b)/m$ into each of the cells of this column. This formula is a re-arrangement of the familiar $y=mx+b$ formula. (When you create the template for your application, use an equation that appropriately describes the data for *your* standard curve.)

When a linear regression is performed on the "NG DNA" and "AVERAGE" columns in this example, the REGRESS CURVE column updates using the "Constant" (bias) and "X Coefficient" (slope) from the linear regression. Using the results, you can see how close each of the points on your standard curve agree with those predicted by the regression analysis.

The Sample Data Section

The final section of an assay template is the Sample Data section. This section is set up to receive the data from the samples. In the example, space is included for triplicate determinations on each of three separate samplings of a DNA preparation. This section will certainly be different for other experiments and/or applications. Thus the beauty of a spreadsheet is that you can change it at will to fill any purpose.

The formulae at the bottom of the Sample Data section calculate the average fluorescence, the standard deviation, the coefficient of variation, and the concentration of DNA. The first two formulae are the Lotus @AVG and @STD functions. The third formula is +(B49*100)/B48. The final formula uses the x coefficient and constant from the regression analysis to determine the DNA concentration for the sample. The formula is +(B48 – $BIAS)/$SLOPE.

CREATING THE TEMPLATE FOR THIS EXAMPLE

Creating the template for this example is easier than it appears. Type the text into the appropriate cells like the previous examples.

To create the box, use a single quote and a series of equal signs for the horizontal line ('===) and a single quote and shift-backslash ('|) for the vertical lines. More instructions are listed below.

1. Into cell F10, type @AVG(B10..E10).
2. Using the / Range Name Create command, create the following ranges:

Range Name	Range of Cells
TESTNO	D4
STATUS	G4
DATERUN	B6
STDSECT	A9 . . E17
STDX	A10 . . A17
STDF	F10 . . F17
STATS	B21
SLOPE	D28
BIAS	E22
DATASECT	A37 . . Z500
REPORT	A1 . . H53

3. Into cell G10, type + (F10 – $BIAS)/$SLOPE. NOTE: Using the dollar sign before a cell name will allow you to "fix" the column and row coordinates. That is, if you were to copy a formula in a down direction without the dollar sign, the row coordinates for SLOPE and BIAS would change by the number of rows that offsets the final cell from the original cell. However, because of the dollar sign in front of $SLOPE and $BIAS they will be copied with no change in the cell designation.
4. Issue the / Copy command to copy from cells F10 . . . G10 to STDF. (i.e., When prompted for the range to copy to, type in STDF and press return.) The formulae will be copied to cells F10..G17.
5. Into cell B48, type @AVG(B37..B45).
6. Into cell B49, type @STD(B37..B45).
7. Into cell B50, type +(100*B49)/B48.
8. Into cell B52, type +(B48 – $BIAS)/$SLOPE.

9. Format the DATERUN cell using the / Range Format Date 4 command for Lotus 1-2-3, or / Format Date 4 command for Symphony.
10. Set the source for the print function to REPORT. In Lotus 1-2-3, issue the / Print Printer Range command. In Symphony, issue the Services Print Settings Source Range command. When prompted for the range, type REPORT and press return.

Before moving on to the macro program that runs this spreadsheet, review some of the goals that you have achieved in anticipation of the macro program. The most obvious goal that you have accomplished was to create named cells and ranges to receive data from the macro program. These cells and ranges allowed you to use names in the commands of the program; the program is therefore easier to read. These named cells and ranges also serve to "marry" the template to the macro program.

You have also initialized the STDX and STDF ranges. These ranges contain the X and Y values that will be used in the regression analysis. The macro program for the LS-2B will redefine the range of cells to include just the number of standards actually run. In the process, the program will issue the / Range Name Create command and then use the {ESC} command to eliminate the part of the range that exists below the starting cell. If the range had not been initialized to contain at least two cells, the {ESC} would cause the menu to back up one prompt.

The STDSECT and DATASECT ranges illustrate another important concept. As you know, it is important to build safety features into your spreadsheet. The STDSECT and DATASECT ranges not only provide a way to make programming easier, but they specify very defined areas of the spreadsheet where data can be placed. When ranges are used in a {PUT} command in the program, an attempt to put data outside the range results in an error message and the macro aborts. For example, if a user tries to place data for unknown specimens in cell S10, the program prevents it. This measure provides a margin of safety because it keeps important regions of the spreadsheet from being overwritten. Thus, a program cannot be accidentally destroyed by placing data into cells that contain a macro.

TEST THE TEMPLATE

Take a moment to test your template to make sure that it is working correctly.

Testing a template before you start programming is very important. If a template does not perform correctly alone, you cannot expect it to perform correctly when it is interacting with a program.

ESTABLISH COMMUNICATIONS

As outlined in Chapter 4 ("Where to Start"), you should create a very simple program that writes a string of characters to your instrument's display, returns the instrument's software version, etc. Figure 4-2 is an example of this type of program.

If the program in Figure 4–2 is modified using "$RE 0" instead of "@CPU", the "Processor Control" light on the LS-2B's display will illuminate. This modification provides a convenient way to test to see if communications are successful for the example.

Additionally, as discussed in "The Configuration Section" of this chapter, when the LS-2B receives a command it responds by transmitting a string that contains an error code. This response provides a convenient test that ensures that data communications coming into the computer are correct. You can implement this test by adding a {READLN} command just below the {WRITELN} command in Figure 4–2 and creating a cell called BUFFER to receive the data.

Once you have modified Figure 4–2 for *your* instrument and established communications, modify the program further to try to get some data from your instrument and place it into your template.

THE MACRO PROGRAM

Once you have established communications, your next step is to create your program. Figures 7–3 through 7–12 show a program that contains all of the elements needed to control a motorized instrument.

The program is similar to the one used for the ACRO-900™, but adds some intriguing new features and programming techniques. Each section is described, in turn, below.

THE AUTO-EXECUTING MACROS

The auto-executing macros shown in Figure 7–3 are similar to those in Chapter 6. They execute according to the Lotus spreadsheet that you are using and set the value of VERSION accordingly. This information will be used later to determine which set of menu commands to issue.

The auto-executing macro for Symphony will show you how to check to see if STAT.APP has been attached. This program needs to be attached before your program can perform regression analyses in Symphony. (In Lotus 1-2-3, the regression routine is permanently attached.) If STAT.APP has not been previously "added-in", the Symphony auto-executing macro (AUTOEXEC) will attach it.

At the end of the appropriate auto-executing macro, the program branches to the main menu. Figure 7–4 is a "wide view" version of the main menu shown in Figure 7–3. It allows you to see the lines of text easier.

INSTRUMENT INITIALIZATION MACROS

Most instruments need to be initialized (configured) before they give meaningful data.

Review your User's Manual to see which parameters need to be initialized and

```
--------P--------Q--------R--------S--------T--------U--------V--------W-----
 1                  /*PROGRAM THAT RUNS THE LS-2B FLUORIMETER*/
 2
 3                /*AUTOEXECUTING MACRO FOR LOTUS 1-2-3*/
 4 \0        {LET VERSION,0}   /*TELLS PROGRAM IT'S RUNNING UNDER 1-2-3*/
 5           /S        /*GOES TO SYSTEM SO COMMUNICATION PARAM'S CAN BE SET*/
 6           {MENUBRANCH MAIN_MU}        /*PUT UP THE MAIN MENU*/
 7
 8                  /*AUTOEXECUTING MACRO FOR SYMPHONY*/
 9 AUTOEXEC  {LET VERSION,1}    /*TELLS PROGRAM IT'S RUNNING UNDER SYMPHONY*/
10           {IF @ISERR(@APP("DOS",""))}{ADD_DOS}      |*CHECK FOR DOS.APP*|
11           {IF @ISERR(@APP("STAT",""))}{ADD_STATS}   |*AND STAT.APP*|
12           {SERVICES}AIDOS~MODE COM1:4800,E,7,1~
13           {MENUBRANCH MAIN_MU}        /*PUT UP THE MAIN MENU*/
14
15                  /*ATTACH DOS.APP TO SYMPHONY*/
16 ADD_DOS   {INDICATE ADDIN}            /*INDICATE THIS SUBROUTINE*/
17           {SERVICES}AADOS~Q           /*ATTACH DOS APPLICATION PROGRAM*/
18           {INDICATE}                  /*RETURN INDICATOR TO LOTUS*/
19
20                  /*ATTACH STAT.APP TO SYMPHONY*/
21 ADD_STATS{INDICATE ADDIN}             /*INDICATE THIS SUBROUTINE*/
22           {SERVICES}AASTAT~Q          /*ATTACH STATISTIC APPLICATION PGM*/
23           {INDICATE}                  /*RETURN INDICATOR TO LOTUS*/
24
25                  /*THIS IS THE MAIN MENU*/
26 MAIN_MU   STANDARDSUNKNOWNS REGRESSIOINITIALIZPRINT    QUIT
27           RUN STD CRUN UNKNORUN REGREINITIALIZPRINT THEQUIT THE MACRO
28           {DO_STDS}{SAMPLES}{DO_STATS{INIT}    {PRINT}  {QUIT}
29           {MENUBRAN{MENUBRAN{MENUBRAN{MENUBRAN{MENUBRANCH MAIN_MU}
30
31
```

Figure 7-3

the commands that you need to send to the instrument to set them. Sometimes, this information is supplied in a separate "Programmer's Manual".

Also, look to see if a status string was returned from the instrument. If so, examine the format of the status string. After you have obtained this information, you can begin to plan your initialization macro.

The initialization process is generally performed by sending the instrument string commands the same way that string commands were sent to the ACRO-900. Usually one command needs to be sent for each parameter. Typically, the command starts with a special character (such as a $, @, %, &, *, etc.).

```
--------P--------Q------------- ----R-----------------S---------------------T--------------------U----------------V----------
22
23
24
25                         /*THIS IS THE MAIN MENU*/
26 MAIN_MU  STANDARDS      UNKNOWNS           REGRESSION               INITIALIZE            PRINT             QUIT
27          RUN STD CURVE  RUN UNKNOWN SAMPLES RUN REGRESSION ANALYSIS INITIALIZE FLUORIMETER PRINT THE RESULTS QUIT THE MACRO
28          {DO_STDS}      {SAMPLES}          {DO_STATS}               {INIT}                {PRINT}           {QUIT}
29          {MENUBRANCH MAIN_MU}{MENUBRANCH MAIN_MU}{MENUBRANCH MAIN_MU}   {MENUBRANCH MAIN_MU}  {MENUBRANCH MAIN_MU}
30
31
```

Figure 7-4

Next, a string of characters specify which parameter is to be set. This string is followed by the value(s) for the parameter. The string is usually terminated with a carriage return (ASCII 13) or a carriage return/line feed (ASCII 13 10) combination.

The sequence of a "header" character, a command mnemonic, a value (or values), and a terminating carriage return/line feed is a VERY common, if not universal, format for command strings that control instruments. Sometimes however, the "header" character is not one of the usual keyboard characters, but is a non-printing control character (such as a CONTROL-B; which is also called STX, or ASCII 002). When such is the case, the terminating character is usually also a non-printing control character (usually a CONTROL-C; which is also called ETX, or ASCII 003). Chapter 4 contains a comprehensive discussion of how to assemble string commands using ASCII control characters.

DESCRIPTION OF THE EXAMPLE INITIALIZATION MACRO

The example initialization macros (INIT and INITLOOP) in Figure 7-5 set the wavelength, fixed scale, photomultiplier voltage, etc., on the LS-2B. The LS-2B, like most other instruments is accompanied by a "Programming Guide". In the guide, it states that a command consists of a preceding dollar sign ($), a two-letter command mnemonic, an optional argument, and a terminating carriage return/line feed.

The INITSTRNG range contains the appropriate strings. This range is in the scratch-pad region of the program. The INITSTRNG named range consists of cells BY5..BY16. The strings placed into the cells of this table can be seen in Figure 7-11.

```
--------Y--------Z--------AA-------AB-------AC-------AD-------AE-------AF----
 1
 2                        /*LS-2B FLUORIMETER INITIALIZATION MACRO*/
 3 INIT       {WINDOWSOFF}{PANELOFF}      /*FREEZE THE DISPLAY*/
 4            {OPEN "COM1",M}             /*OPEN THE COM PORT*/
 5            {LET TESTNO,1}              /*INITIALIZE TEST STATUS NUMBER*/
 6            {FOR INC,0,11,1,INITLOOP}   /*SEND OUT INITIALIZATION STRINGS*/
 7            {WINDOWSON}{PANELON}        /*UNFREEZE THE DISPLAY*/
 8            {CLOSE}                     /*CLOSE THE COM PORT*/
 9
10
11                    /*SENDS AN INITIALIZATION COMMAND TO LS-2B*/
12                /*GETS STATUS STRING BACK AND TESTS TO MAKE SURE*/
13                    /*COMMAND WAS SUCCESSFULLY COMPLETED*/
14 INITLOOP {WINDOWSON}{WINDOWSOFF}      /*UPDATE DISPLAY; PAUSE*/
15            {WRITELN @INDEX(INITSTRNG,0,INC)}    /*SEND OUT COMMAND STRING*/
16            {READLN STATUS}             /*READ STATUS FROM FLUORIMETER*/
17            {IF @RIGHT(@CLEAN(STATUS),4)<>"0000"}{QUIT}
18            {LET TESTNO,TESTNO+1}       /*DISPLAY TEST/INIT NUMBER*/
19
20
21
```

Figure 7-5

The initialization string commands are chosen sequentially using the loop counter and the @INDEX funtion in the {WRITELN} command of INITLOOP. For example, at the start, INC is zero and the @INDEX takes on the value in the first cell of INITSTRNG ($RE 0). {WRITELN} sends this string out to the port. Next, INC increments to one, @INDEX becomes $AC 0, and {WRITELN} sends it out to the port. This process continues until all initialization strings have been transmitted.

TESTING FOR SUCCESSFUL INITIALIZATION

When an initialization string is received by an instrument, the instrument deciphers it, carries out the tasks specified by the command (by moving motors, etc.), and then usually returns a string to the personal computer. This string is typically a report on whether the command was understood and how successful the operation was. It is very common for instruments to return error codes. These error codes can be put to good use in your program. The INITLOOP macro will show you two important ways to use this information.

The INITLOOP macro uses a {LET} command to display returned error codes in a cell called STATUS for the user to view. This display provides positive feedback to the user.

Perhaps more importantly, a macro like INITLOOP can be used to provide a margin of safety to your program. That is, the macro can use an {IF} test to check a returned string (after it has been parsed to remove characters not related to the status) to see if the string was equal to the one used to signal the successful completion of a task (0000 in this example). If the two strings are not equal, the program would know that the initialization was not successful and the macro would abort. The test would therefore ensure that the instrument is not used until it is properly initialized.

The parser for the test in this example removes the terminating carriage return/line feed combination and takes the last four characters of the string. Under certain circumstances, the LS-2B (like many other instruments) will return a character(s) before the four-digit error code. The carriage return/line feed and the character(s) are of no use and are therefore automatically discarded if an @RIGHT and @CLEAN are used.

The < > in the {IF} test means "not equal to".

TIMING DURING INITIALIZATION

One final feature of the initialization macro deserves explanation because it is a feature you may need to use often. The {WINDOWSON} {WINDOWSOFF} commands at the beginning of the INITLOOP serves two functions. First, it updates the display so that the current test number and status are displayed. Second, the pair performs an even more important function. As stated in previous chapters, timing is often a critical part of communicating with an instrument. Instrument initialization is no exception. In fact, timing during initialization is every bit as important as it is during sampling.

The LS-2B provides a good example of the use of this pair of commands. If the {WINDOWSON} {WINDOWSOFF} commands are omitted from the loop, the communications between the personal computer and the instrument become confused, the spreadsheet issues an error, and the macro aborts. With the {WINDOWSON} {WINDOWSOFF} in place, the two commands allow just enough time for the sequence to work flawlessly.

If your instrument has timing problems, you can also use this combination to slow your macro down. Alternatively, you can use a {WAIT @NOW+@TIME(0,0,1)} to pause a program long enough for an instrument to catch up to it. The amount of time to pause can only be determined experimentally after you have written your initialization program.

One other word of advice is: If your instrument gets confused, hangs, or does not respond to string commands, turn the instrument off momentarily and then turn it on again. This act resets the instrument to its power-on state.

STANDARDIZATION MACROS

Most experiments are calibrated with a set of standards that have known values. The results from this calibration portion of an experiment are usually retained for the assignment of values to unknown samples. If you perform experiments of this nature, your program will need a macro that handles automatic calibration procedures. This type of macro is called a ''standardization macro''.

A standardization macro is usually divided into three parts. The first part usually calls an initialization subroutine. The function and format of initialization macros were described in the last few sections. Calling an initialization macro just before assaying standards will ensure that an instrument is properly configured before any testing is performed.

The second part of a standardization macro runs an instrument through a sequence of motions that emulate the samples. These motions provide an assay that results in data from the calibration samples. The macro retrieves this data from the instrument and places the results into the standardization section of the template.

The third part of a standardization macro usually performs some sort of regression analysis on the data to determine a ''standard curve''. This standard curve is set up for later use in an experiment (to calculate values for the unknown samples).

The standardization macros in Figures 7-6 and 7-7 illustrate some of the features to include in your program. The following describes each of these features.

A standardization macro begins by using the @NOW function to place a date-

```
--------AH-------AI-------AJ-------AK-------AL-------AM-------AN-------AO-----
 1
 2            /*THIS IS THE MACRO FOR THE STDS AND LINEAR REGRESSION*/
 3 DO_STDS    {INDICATE INIT}            /*SIGNAL INITIALIZATION PROCEDURE*/
 4            {INIT}                     /*INITIALIZE THE FLUORIMETER*/
 5 \V         {LET DATERUN,@NOW}         /*DATE STAMP THE SPREADSHEET*/
 6            {GETNUMBER "How Many Standards In This Batch? ",COUNT}
 7            {GETNUMBER "How Many Replicates / Standard? ",REPS}
 8            {WINDOWSOFF}               /*FREEZE THE DISPLAY*/
 9            {IF VERSION=0}/REB10..E17~ |*CLEAR DATA FROM STANDARDS SECTION*|
10            {IF VERSION=1}/EB10..E17~  |*0 => LOTUS 1-2-3;1 => SYMPHONY   *|
11            {INDICATE I/O}             /*STATUS = I/O PHASE*/
12            {PANELON}{PANELOFF}        /*UPDATE THE MODE INDICATOR*/
13            {OPEN "COM1",M}            /*OPEN THE COM PORT*/
14            {FOR INC,1,COUNT,1,GET_STD}/*READ STANDARDS AND RECORD*/
15            {CALC}                     /*FORCE RECALC OF SPREADSHEET*/
16            {CLOSE}                    /*CLOSE THE COM PORT*/
17 DO_STATS   {INDICATE STATS}           /*MACRO FOR LINEAR REGRESSION*/
18            {PANELON}{PANELOFF}        /*UPDATE THE MODE INDICATOR*/
19            {WINDOWSOFF}               /*FREEZE THE DISPLAY*/
20            /RNCSTDX~{ESC}.{RIGHT}{END}{DOWN}{LEFT}~      |*REDEFINE*|
21            /RNCSTDF~{ESC}.{LEFT 4}{END}{DOWN}{RIGHT 4}~  |*RANGES  *|
22            {IF VERSION=0}{LET STATVERS,"/DR"}  /*USE 1-2-3 MENU*/
23            {IF VERSION=1}{LET STATVERS,"/RR"}  /*USE SYMPHONY MENU*/
24 STATVERS   /RR                        /*SELF MODIFYING MACRO TARGET CELL*/
25            XSTDX~                     /*SET X RANGE TO STD VALUES*/
26            YSTDF~                     /*SET Y TO CORRECTED FLUOR VALUES*/
27            OSTATS~G                   /*SET OUTPUT RANGE FOR STATS*/
28            {CALC}                     /*FORCE RECALC OF SPREADSHEET*/
29            {INDICATE}                 /*SIGNAL COMPLETION*/
30            {WINDOWSON}                /*REFRESH DISPLAY*/
31            {PANELON}                  /*TURN PANEL BACK ON*/
32
33
34
```

Figure 7-6

```
--------AQ-------AR-------AS-------AT-------AU-------AV-------AW--------AX----
 1
 2                      /*LOOPS FOR THE STANDARD CURVE*/
 3
 4               /*READ THE RESULTS FOR ONE COLUMN (ONE SAMPLE)*/
 5 GET_STD    {FOR CNT,1,REPS,1,STD_REP} /*READ STANDARD REPLICATES*/
 6            {DOWN}                     /*RESET CELL-PNTR FOR NEXT SAMPLE*/
 7
 8                      /*READ AN INDIVIDUAL DATA POINT*/
 9 STD_REP    {WINDOWSON}                /*REFRESH THE SCREEN*/
10            {WRITELN GO}               /*SEND GO COMMAND TO LS-2B*/
11            {WINDOWSOFF}               /*FREEZE DISPLAY; DELAY FOR LS-2B*/
12            {READLN BUFFER}            /*CLEAR OUT LS-2 ERROR CODE*/
13            {IF @RIGHT(@CLEAN(BUFFER),4)<>"0000"}{QUIT}  /*TEST SUCCESS*/
14            {READLN BUFFER}            /*PULL IN THE NEW STRING OF DATA*/
15            {LET NUM,@VALUE(@MID((BUFFER),@LENGTH(BUFFER)-10,4))}
16            {LET POWER,@VALUE(@MID((BUFFER),@LENGTH(BUFFER)-11,1))}
17            {PUT STDSECT,CNT,INC,NUM/10^POWER}
18            {WINDOWSON}                /*REFRESH SCREEN WITH NEW INFO*/
19            {WINDOWSOFF}               /*FREEZE DISPLAY*/
20            {WAIT @NOW+@TIME(0,0,3)}   /*WAIT 3 SECONDS FOR AUTOCHANGER*/
21
22
23
```

Figure 7-7

stamp into the DATERUN cell of the template. This function is vital for you to include in your macro because it ensures that the template bears the date that an experiment was actually performed.

The next step ensures that old data is removed from the template. Erasing data in a template before new data is added is desirable. If a previous experiment had more replicates than the current one, all of the old data would not be overwritten. By erasing all of the old data in the range first, you can ensure that old data will not be used in @functions and equations that use the range. Also, because data is added one point at a time, it is much less confusing to a user to observe the data as it is being added to the template if the template does not contain a mixture of old and new data.

The process of erasing old data is performed by the two {IF VERSION} tests in DO_STDS. These tests check to see if the program is running under Lotus 1-2-3 or Symphony and issue the appropriate set of commands.

The next section of the DO_STDS macro performs replicate assays on several different standards. The process is controlled by two nested {FOR} loops. The first {FOR} loop is in the DO_STDS macro and repeats the GET_STD macro for each standard concentration being tested. The second {FOR} loop is in GET_STD and executes STD_REP for each replicate of the standard being tested. STD_REP issues the GO command (which is the string $RD 1, as shown in Figure 7-11) each time STD_REP is called from the {FOR} command in GET_STD.

The GO commands for most instruments are relatively simple at this point in the control because the initiation macro has normally already performed the majority of the work. That is, the instrument already knows what to do at this point and just needs for you to tell it when to start.

On some instruments, however, you may need to control each step of the sampling process with a set of string commands. If your instrument is in this category, use a series of {WRITELN} or {WRITE} commands to send each string to your instrument.

You may also need a series of {WAIT} commands between {WRITELN} commands to provide enough time for an instrument to complete a specified task(s) before sending the next command. That is, when you create and evaluate your program, you should be constantly aware that timing is usually the most critical part of all programs involved in instrument control.

Next, STD_REP gets the data, parses it, and places it in the template. This process is described more fully in a subsequent section of this chapter.

LINEAR REGRESSION MACROS

The example linear regression macro is part of the DO_STDS macro in Figure 7-6.

Although this macro is part of the standards macro, it can be executed separately. Notice that a cell called DO_STATS was placed at the appropriate point in the macro. This line initiates the regression analysis. This section of the macro is also used elsewhere in the program. That is, if you were to choose the Regression menu choice, it would break into this macro at DO_STATS, execute the macro at that point, and then return to the menu.

This macro contains all of the basics that you will need to know to perform simple linear regressions on standard curves, correlation studies, etc. The following looks at each element separately.

The /RNC commands in the DO_STATS macro are necessary to redefine the X and Y ranges used for the regression analysis. Take a closer look at the reasons for using this programming technique.

Earlier in the DO_STDS macro, a user is allowed to specify how many standards are to be assayed. If the number of standards is less than eight, some cells in the STDF range would contain ERR because ERRs appear in cells containing @AVGs that have no values in their ranges. If any cells in the regression ranges contain ERR, the Lotus regression program aborts. On the other hand, if a user would want more than eight standards, some standards would not be included in the analysis.

For these reasons, use named ranges for the X and Y input ranges of the regression program. Then, redefine the ranges and "feed" the names back into the program to get the appropriate number of data points.

One method that you can use to reset ranges makes use of the fact that the first column of the standard data section always contains a filled cell for each sample. You can make use of this method and it can serve as a "yard stick" to specify how far down the STDX and STDF columns should go when setting their ranges.

More specifically, your program can use the {ESC} command to delete the old range, "anchor" the range with a period (.), move the cell-pointer to the first column of the standard data section, and then use the {END} {DOWN} combination to move the highlight to the end of the column. Your program can then reset the cell-pointer back to the appropriate column and issue the return command (~).

If the method described above seems complicated, go through the keystrokes as they appear in DO_STATS manually. The method will immediately become apparent.

The next section of the program assigns the STDX and STDF ranges as the X and Y ranges for the regression analysis. The two {IF} statements are used to modify one of the lines of the program (range name STATVERS). The result is that the macro is customized to issue the Lotus 1-2-3 or the Symphony command for regression. This technique is referred to as the "self-modifying macro technique" because it places a command into the path of a program. The command is then executed as the spreadsheet encounters it.

This method of modifying a macro "on the fly" is very powerful and useful; it is a rare capability in the world of programming languages. Chapters 8 and 9 will show you some situations where self-modifying macros can be invaluable.

After the STATVERS cell is modified, commands are issued to activate the Regression menu and redefine the X, Y, and Output ranges. The macro concludes with the GO command and regression is calculated and placed into the appropriate cells.

MACROS FOR UNKNOWN SAMPLES

Typical macros for performing automated assays on unknown samples and retrieving the data are shown in Figures 7-8 and 7-9. These macros are similar to the standardization macros.

These macros contain some interesting new programming techniques. The next few sections of this chapter are devoted to explaining the techniques.

HELP WINDOWS

The first technique appears in the GETCUR subroutine. When a user wants to perform analyses on unknown samples, it is desirable to allow him or her to be able to set the cell-pointer to any position within the DATASECT range of the template.

```
--------AY-------AZ-------BA-------BB-------BC-------BD----------BF-------BG----
 1
 2                   /*THIS IS THE MACRO FOR THE SAMPLES*/
 3
 4 SAMPLES   {GETCUR}        /*HAVE USER MOVE CELL-PNTR TO STARTING POSITION*/
 5           {GETNUMBER "How Many Samples In This Batch?",COUNT}
 6           {GETNUMBER "How Many Replicates / Sample?",REPS}
 7           {WINDOWSOFF}                 /*FREEZE THE DISPLAY*/
 8           {INDICATE I/O}               /*CHANGE STATUS TO I/O MODE*/
 9           {PANELOFF}                   /*FREEZE THE PANEL*/
10           {OPEN "COM1",M}              /*OPEN THE COM PORT*/
11           {SETUP}                      /*SETUP LOOP OFFSETS FOR CELL COORDS*/
12           {FOR INC,X_VALUE,COUNT,1,GET_SAMP}  /*READ SAMPLES AND RECORD*/
13           {WINDOWSON}{PANELON}         /*TURN DISPLAY BACK ON*/
14           {CALC}                       /*FORCE RECALC OF SPREADSHEET*/
15           {INDICATE}                   /*SIGNAL COMPLETION*/
16           {CLOSE}                      /*CLOSE THE COM PORT*/
17
18                   /*LOOPS FOR THE SAMPLES*/
19 GET_SAMP {FOR CNT,Y_VALUE,REPS,1,GET_REP}     /*READ SAMPLE REPLICATES*/
20           {RIGHT}                      /*RESET CELL-PNTR FOR NEXT SAMPLE*/
21
22 GET_REP   {WINDOWSON}                  /*REFRESH THE SCREEN*/
23           {WRITELN GO}                 /*SEND GO COMMAND TO LS-2*/
24           {WINDOWSOFF}                 /*FREEZE DISPLAY; DELAY FOR LS-2*/
25           {READLN BUFFER}              /*CLEAR OUT LS-2B ERROR CODE*/
26           {IF @RIGHT(@CLEAN(BUFFER),4)<>"0000"}{QUIT}   /*TEST SUCCESS*/
27           {READLN BUFFER}              /*PULL IN THE NEW STRING (DATA)*/
28           {LET NUM,@VALUE(@MID((BUFFER),@LENGTH(BUFFER)-10,4))}
29           {LET POWER,@VALUE(@MID((BUFFER),@LENGTH(BUFFER)-11,1))}
30           {PUT DATASECT,INC,CNT,NUM/10^POWER}
31           {WINDOWSON}                  /*REFRESH SCREEN WITH NEW INFO*/
32           {WINDOWSOFF}                 /*FREEZE DISPLAY*/
33           {WAIT @NOW+@TIME(0,0,3)}     /*WAIT 3 SECONDS FOR AUTOCHANGER*/
34
35
```

Figure 7-8

```
--------BI-------BJ-------BK-------BL-------BM-------BN-------BO-------BP----
 1
 2     /*MACRO SUBROUTINES THAT SUPPORT SAMPLE MACRO AND ALSO PRINT*/
 3
 4            /*MACRO THAT ALLOWS USER TO POSITION THE CURSOR*/
 5            /*IF IN SYMPHONY, DISPLAYS A MESSAGE IN ALTERNATE WINDOW*/
 6 GETCUR     {IF VERSION=1}{WINDOW}{WAIT @NOW+@TIME(0,0,3)}{WINDOW}
 7            {?}       /*LET USER MOVE CURSOR TO STARTING*/
 8                      /*CELL AND PRESS RETURN*/
 9
10
11            /*MACRO THAT GETS THE CURSOR COORDINATES AND*/
12            /*ADJUSTS THE LOOP COUNTERS SO {PUT} RANGES*/
13            /*PLACE DATA INTO THE APPROPRIATE CELLS*/
14 SETUP      {LET X_VALUE,@CELLPOINTER("COL")-2}  |*ADJUST INITIAL VALUE OF*|
15            {LET Y_VALUE,@CELLPOINTER("ROW")-37}|*LOOP FOR CELL-POINTER  *|
16            {LET COUNT,COUNT+X_VALUE-1}|*ADJUST UPPER COUNTERS*|
17            {LET REPS,REPS+Y_VALUE-1}  |*FOR CELL-POINTER     *|
18
19            /*PRINT MACRO: PRINT; RETURN TO MENU1*/
20 PRINT      {CALC}                     /*FORCE RECALCULATION OF TEMPLATE*/
21            {WINDOWSOFF}{PANELOFF}     /*FREEZE THE DISPLAY*/
22            {IF VERSION=0}/PPRREPORT~AGPQ                 /*LOTUS 1-2-3*/
23            {IF VERSION=1}{SERVICES}PSSRREPORT~QAGPQ      /*SYMPHONY*/
24            {WINDOWSON}{PANELON}       /*THAW THE DISPLAY*/
25
26
```

Figure 7-9

With this feature, batches of samples can be run without having to worry about always starting at column B or row 37, etc.

After the cell-pointer has been set to a starting cell position, the user needs only to press return to start the assay. Once the assay is started, the data will be placed into the template, beginning at the specified cell.

This information is not apparent to the user. You can use a window that you have prepared off to the side of the spreadsheet to give a user instructions. This window will contain a prompt. The ability to overlay windows is only available in Symphony.

Using a Symphony "window" is similar to having a television studio with several video cameras: each of the cameras being aimed at different portions of the set. When a producer wants to view something different on the set, he switches to the appropriate camera. In Symphony, once you have windows created and "aimed" at the appropriate sections of a spreadsheet, you can switch between them and view a section without moving the cell-pointer. Switching between windows is much faster than moving the cell-pointer. You can also decrease the size of a window and overlay it onto a bigger window (thereby viewing both windows simultaneously).

The ability to create windows in Symphony is a very useful feature. It allows you, as a programmer, to quickly display sections of a spreadsheet so that data can be reviewed. It also allows you to display warning messages, error messages, help messages, and prompts.

To implement this feature: The {IF} command tests whether a program is running under Symphony. If it is, the macro uses the {WINDOW} command to overlay a help window that was prepared during programming. The {WAIT} command leaves the help window on the display and the second {WINDOW} command removes it. A delay of about 3 seconds is both tolerable and long enough for even a slow reader to comprehend.

The text that appears in the example help window is shown in Figure 7–10.

The instructions on how to create the window are provided in a subsequent section of this chapter.

FINDING THE CELL-POINTER

The second new technique can be seen in the SETUP subroutine. This macro redefines starting and ending values for the loop counters so that they become aligned with the cell-pointer position specified by the user. By adjusting the loop counters, the column and row offsets in the {PUT} command of GET_REP also become aligned with the starting cell-pointer coordinates. When called upon, the {PUT} command positions data below, and to the right of, these starting coordinates.

The critical component in this macro is the @CELLPOINTER function. The @CELLPOINTER function tests the cell that is currently highlighted by the cell-pointer. The syntax of this function is

```
@CELLPOINTER(attribute)
```

The attribute can be "address", "row", "col", "contents", "type" (blank, number, or label), "prefix" (left-center-right label positioning), "protect", "width", or "format". You can call upon this useful function whenever you need information about any cell in the worksheet.

The @CELLPOINTERs in SETUP use the "col" and "row" attributes. Thus, the @CELLPOINTER functions return the numeric equivalents of the cell-pointer's current column and row. As a point of reference, column A is 1, column B is 2, and so on. The row numbers correspond one to one with the row; row one is the top row of the spreadsheet.

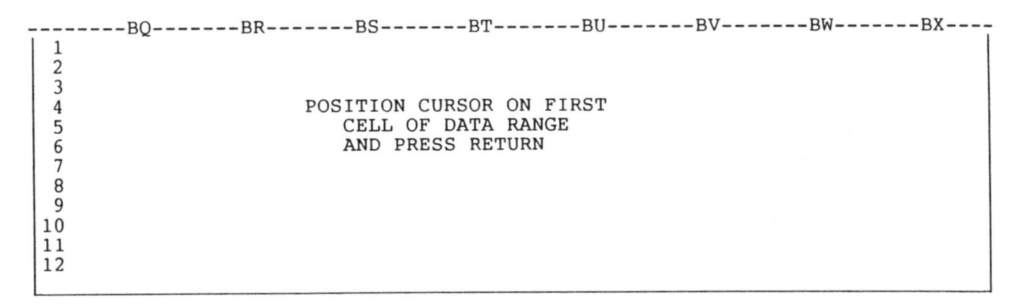

Figure 7–10

If you recall, {PUT} commands use a range, and column and row offsets within that range, to pinpoint the position of the cell that is to receive data. The top-left corner of the range is at offset (0,0). In the example program, DATASECT is defined as cells B37..Z500. Therefore, the cell that is at position (0,0) is cell B37.

Column and row @CELLPOINTER functions would return a position of (2,37) for cell B37. For this reason, you must apply a correction to the values returned by @CELLPOINTER whenever values are used to specify offsets for {PUT} commands.

To correctly specify the offsets, you would subtract the column and row coordinates (relative to cell A1) of the starting (top-left) cell of the named range from the values returned by the @CELLPOINTER functions. (Remember that columns begin at 1, not zero; e.g., column B = 2, not 1.)

For example, suppose a {PUT} command specified DATASECT as its range and that DATASECT was defined as B37..Z500. Then, the coordinates of B37 (2,37) would need to be subtracted from the values returned by @CELLPOINTER functions before the coordinates could be used by the {PUT} command. The following equations would result:

```
@CELLPOINTER("col")-2
@CELLPOINTER("row")-37
```

If you apply a similar correction, you can use the column and row offsets from @CELLPOINTER in {PUT} commands.

The {PUT} command in GET_REP uses loop counters in the SAMPLES and GET_SAMP macros as column and row offsets. Therefore, if you change the values for the two loop counters based on information from @CELLPOINTER, {PUT} can use the information to position data.

The equations in the {LET} commands of SETUP apply these corrections. The first two {LET} commands calculate X_VALUE and Y_VALUE. These two values become the initial values for the {FOR} loops used later in SAMPLES and GET_SAMP.

The upper limits for the loops, COUNT and REPS, are also re-adjusted. The calculations in this set of {LET} commands take the values for the number of samples and the number of replicates per sample that are input by the user, and add the X_VALUE and Y_VALUE correction factors. This correction is necessary because if you are adding an amount to the initial value for the loop, you must add the same amount to the upper limits of the loop so that the loop will execute the same number of times. One is subtracted from each value to account for a quirk in the way that the loop counters are tested. (i.e., The upper limit test is made AFTER the loop has been executed, NOT BEFORE.)

The remainder of the macro is almost identical to the one that gets fluorescent readings for the standards. Now that the loop counters have been adjusted, the {PUT} command will place the values into their appropriate cells.

The @CELLPOINTER function has one caveat that you need to be aware of. When @CELLPOINTER is used in a macro, it may return information that is not current, because the worksheet is not necessarily updated during macro execution. In the example program, return is pressed after the cell-pointer is set. This action recalculates the spreadsheet, thereby eliminating the problem. If you are using this function without having a user press return, then you should use a {CALC} command to ensure that the cell's attribute and value are up-to-date when they are tested.

TESTING ERROR CODES

The sequence of actions that the LS-2B undertakes after receiving the GO command is typical of many instruments, although the particulars may be quite different. For example, when the LS-2B receives the GO command, it returns an error code (relating to whether it understood the command), moves the Autochanger to the first position, aspirates a sample, waits a period of time, reads, and integrates the results. The instrument, like many other instruments, completes the cycle by sending a second string that contains the results and an error code (that relates to the success of the cycle). The results in this case are in "fluorescence units".

The pairs of {READLN}s in STD_REP (Figure 7-7) and in GET_REP (Figure 7-8) illustrate how to structure a macro to respond to this scenario.

The first {READLN} in these macros is required to clear the first error code out of the "pipeline" so that the fluorimetric results can be received. The first error code is handled in the same way that the error codes in the initialization macro were handled. That is, the string is parsed and tested for successful completion of the command. If the instrument has sent an error code to indicate a problem, the macro is aborted. If the instrument has sent a 0000 to signal success, the result of the assay is read in as part of the second string and the second string is parsed according to the formulae contained in the two {LET} commands. These macros could parse the second string to obtain and test the error code after the second {READLN} command.

STRINGS THAT CONTAIN MULTIPLE PIECES OF INFORMATION

Data strings that are transmitted from instruments can be complex and require substantial mathematical manipulation before they can be placed into a template. The data string that is output from the LS-2B represents three numbers, separated by commas and ending in a carriage return. The following is an example of a string that originated from the LS-2B:

```
10818,0,00<CR>
```

The first number has five digits and represents the relative fluorescence. The first digit of this number is the power of ten that the next four digits are divided by. Thus, the 10818 would translate to 81.8 fluorescence units (0818/10E1).

The second number is a single-digit error flag. It is 1 if an error has occurred during integration and 0 if not.

The third number has two digits and indicates the number of repeat integration cycles remaining. That is, the LS-2B allows you to integrate a sample several times. This number is 0 if only one integration was called for or if the data is from the last integration.

Whenever a string contains more than one piece of information, you can use a series of @MID functions to parse out each individual piece as it is needed. For example, in the string shown above, you could use an @MID function to get the 10818 (fluorescence units), an @MID to get the 0 (error code) and a third @MID function to get the 00 (number of integration cycles remaining).

The following section will show you how to get the fluorescence units out of the string. The error code and integration cycles work in identical fashions.

PARSING VARIABLE LENGTH STRINGS

The parsing functions in STD_REP (Figure 7-7) and GET_REP (Figure 7-8) will introduce you to some new concepts. As stated in the last section, the numerator is the second through the fifth digit of the string. However, the example is misleading. The actual string is much different than the one shown.

First, the string begins with a line feed that is left over from reading in the four-digit error code in the first {READLN}. That is, {READLN}s do *NOT* always clear line feeds out of DOS's port buffer. You may often find a line feed (ASCII 10) at the beginning of the strings containing your data. The preceding line feed will need to be removed before you can obtain the value of your data.

Second, the line feed is followed by several characters that are not sent by the LS-2B. These characters are ASCII 153s and are placed into the buffer cell of the spreadsheet when {READLN} waits for characters from an idling communications port. These characters are common when delays occur in data transmission. That is, they are issued whenever an RS-232 port is "idling".

The problem that they present is that the number of character 153s can be variable, depending on the length of the delay. The next two chapters will show you several ways to program around these characters if they represent a problem; but for most programs, you can just parse them out of the string. That is, you can count backward from the carriage return at the end of the string. By using the end of the string as a reference, you can be sure to obtain the correct value when you convert the string. The following two {LET} commands exemplify this concept:

```
{LET NUM,@VALUE(@MID((BUFFER),@LENGTH(BUFFER)-10,4))}
{LET POWER,@VALUE(@MID((BUFFER),@LENGTH(BUFFER)-11,1))}
```

The section titled "Cutting A Piece Of The Pie" in Chapter 4 explains how to obtain this information from your data strings.

To summarize, read some data from your instrument into a BUFFER cell. Then use the {EDIT} key (F2) to expose the true contents of the cell. By moving the cursor back and forth through the cell, you can expose any line feeds, "non-printing" character 153s (which will appear as filled-in boxes) before the string, and the carriage return after the string.

Next, temporarily place an @LENGTH(BUFFER) function into a nearby cell to count the total number of characters in the string and then count backward from this number to the string(s) that your are interested in. Then, prepare a second temporary cell near the BUFFER cell to test the formulae and "fine tune" them until you obtain correct results.

As always, more than one way to parse a string exists. For example, the following set of equations can be used to parse the LS-2B's string:

```
{LET NUM,@VALUE(@LEFT(@RIGHT(@CLEAN(BUFFER),9),4))}
{LET POWER,@VALUE(@LEFT(@RIGHT(@CLEAN(BUFFER),10),1))}
```

Both of these functions clear out the terminating carriage return and use the end of the string as a reference. They then count backward 9 (10) characters and mark the position as the starting point for a new string. The following are the components of the equations that perform this task:

```
@RIGHT(@CLEAN(BUFFER),9)
@RIGHT(@CLEAN(BUFFER),10)
```

The first 4 (1) characters in the new string are taken. The @LEFT function performs this function in the equation. That is, the @LEFT function picks the first 4 (1) characters from the @RIGHT's starting point. Below is the pertinent part of the equation:

```
@LEFT(@RIGHT(@CLEAN(BUFFER),9),4)
@LEFT(@RIGHT(@CLEAN(BUFFER),10),1)
```

The chosen characters are used by the @VALUE function to calculate the value of NUM and POWER. This combination of @RIGHT and @LEFT functions is powerful and you will find that it can be used in many of the applications that you develop.

EXPONENTS IN STRINGS

The {PUT} command in STD_REP (Figure 7-7) and GET_REP (Figure 7-8) that assembles the final equation and places the value into the appropriate cell of the spreadsheet uses the following formula:

NUM/10^POWER

In this equation ^ means "to the power of".

PROGRAM TIMING

After the results are {PUT} into the spreadsheet, the STD_REP and GET_REP programs encounter a {WAIT} command. This time delay is necessary to pause the program until the Autochanger completes its action. If this {WAIT} had not been placed into the program, the LS-2B would have been "busy" and would have ignored any commands that the program tried to sent to it.

If your instrument sends its data out separately, but immediately after it sends out its error code string, AND you are using a slow personal computer (such as an IBM PC or XT), you may get a System Error message if an {IF} command is placed between two {READLN} commands. {IF} tests take time. If the personal computer is not ready to receive a data string, then Lotus will abort the macro and give a System Error message.

If this description fits your instrument, move the {IF} test to a position just below the second {READLN}. This change will test STATUS after the second {READLN} during a period when timing is not vital. This solution should resolve any timing problems, but is an "after the fact" test: You will know that a problem has occurred, but after the data has been read in.

THE "SCRATCH-PAD" AND STRING COMMANDS

When you control an instrument that takes several string commands, the most convenient way to organize the strings is to place them in a table in your scratch-pad.

Figure 7-11 shows the range that contains the initialization strings that are sent to the LS-2B. They are contained in the range named INITSTRNG (BY5..BY16).

The format of string commands was discussed previously in this chapter. As you can see from Figure 7-11, commands that are sent to the LS-2B begin with a dollar sign ($). The $ is followed by a two-letter mnemonic that specifies the task that the LS-2B should perform. The third argument specifies the particulars for the task to be performed. For example, the 750 in the $FL command specifies a voltage of 750 for the photomultiplier tube and the 458 in the $GM command specifies an emission wavelength of 458 nanometers. The final part of the command is a carriage

```
--------BX-------BY-------BZ-------CA-------CB-------CC-------CD-------CE----
 1
 2         /*NAMED RANGES FOR STRING COMMANDS AND "SCRATCH PAD"*/
 3
 4         /*TABLE OF INITIALIZATION STRINGS FOR THE LS-2B*/
 5 INITSTRNG$RE 0              /*SET PROMPT MODE TO "FREE RUNNING"*/
 6         $AC 0              /*SET AUTOCONC TO OFF*/
 7         $AZ 0              /*SET AUTOZERO TO OFF*/
 8         $FR 4              /*SET FILTER RESPONSE TO 4*/
 9         $FS 3              /*SET FIXED SCALE TO 3*/
10         $FL 100            /*SET UP NORMAL FLUORESCENCE MODE*/
11         $FL 750            /*SET HT VOLTAGE TO 750*/
12         $GM 458            /*SET WAVELENGTH TO 458*/
13         $ZM 2              /*SELECT AUTOCHANGER; PUMP FWD.*/
14         $ZT 10057          /*SET SAMPLE TIME TO 5.7 SECS*/
15         $ZT 20015          /*SET DELAY TIME TO 15 SECS*/
16         $ZT 40000          /*SET PURGE TIME TO ZERO SECS*/
17
18
19                  /*STRING CORRESPONDING TO A "GO" COMMAND*/
20 GO      $RD 1                       /*SEND OUT A "READ" COMMAND*/
21
22
23                  /*SCRATCHPAD OF NAMED RANGES*/
24 VERSION                    /*0=>LOTUS 1-2-3; 1=>SYMPHONY*/
25 INC                        /*COUNTER FOR LOOP1*/
26 CNT                        /*COUNTER FOR LOOP2*/
27 COUNT                      /*TOTAL NUMBER OF SAMPLES*/
28 REPS                       /*NUMBER OF REPLICATES/SAMPLE*/
29 BUFFER                     /*INPUT BUFFER FOR FLUORIMETER DATA*/
30 NUM                        /*NUMERATOR FOR FLUORIMETRIC UNITS*/
31 POWER                      /*POWER OF 10 FOR FLUORIMETRIC UNITS*/
32 X_VALUE                    /*COLUMN OFFSET FOR THE CELL-POINTER*/
33 Y_VALUE                    /*ROW OFFSET FOR THE CELL-POINTER*/
34
35
36
```

Figure 7-11

return/line feed combination. This combination is automatically added by the {WRITELN} command in the macro.

Below the INITSTRNG range is the range called GO. Your programs will be easier to read if you use descriptive commands (such as {WRITELN GO}), instead of commands like {WRITELN "$RD 1"}.

Gathering your string commands in the scratch-pad and using tables from which your macro program can select the strings present many benefits. Tables make your program more organized, more "readable", easier to follow, and easier to change if the need arises. These benefits provide powerful incentive to use tables routinely.

CREATING THE PROGRAM

This section contains the details on how to create the program that runs the LS-2B. Although your program may be different than the one presented in Figures 7-3 through 7-11, you will need to perform the same general tasks.

```
--------CG-------CH-------CI---------CJ--------CK-------CL-------CM-------CN----
 1
 2                           NAMED RANGES
 3                           ===================
 4               ADD_DOS     Q16        NUM      BY30
 5               ADD_STATS   Q21        POWER    BY31
 6               AUTOEXEC    Q9         PRINT    BJ20
 7               BIAS        E22        REPORT   A1..H53
 8               BUFFER      BY29       REPS     BY28
 9               CNT         BY26       SAMPLES  AZ4
10               COUNT       BY27       SETUP    BJ14
11               DATASECT    B37..Z500  SLOPE    D28
12               DATERUN     B6         STATS    B21
13               DO_STATS    AI17       STATUS   G4
14               DO_STDS     AI3        STATVERS AI24
15               GETCUR      BJ6        STDF     F10..F17
16               GET_REP     AZ22       STDSECT  A9..E17
17               GET_SAMP    AZ19       STDX     A10..A17
18               GET_STD     AR5        STD_REP  AR9
19               GO          BY20       TESTNO   D4
20               INC         BY25       VERSION  BY24
21               INIT        Z3         X_VALUE  BY32
22               INITLOOP    Z14        Y_VALUE  BY33
23               INITSTRNG   BY5..BY16  \0       Q4
24               MAIN_MU     Q26        \V       AI5
25
26
27
28
29
30
```

Figure 7–12

Your first task is to decide on the positioning of the program.

Notice that the example program is positioned horizontally across the spreadsheet rather than vertically. This orientation provides clearance for the data. That is, if the program had been placed vertically down column J, the amount of data that could be placed in the spreadsheet would have been limited and it might have overlapped the program.

Your next task is to enter the text for your program. You then need to activate the program. You can activate your program by performing tasks similar to the following, which are those needed for the LS-2B program:

1. Use the / Range Name Label Right (/RNLR) command on columns P, Y, AH, AQ, AY, BI, and BX to create named ranges for each macro or named cell.
2. Use the / Range Name Create (/RNC) command to create the range named INITSTRNG for cells BY5..BY16.
3. Use the / Range Name Table (/RNT) command to create a table of range names in the cells starting at CI4. (See Figure 7–12).
4. Check the table to determine if any ranges are missing. If a range is missing, add it.
5. If you are using Lotus Symphony, issue the {SERVICES} Settings Auto-

execute Set (F9, SAS) command and type AUTOEXEC. Press return. AUTOEXEC will then execute each time the Symphony spreadsheet is retrieved.

6. Store the spreadsheet in a file.

CREATING WINDOWS

If you are using Symphony, you have one additional task to complete. The task is to create a "window" like the one that appears in Figure 7–11. As explained previously in this chapter, this window contains a help message to new users and alerts them to move the cell-pointer to the first cell into which they want data to be placed and to press return.

Creating the example window is simple. Use the following sequence:

1. Move the cell-pointer to cell BQ1.
2. Issue the {SERVICES} Window Create command.
3. When you are prompted for the new window name, type MESSAGE and press return.
4. When prompted for the spreadsheet work environment, choose Sheet.
5. When asked for the Window area, use the up arrow key to move the highlighted area to cell BW10. Use the left arrow key to move to cell BV10. Then press the TAB key and arrow right to cell BR10. Press TAB again and arrow to BR2. Press return when you are satisfied with the location and size of the window.
6. At the next menu, choose Borders, Line, and Quit. These commands eliminate the row and column designators for the window and make the window look more like a message.
7. When back in the spreadsheet environment with the cell-pointer in the MESSAGE window, move the cell-pointer in the window to center the text as best you can. If you cannot center it because the text jumps, move the cell-pointer to the left-hand side of the window and change the column width. (In the example, I used the / Width Set (/WS) command to set the column width to 6 and the text was perfectly centered.)
8. If you are not satisfied with the placement or size of the window, issue the {SERVICES} Window Layout command (F9 WL). This command will return you to the "Identify Window Area" prompt and allow you to repeat the sequence from Step 5 (above) to redefine the window.
9. Press the WINDOW key (F6) to remove the MESSAGE window from the display and return to the MAIN window.
10. Save the spreadsheet.

The LS-2B program is now ready for use. If you have an LS-2B, you could run it at this point. If you have another instrument that you are trying to control, you will find that most of this program can be used "as-is" for that instrument.

TEST IT!

As has become customary, you should thoroughly test your program under actual experimental conditions. However, testing is even more critical in this category than for some of the others. Whenever you are controlling motions in an instrument, timing is critical. Testing the system under actual conditions often uncovers timing problems that cannot be found by just testing individual macro subroutines.

WHAT'S NEXT?

The next chapter will show you how to get data from a different category of instruments—instruments that cannot be prompted for their data from a personal computer.

Although several new concepts will be introduced, you should see a trend arise. The trend is that most common instruments require the same kind of programs. Pay particular attention to the similarities between the ACROSYSTEM® Interface, the Perkin-Elmer LS-2B Fluorimeter, and the Bio-Tek EL-309 plate reader in the next chapter. You should also think about what you will need to know about your instrument to interface it to a spreadsheet.

8

Getting Data from Instruments that You Cannot Prompt

In the last chapter, you learned how to interface a robotic instrument and how to increase the professionalism of your program. This chapter will show you how to interface another category of instruments. This category is comprised of instruments that transmit data without being prompted for it by a string command from a personal computer.

Typically, these instruments send out data after a ''start'' button is pressed on the instrument. That is, they are instruments that are not controlled by a personal computer.

This category also includes bar code readers. These devices are activated when they detect the presence of a bar code label. Once activated, the device reads the bar code and transmits the data from its RS-232 communications port.

Not knowing when data will be received by your program requires some special programming techniques. That is, your program needs to remain in an ''idle'' state until data is received and then it can continue processing. This category of instruments also requires special techniques to open the communications port. It is the primary purpose of this chapter to teach you these techniques.

One variation on this category of instruments is presented in the next chapter. This variation contains instruments that are also not controlled by personal computers; but the instruments *continually* send out data. For example, balances are in this sub-category and send out data non-stop. This method of communications can be difficult to deal with, but once you know some special programming techniques you can be quite successful. Because these programming techniques comprise a superset of the techniques in this chapter, you will need to understand the concepts contained in this chapter before proceeding.

You will also learn a relatively large number of new programming techniques in this chapter. These techniques will help you refine your user interface and add more features to your menus.

Two examples are used in this chapter. The primary example is the Bio-Tek Model EL-309 Microplate Reader. A brief discussion of a Symbol Technologies LS-6000 Bar Code Reader also provides an example because the method of data transfer from the bar code reader is a slight (but very common) variation of the EL-309's.

BRIEF DESCRIPTION OF THE EXAMPLE INSTRUMENT

The Bio-Tek EL-309 Microplate Reader is a good example of an instrument whose data transmission is controlled by the instrument and not by a personal computer. Because you may not have this exact instrument, this section will give a brief overview of the instrument and its use. This overview will give you a better understanding of the example and how you might modify the example to fit your needs.

The EL-309 (and its newer version, the EL-311) is available from Bio-Tek Instruments, Inc., P.O. Box 998, Hyland Industrial Park, Winooski, Vermont, 05404. It is designed to measure the optical density of solutions in 96-well microtitration plates. A microtitration plate is a plate usually made of transparent polystyrene and measures approximately 120 mm long by 80 mm wide by 15 mm deep. The format of the plate is 12 columns of wells across by 8 rows deep. (See Figure 8–1.) Each well has a diameter of about 7 mm and a depth of about 11.25 mm. The small internal volume of the wells (ca. 350 microliters) and the convenient format have made these plates very popular (if not ubiquitous) in Immunology, Cell Biology, Microbiology, Pharmacology, and Biotechnology laboratories.

One of the most common uses of these plates is for *E*nzyme *L*inked *I*mmuno-*S*orbent *A*ssays (ELISA). However, other uses abound for microtitration plates. Some of these other uses are protein, enzyme, cytotoxicity, cytoproliferation, hemagglutination and *M*inimal *I*nhibitory *C*oncentration (MIC) assays. All of these assays produce a color or turbidity that is proportional to the concentration of analyte present.

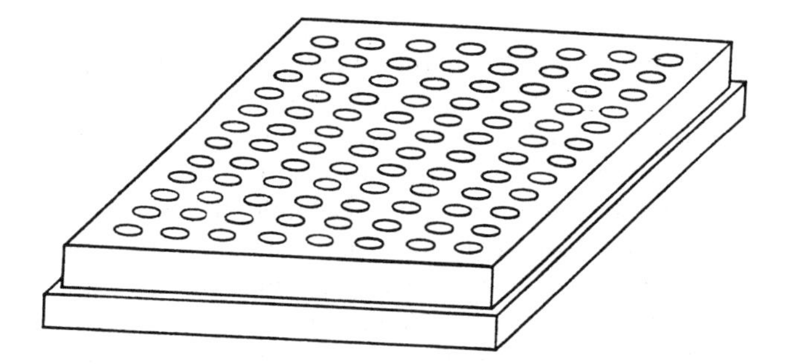

Figure 8–1

The EL-309 automatically reads all wells in the plate and prints the results in a 12 by 8 array. The EL-309 has a parallel port and a serial port for transmission of optical density data to a printer.

The EL-309 has an optional feature that allows a personal computer to control the reader through the serial port. The reader accepts simple string commands (similar to the ones used by the LS-2B in the last chapter) to move the plate carrier and filter wheel, to read the light level at any detector, to send messages to the printer and display, and to perform a variety of other tasks.

Although the EL-309 can be controlled completely from a personal computer, the example uses the EL-309 in a manual mode only. That is, the instrument will not be controlled from the personal computer as was the LS-2B Fluorimeter in the last chapter.

In manual mode, a user presses a start button and the instrument reads a plate. After the plate is read, the data is transmitted out to the RS-232 port for printing.

This example will take the data that would customarily be sent to the printer and capture it in the spreadsheet.

INSTALLATION

Your usual first step is to install the instrument. Most instruments in this category come ready to use, including the EL-309. If your instrument requires an installation procedure, use the procedure to set up and calibrate the instrument. Also, configure your instrument for any parameters needed to collect your data.

The EL-309 illustrates the current state-of-the-art method of configuring an instrument to perform testing according to a user specified protocol. The user specifies the protocol by answering a series of prompts. The answers to these prompts form an organized group of instrument settings that can be stored as a unit in the instrument's "non-volatile" memory. Various manufacturers have different names for this collection of settings. However, "file", "mode", "method", and "procedure" seem to be the most popular.

For example, the EL-309 has nine basic operating setups. These setups are referred to by Bio-Tek as "operating modes". The modes are user selectable from the control panel. Modes 1 and 2 are pre-set at the factory and cannot be changed. Mode 3 is a mode for computer control. Modes 4 through 9 allow a user to enter specific parameters and store them in non-volatile memory for future use. In these modes, a user may define parameters such as filter numbers, single or dual wavelength, blanking patterns, output options, and reading options. The EL-309 prompts the user for the parameters on the display.

The EL-309 Operator's Manual describes how to set up the modes. For example, you press right and left arrow keys to move through a list of parameters. To change a parameter, you use an OPTION key. The following are the parameters used for mode 4 when the example program was developed for this chapter:

```
SKIP HEADING
NO USER ID
PRINT DENSITIES
NO LIMIT SYMBOLS
NO RANGE OUTPUT
TRANSMIT SERIAL
FULL 8x12 FORMAT PRINTOUT
SINGLE WAVELENGTH
WAVELENGTH FILTER = 405
READ MODE IS: READ-AND-EJECT
BLANKING MAP: CONSTANT
USER BLANK: 0.000
```

A number of other features can be programmed in the modes. For example, you can require a user ID, read the plate in a dual wavelength mode, and/or print in a column format. You also have flexible blanking. You can blank an entire plate or divide it into halves, quarters, rows, or columns, with each sector using its own blank. You can also specify which wells in the sectors are to be averaged and used as blanks. To keep the example simple, just the basic features are used in this chapter.

Once you have configured your instrument, take care to ensure that the instrument is working correctly in manual mode before you start programming. Do not try to interface an instrument unless you are absolutely certain that it is working correctly. Personal computer control only complicates matters if problems exist and make it more difficult to pinpoint the cause of these problems.

MAKING THE CONNECTION

Your next step is to determine the kind of cable to use. The RS-232 cable for the EL-309 requires some thought to determine and illustrates some important points to be aware of.

The Operator's Manual lists the instrument as Data Communications Equipment (DCE). That is, it transmits its data on pin 3. Because IBM personal computers and compatibles are DTEs, you would expect a straight-through cable to work. However, it does not.

It does not work because whenever a serial transmission is to be made, the EL-309 first checks pin 20 (Data Terminal Ready, DTR) to see if it is active (HIGH). If DTR is inactive, no transmission occurs. If DTR remains inactive for more than one second, the entire attempt to transmit data is abandoned. If the DTR is active, the EL-309 checks pin 4 (Request To Send, RTS) to see if it is also active (HIGH). If it is, the EL-309 will transmit its data. Thus, no data will be sent from the EL-309 unless pins 4 and 20 are held active (HIGH) at all times.

The IBM and its compatibles do not normally activate pin 4 when you power-up the computer or reboot. Therefore, you need a special cable. The cable is one that is straight-through, but the wires between the two pin 4's and two pin 6's are not connected. On the side that plugs into the EL-309, pin 4 is tied to pin 6, which has the correct voltage. Therefore, the cable looks like the following:

```
EL-309          IBM
  1---------------1
  2---------------2
  3---------------3
  4\              4
  6/              6
  5---------------5
  7---------------7


 20--------------20
```

Some alternative cables also work. The following is an illustration of one of them:

```
EL-309          IBM
  1---------------1
  2---------------2
  3---------------3
  4\            / 4
  5/            \ 5
  6---------------6
  7---------------7
  8---------------8
 20--------------20
```

Another is:

```
EL-390          IBM
  1---------------1
  2---------------2
  3---------------3

  4\
    >-------------8
  5/

                / 4
  8------------<
                \ 8

  6---------------6
  7---------------7
 20--------------20
```

Note that these configurations are NOT null-modems because the pin 2 at one end is connected to pin 2 at the other end and the pin 3 at one end is connected to pin 3 at the other end.

Appendix A contains additional information to help you figure out the cable.

Having to tie together combinations of pins 4, 5, 6, and/or 8 in both straight-through and null-modem cables is VERY commonly required when you use the Device File Name method of obtaining data from instruments. Tying these pins together disables the normal handshaking (data flow control) process that prevents many instruments from transmitting their data.

SETTING THE COMMUNICATIONS PARAMETERS

The next step is to set up the communications port.

The instruments in this category usually require slower Baud rates than the instruments in the categories that are prompted for data. This slower rate is owing to the large amounts of data that are usually transmitted without a break in the stream of data. The DOS buffer that receives this data is only 512 bytes. If the number of characters that the instrument transmits exceeds 512 bytes, it may overrun the DOS buffer. Therefore, your program needs the capability to quickly remove some of the received data from the DOS buffer or it will be overrun.

If the DOS buffer is overrun, Lotus aborts your program and issues a "System Error" message. For this reason, you will usually need to start with a slow Baud rate, program efficiently and use a personal computer with a speed at least as fast as an 8 megahertz IBM AT. Then, after your program is running successfully, you can increase the Baud rate in a stepwise fashion until you no longer get reliable communications. Complete the process by returning to the last Baud rate that gave acceptable performance.

The EL-309 is a perfect example of an instrument in this category. The factory default for the serial port is a Baud rate of 300, 7 data bits, no parity, two stop bits, and an end of line carriage return/line feed. The factory settings can be changed, but after the microplate is read, the instrument sends out the entire plate's worth of data without a break. In other words, a burst of 594 bytes is transmitted, without interruption, from the RS-232 port. Thus, your program needs to quickly remove at least 82 bytes from the DOS buffer or it will be overrun. Additionally, it means that you should start with a slow Baud rate.

If you own an EL-309 and have been using it at a Baud rate other than 300, return the serial port parameters to that speed. The EL-309 represents one of the new ways that are becoming popular for setting communications parameters. If you recall, in previous chapters, these parameters were changed by positioning switches or moving jumpers. The current state-of-the-art method is for the user to place the instrument in a special mode and then change the parameters via the instrument's keyboard. This procedure places the parameters into a type of memory that can retain its contents when the power is removed from it. This type of memory is referred to as "non-volatile" memory.

The user then returns the instrument to test mode and resets the instrument.

As in the case of switch and jumper settings, resetting the instrument is crucial to ensure that it is using the new communications parameters.

The procedure for the EL-309 is typical of this method of setting communications parameters. A user begins by moving a NORM/CONTROL switch on the rear panel toward the right (as viewed from the rear). The switch is now in a CONTROL position. Next, the user presses a reset button next to the switch. After the EL-309 resets, the user presses the Enter key on the front panel and then the Serial Out key. The instrument will prompt for the communications parameters. Left and right arrow keys move through the parameters. An option key changes the parameters. At the end of the list of communications parameters, the user slides the NORM/CONTROL switch on the rear panel back to the NORM position and presses the reset button.

REPORT TEMPLATES

Your next task is to plan the template that will receive data from your instrument. Figure 8–2 shows the example template created for this application.

At the top of the template is the now familiar date-stamp cell and a cell that is used for an identification number for the plate. It can be a barcode number or just a number written on the cover or side of the plate. A subsequent section will show you how to use this number to store the results of a current plate in a file. Using the plate number makes it easier to identify which file to retrieve when you want to review results.

The remainder of the example template is divided into three sections. The first section is set up to receive data from the EL-309. Unlike an LS-2B Fluorimeter, where samples are read and the results are reported upon individually, the EL-309 reads an entire plate and sends the data for all 96 wells at once. This section of the template facilitates the viewing of standards and unknowns in the same format as the plate (12 by 8). In the example, column 1 contains the standards, which are run singly. To use this example in your laboratory, you may want to change this format to suit your needs.

The second section of the template is another 12 by 8 matrix and it is set up to receive and display final concentration results. These results are calculated using the data received from the EL-309 and the results of the linear regression in the bottom section of the template.

The linear regression section is similar to the one used in the last chapter. However, a table has been added for the concentration of each standard and its absorbance. The macro program provides the ability to manually enter standard concentrations. Absorbance is automatically copied to the table after the plate is read.

As you can see, the template could be used as a report for your lab notebook.

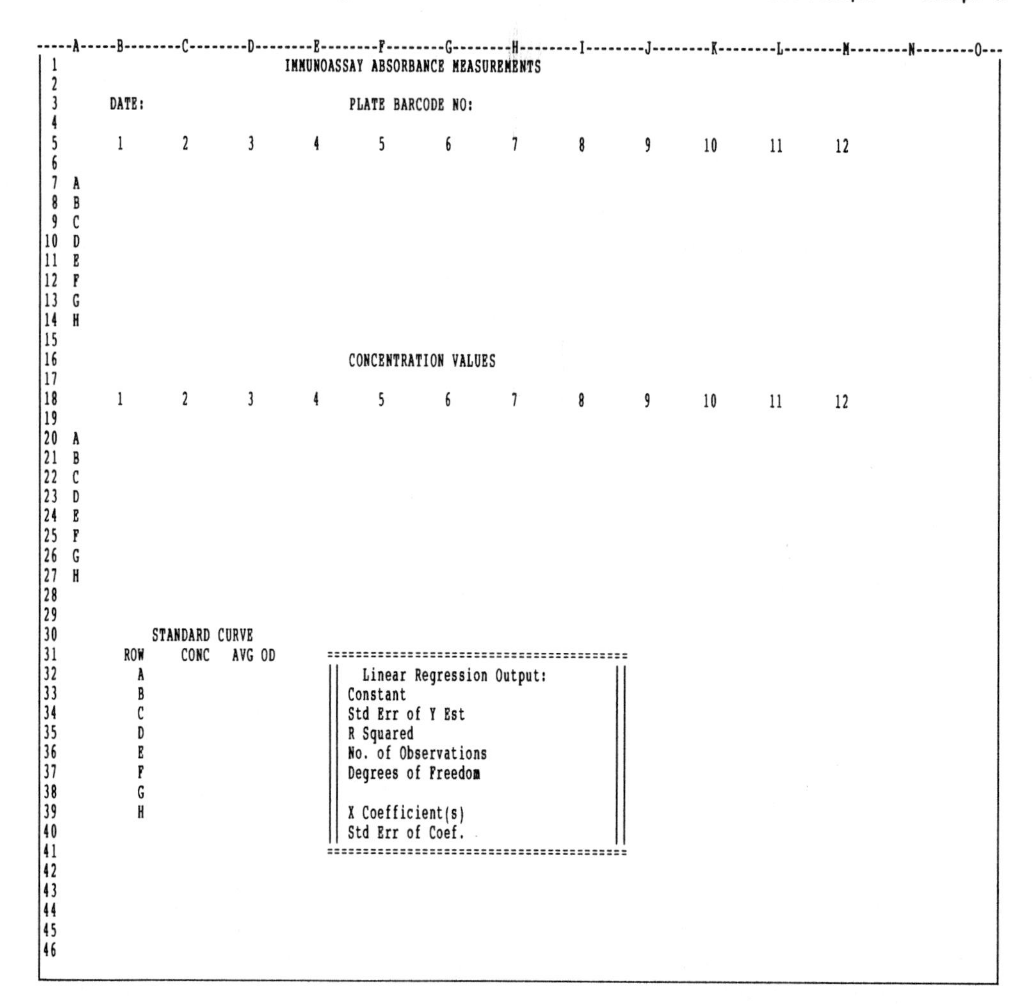

Figure 8–2

However, it is too wide to print on one 8-½- by 11-inch piece of computer paper. Most dot matrix printers allow you to print in compressed mode. In this mode, characters are smaller and you can print more of them on the same line. A subsequent section of this chapter will show you how to send a compressed mode command to your printer.

CREATING THE TEMPLATE

Create a template like the one shown in Figure 8–2 by entering text into the appropriate cells. Note that unlike the LS-2B template, this template does not contain calculations. Formulae in cells slow the speed of macro program execution and the spreadsheet's recalculation. Because the communications part of this application requires maximal execution speed, calculations will be made by a macro subroutine and NOT by formulae in cells.

You must create several named cells and ranges so that the template "talks" to the program. For example, you need to create the ranges called ABS and CONC to receive data from {PUT} commands in the program. These ranges also serve to place boundaries around areas wherein data can be placed. They therefore provide a margin of safety by ensuring that data cannot overwrite part of the template.

Use the / Range Name Create command to create the following ranges:

DATERUN	C3
ABS	B7 . . M14
CONC	B20 . . M27
STATS	F33
SLOPE	H39
BIAS	I33
STDCONC	C32 . . C39
STDABS	D32 . . D39
PLBCODE	I3
REPORT	A1 . . M44
ABSREP	A1 . . M14
STDCOL	B7 . . B14

You can use a number of improvements to make a template look more organized and polished. One improvement is to set the width of each column in the template to fit the text in the column. For the example template, you would set the width of column 1 to a value of 3. Start by moving the cell-pointer to column one. In Lotus 1-2-3, issue the / Worksheet Column Set-Width 3 command; in Symphony, issue the / Width Set 3 command.

Another improvement you can use is to format special cells. Using the example template, note that the DATERUN cell needs special formatting. Set the format of DATERUN by using the / Range Format Date 4 command for Lotus 1-2-3, or / Format Date 4 command for Symphony.

A third improvement would be to set cells so that they display the appropriate number of significant figures. That is, if the data that comes from your instrument is only accurate to a certain number of decimal digits, you would not want to display the data past that number of digits. On the other hand, you also want to consistently display all numbers with the same number of decimal digits. The Lotus default display does not hold true for decimals that end in zeros, unless the cell has been specially formatted (e.g., 0.300 would be displayed as .3 if the cell has not been specially formatted).

In this example, set the ABS and CONC ranges to three decimal places. In Lotus 1-2-3, issue the / Range Format Fixed 3 command; in Symphony, issue the / Format Fixed 3 command. Specify ABS as the range of cells. Then repeat for CONC.

THE MACRO PROGRAM

After testing your template, plan your program. Figures 8-3 through 8-6 and 8-9 through 8-14 show a typical program for this category of instruments. The figures contain all of the elements that you will need to know to get data from an instrument that cannot be prompted for its data.

The program is similar to the one used for the LS-2B, but adds many new features and programming techniques, which are explained below.

STARTING WITH A SLOW BAUD RATE

The auto-executing macros in Figure 8-3 are essentially the same as the ones used for the LS-2B program. The one exception is the communications set-up. In this case, we use a Baud rate of only 300 bits per second.

As you know, this instrument sends out a burst of 594 bytes to the personal computer without a break; the DOS buffer that receives the data is only 512 bytes. Therefore, your program needs to remove at least 82 bytes from the DOS buffer or

```
--------S--------T--------U--------V--------W--------X--------Y--------Z-----
 1              /***INITIALIZATION PROGRAMS FOR BIO-TEK EL309***/
 2
 3                  /*AUTOEXECUTING MACRO FOR LOTUS 1-2-3*/
 4 \0        {LET VERSION,0}    /*TELLS PROGRAM IT'S RUNNING UNDER 1-2-3*/
 5           /S         /*GOES TO SYSTEM SO COMMUNICATION PARAM'S CAN BE SET*/
 6           {MENUBRANCH MAINMENU}       /*PUT UP THE MENU*/
 7
 8                  /*AUTOEXECUTING MACRO FOR SYMPHONY*/
 9 AUTOEXEC  {LET VERSION,1}    /*TELLS PROGRAM IT'S RUNNING UNDER SYMPHONY*/
10           {IF @ISERR(@APP("DOS",""))}{ADD_DOS}      |*CHECK FOR DOS.APP*|
11           {IF @ISERR(@APP("STAT",""))}{ADD_STATS}   |*AND STAT.APP     *|
12           {INDICATE DOS}                |*USE SLOWEST BAUD RATE:  PER *|
13           {SERVICES}AIDOS~              |*BIOTEK MANUAL, IT IS EASY TO*|
14           MODE COM2:300,N,7,2~          |*OVERRUN DOS BUFFER, SINCE IT*|
15           {INDICATE}                    |*DUMPS 594 BYTES ALL AT ONCE *|
16           {CLOSE}            /*CLEAR THE PORT*/
17           {S}WUMAIN~         /*SWITCH TO MAIN WINDOW*/
18           {MENUBRANCH MAINMENU}       /*PUT UP THE MENU*/
19
20                  /*ATTACH DOS.APP TO SYMPHONY*/
21 ADD_DOS   {INDICATE ADDIN}            /*INDICATE THIS SUBROUTINE*/
22           {SERVICES}AADOS~Q           /*ATTACH DOS APPLICATION PROGRAM*/
23           {INDICATE}                  /*RETURN INDICATOR TO LOTUS*/
24
25                  /*ATTACH STAT.APP TO SYMPHONY*/
26 ADD_STATS{INDICATE ADDIN}             /*INDICATE THIS SUBROUTINE*/
27           {SERVICES}AASTAT~Q          /*ATTACH STATISTIC APPLICATION PGM*/
28           {INDICATE}                  /*RETURN INDICATOR TO LOTUS*/
29
30
31 \M        {MENUBRANCH MAINMENU}       |*CALL THE MAIN MENU.  THIS IS A  *|
32                                       |*WAY TO INITIATE PROGRAM AFTER   *|
33                                       |*QUIT HAS BEEN CHOSEN IN MAINMENU*|
34
```

Figure 8-3

the buffer will be overrun and Lotus will abort your program and issue a "System Error" message. For this reason, you must start with a slow Baud rate and then increase it after the program is running reliably.

Remember, if you change the Baud rate in an instrument, you must also change the rate in the personal computer. If the two Baud rates are not in sync, you will be misled into concluding that the new Baud rate does not work correctly.

"LINKED" MENUS

Figure 8-4 shows you how to create a system that contains a "tree" of menus. Each of these menus has a format that is exactly the same as the one created for the LS-2B. However, the menus are different because they are "linked" together. That is, when a user chooses from one menu, a {MENUBRANCH} command jumps to the next menu. This process continues down the hierarchy of menus until the user "zeros" in on the task to be performed, at which time the appropriate macro is called upon to do the work.

For example, suppose a user wanted to print a copy of the absorbances that were obtained for a plate. Choosing the Utilities option from the MAINMENU branches to the UTILITIES menu. Choosing Print from the UTILITIES menu branches to the PRNMENU. Choosing Absorbance from the PRNMENU menu first sets the print range definition and then calls the PRINT subroutine, which prints the report. As you can see, hierarchical menus are user friendly because they allow a user to know what a program can do and what options are available at each point in a program.

SELF-MODIFIED RANGE SPECIFICATIONS

The PRNMENU menu has some interesting programming techniques under the Report and Absorbance choices. Both of these menu choices call the same subroutine, PRINT. {LET} commands in these two choices modify one line of the program in the PRINT subroutine so that the proper range is printed. This line of the program is called REP_SPEC in Figure 8-11. Thus, when Report is chosen, the REP_SPEC cell reads REPORT~ and when Absorbance is chosen, the cell reads ABSREP~. When the spreadsheet reads down the lines of the program during the macro's execution, it reads the appropriate range name as a command, and executes it.

CENTERING DATA ON THE DISPLAY

You probably noticed the {HOME} command throughout the menus. The {HOME} command moves the cell-pointer to cell A1. Using the {HOME} command in the macros serves two purposes. First, the command scrolls the screen to the top left

```
--------AD-------AE----------------------AF-----------------------AG-----------------------AH--------------------AI----------AJ----
 1                                /***BIOTEK PLATE READER MENUS***/
 2
 3 MAINMENU RUN                      UTILITIES                 QUIT
 4          RUN PLATE READER         OTHER UTILITIES           QUIT THE MACRO
 5          {MENUBRANCH RUNMENU}     {MENUBRANCH UTILITIES}    {QUIT}
 6
 7 RUNMENU  BIOTEK                   STATS                     QUIT
 8          READ A PLATE             STD CURVE/REGRESSION      RETURN TO THE MAIN MENU
 9          {INDICATE WAIT}          {MENUBRANCH STATMENU}     {MENUBRANCH MAINMENU}
10          {CALC}
11          {BRANCH BIOTEK}
12
13 UTILITIESVIEW                     FILE                      PRINT                     REGRESSION            QUIT
14          VIEW RESULTS             SAVE RESULTS              PRINT THE RESULTS         REGRESSION ANALYSIS   RETURN TO MAIN MENU
15          {MENUBRANCH VIEWMENU}    {MENUBRANCH FILEMENU}     {MENUBRANCH PRNMENU}      {MENUBRANCH STATMENU} {MENUBRANCH MAINMENU}
16
17 VIEWMENU ABSORBANCE               CONCENTRATION             REGRESSION                QUIT
18          VIEW ABSORBANCE DATA     VIEW CONCENTRATIONS       VIEW REGRESSION RESULTS   RETURN TO UTILITIES MENU
19          {IF VERSION=1}{S}WUMAIN~ {IF VERSION=1}{S}WUMAIN~  {IF VERSION=1}{S}WUSTATS~ {MENUBRANCH UTILITIES}
20          {GOTO}ABS~               {GOTO}CONC~               {HOME}
21          {?}~                     {DOWN 7}                  {GOTO}STATS~
22          {HOME}                   {?}~                      {LEFT 4}{UP 2}
23          {MENUBRANCH VIEWMENU}    {HOME}                    {?}~
24                                   {MENUBRANCH VIEWMENU}     {IF VERSION=1}{S}WUMAIN~
25                                                             {MENUBRANCH VIEWMENU}
26
27          /*THIS IS A MENU TO CHOOSE THE SOURCE OF THE PRINTOUT*/
28 PRNMENU  REPORT                   ABSORBANCE                QUIT
29          PRINT ENTIRE REPORT      PRINT ABSORBANCE DATA     RETURN TO UTILITIES MENU
30          {LET REP_SPEC,"REPORT~"} {LET REP_SPEC,"ABSREP~"}  {MENUBRANCH UTILITIES}
31          {PRINT}                  {PRINT}
32          {MENUBRANCH PRNMENU}     {MENUBRANCH PRNMENU}
33
34          /*THIS IS A MENU FOR CREATING STD CURVES AND CALCULATING CONCENTRATIONS*/
35 STATMENU CURVE                    STATS                     VIEW                      GRAPH                 QUIT
36          ENTER THE STD CONCS      PERFORM LINEAR REGRESSION VIEW REGRESSION RESULTS   VIEW STANDARD CURVE   RETURN TO RUN MENU
37          {ENTER}                  {DO_STATS}                {IF VERSION=1}{S}WUSTATS~ /G                    {MENUBRANCH RUNMENU}
38          {MENUBRANCH STATMENU}    {MENUBRANCH STATMENU}     {HOME}                    {IF VERSION=0}NUSTDCURVE~Q
39                                                             {GOTO}STATS~              {IF VERSION=1}1NUSTDCURVE~QPQ
40                                                             {LEFT 4}{UP 2}            {MENUBRANCH STATMENU}
41                                                             {?}~
42                                                             {IF VERSION=1}{S}WUMAIN~
43                                                             {MENUBRANCH STATMENU}
44
45                                   /*THIS MENU TAKES CARE OF FILE OPERATIONS*/
46 FILEMENU ARCHIVE                  RETRIEVE                  BACKUP                    QUIT
47          STORE RESULTS            RETRIEVE FILED RESULTS    FILE ENTIRE SHEET         RETURN TO UTILITIES MENU
48          {ARCHIVE}                {RETRIEVE}                {IF VERSION=1}{S}WUMAIN~  {MENUBRANCH UTILITIES}
49          {MENUBRANCH UTILITIES}   {MENUBRANCH UTILITIES}    {IF VERSION=0}/FS{?}~
50                                                             {IF VERSION=1}{S}FS{?}~
51                                                             {RIGHT}~{HOME}
52                                                             {MENUBRANCH UTILITIES}
53
54
```

Figure 8-4

corner of the spreadsheet. Then, when {GOTO} commands are issued to display a particular section of the spreadsheet (e.g., the regression output), the section is centered on the display.

The second purpose of {HOME} is to reset the cell-pointer to cell A1 to ready the spreadsheet for the next set of commands.

DATA CAPTURE MACROS FOR NON-PROMPTED INSTRUMENTS

The macro program shown in Figure 8–5 is an example that shows how to get data from instruments that send out data without being prompted for it by a string command from a personal computer.

Not knowing when data will be received by your program requires some special programming techniques. Your program needs to remain in an "idle" state until data is received. Then, it can continue processing.

This category of instruments also requires special techniques to open the communications port. These techniques are shown in Figure 8–5. Each technique is explained in detail in the following sections.

```
--------AM-------AN-------AO-------AP-------AQ-------AR-------AS-------AT----
 1                       /***BIOTEK EL-309 PLATE READER DRIVER***/
 2
 3          /*CABLE IS STRAIGHT THRU WITH OPEN PIN 4 (RTS) AND RTS/DSR*/
 4                      /*(4/6) JUMPED ON THE PLATE READER SIDE.*/
 5
 6 /*BIOTEK PARAMETERS ARE: SKIP HEADING; NO ID; PRINT DENSITIES;*/
 7 /*NO LIMITS; NO RANGES; TRANSMIT SERIAL; 8X12 FORMAT; SINGLE*/
 8 /*WAVELENGTH; WAVELENGTH FILTER WL=405; READ AND EJECT; CONSTANT*/
 9 /*BLANK; USER BLANK 0.000*/
10
11                      /*GET THE DATA FROM THE PLATE READER*/
12 BIOTEK    {LET DATERUN,@NOW}          /*DATE STAMP THE SPREADSHEET*/
13           {WINDOWSOFF}{PANELOFF}      /*FREEZE THE DISPLAY & PANEL*/
14           {BLANK ABS}                 |*CLEAR OUT THE OLD DATA*|
15           {BLANK CONC}                |*                      *|
16           {WINDOWSON}{WINDOWSOFF}     /*UPDATE THE CLEARED DISPLAY*/
17           {GETLABEL "PLEASE ENTER BARCODE LABEL: ",PLBCODE}
18           {IF VERSION=1}{PROMPT}      /*PUT UP PROMPT TO START BIOTEK*/
19 RTRY      {ONERROR RESET,ERRMSG}      /*(THIS ALLOWS [BREAK] TO BE USED)*/
20           {OPEN "COM2",M}      |*DUE TO FLAW IN DOS, SOMETIMES DOESN'T*|
21           {INDICATE READ}      |*RELIABLY OPEN FIRST TIME THRU.       *|
22           {PANELON}                   /*UPDATE STATUS*/
23           {PANELOFF}{WINDOWSOFF}      /*FREEZE THE DISPLAY*/
24           {READLN BUFFERA}     /*INITIALIZE THE TRAP LOOP*/
25           {TRPLOOP}            /*TRAP LOOP TO CLEAR OUT PORT*/
26           {ONERROR \M}         /*RE-DIRECT ERROR TRAP*/
27           {READLN BUFFERB}     -------------------------
28           {READLN BUFFERC}     |*READ THE OTHER ROWS: *|
29           {READLN BUFFERD}     |*MUST BE THIS WAY     *|
30           {READLN BUFFERE}     |*BECAUSE OF THE SPEED *|
31           {READLN BUFFERF}     |*THAT THE DATA IS     *|
32           {READLN BUFFERG}     |*COMING IN.           *|
33           {READLN BUFFERH}     -------------------------
34           {INDICATE PARSE}     /*SIGNAL PARSING PHASE*/
35           {CALC}               /*FORCE RECALC OF THE SPREADSHEET*/
36           {CLOSE}              /*CLOSE THE PORT*/
37           {PARSEDATA}          /*PARSE OUT THE 8 STRINGS*/
38           {INDICATE}           /*SIGNAL MACRO COMPLETION*/
39           {WINDOWSON}{PANELON}        /*THAW DISPLAY*/
40           /CSTDCOL~STDABS~     /*COPY STD ABS' FROM PLATE TO STATS SECTION*/
41           {ARCHIVE}            /*CREATE A FILE AND STORE REPORT IN IT*/
42           {MENUBRANCH RUNMENU}        |*USE BRANCH BECAUSE YOU CAN ONLY*|
43
44
45
```

Figure 8–5

HANDLING ERRORS

The {ONERROR} command in Figure 8-5 is an important command for working with un-prompted instruments. The {ONERROR} command is also useful for prompted instruments and for general macro programs that do not interface with instruments.

This Lotus command establishes a procedure to be followed if an error occurs during macro execution. A 1-2-3 or Symphony error is any action that causes a beep, a macro to abort, and an error message to be displayed in the lower left corner of the screen. Typically, such an error occurs when a spreadsheet cannot successfully complete a command with the specifications provided.

The {ONERROR} command provides a way to "trap" an error and keep the macro executing. One such error can occur when you open the communications port. Some instruments, such as the EL-309, can, under certain circumstances, cause a System Error when the port is *re*opened. This System Error is harmless and can be corrected by reissuing an {OPEN} command. The {ONERROR} command can implement corrective action automatically. The syntax of the {ONERROR} command is

```
{ONERROR branch_location,message_location}
```

The branch_location part of this command is the name of the macro to branch to if an error occurs. The message_location is the cell to which the {ONERROR} command should transfer the error message.

Using message_location is essential to your programming because it allows you to test the message and differentiate between error causes. For example, when a user presses CONTROL-BREAK, it interrupts a program by causing an error. If message_location were not used, you could not test whether the error was due to a System Error (caused by the instrument) or due to [CONTROL][BREAK] (issued by a user who just wanted to abort the plate reading macro). This omission could lead to an "infinite loop" that could only be stopped by the user rebooting the personal computer.

By planning ahead, you can use the {ONERROR} command to design an error-handling macro that checks error messages and deals with each error category in a specific way.

Examine how {ONERROR} works in the example program. First, the {ONERROR} command tells the program to place any error messages into the cell labeled ERRMSG. This cell is in the scratch-pad portion of the program (see Figure 8-12). {ONERROR} also tells the program to go to the macro called RESET whenever an error occurs (see Figure 8-6). After these specifications have been set, the macro proceeds to open the port.

If an error occurs, the error message is placed in ERRMSG and program execution is transferred to RESET. The RESET macro tests the error message. If the

```
--------AY-------AZ-------BA-------BB-------BC-------BD-------BE-------BF----
 1                /***MACROS USED BY BIOTEK READER DRIVER***/
 2
 3 PROMPT    {WINDOWSON}                   |*BRIEFLY CHANGE WINDOW        *|
 4           {SERVICES}WUMESSAGE~          |*TO THE ONE THAT HAS THE      *|
 5           {WAIT @NOW+@TIME(0,0,3)}      |*MESSAGE PROMPTING THE USER TO*|
 6           {SERVICES}WUMAIN~             |*PRESS THE START ON THE       *|
 7           {WINDOWSOFF}                  |*BIOTEK READER                *|
 8
 9           /*DOES ERROR HANDLING--CHECKS FOR SYSTEM ERROR AND RECYCLES*/
10           /*NOTE: THIS WAY ALLOWS <CTRL> BREAK TO STILL BE OPERATIONAL*/
11 RESET     {IF @EXACT(ERRMSG,"System error -- Press [HELP]")}{BRANCH RTRY}
12           {BRANCH MU_RESET} |*SOME OTHER ERROR: RECYCLE AND*|
13                             |*RETURN TO NORMAL OPERATION   *|
14
15           /*THIS FUNCTION RESETS THE SYSTEM AND BRANCHES TO MAIN MENU*/
16 MU_RESET {INDICATE}                     /*RETURN CONTROL TO LOTUS*/
17           {CALC}                        /*FORCE RECALCULATION OF SHEET*/
18           {ONERROR \M}                  /*RESET-REDIRECT ERROR TRAP*/
19           {MENUBRANCH MAINMENU}         /*RETURN TO MAIN MENU*/
20
21                    /*LOOP FOR IDLING BIOTEK READER*/
22 TRPLOOP   {IF @CODE(BUFFERA)<>153}{RETURN}
23           {READLN BUFFERA}              |*BEFORE DATA ARRIVES, LOTUS   *|
24           {BRANCH TRPLOOP}              |*PUTS CHAR 153'S INTO BUFFERA.*|
25                                         |*THE DATA BURST STARTS        *|
26                                         |*WITH A <CR>.  THIS LOOP GETS *|
27                                         |*RID OF MEANINGLESS (153) DATA*|
28                                         |*AND ALSO GUARDS FOR THE CASE *|
29                                         |*WHERE THERE ARE MORE THAN 240*|
30                                         |*CHARACTER 153'S (WHICH WOULD *|
31                                         |*ABORT THE {READLN})          *|
32
33                    /*STRING PARSING FUNCTIONS*/
34 PARSEDATA{FOR ROW,0,7,1,ROWPARSE}       /*PARSE THE STRINGS*/
35           {RECALC ABS}                  /*FORCE RECALCULATION OF SHEET*/
36           /CSTDCOL~STDABS~              |*COPY STANDARDS TO REGRESS*|
37           {INDICATE}                    |*SECTION.                 *|
38           {CALC}
39
40 ROWPARSE {LET CURNT,@INDEX(BUFBLK,0,ROW)}     |*MOVE ROW'S STRING INTO*|
41           {FOR COL,0,11,1,CELLPARSE}          |*STAGING AREA; PARSE   *|
42
43                    /*PARSE ONE CELL FROM THE STRING*/
44 CELLPARSE{PUT ABS,COL,ROW,@VALUE(@LEFT(@RIGHT(CURNT,72-6*COL),5))/1000}
45
46
```

Figure 8-6

message is equal to the one that is displayed when a port problem occurs, the program branches back to the main program and tries again to open the port. If some other error occurs or if a [CONTROL][BREAK] is pressed, the RESET macro branches to MU_RESET. MU_RESET forces a recalculation of the spreadsheet and branches back to the MAINMENU. This procedure resets the program so that the user can continue.

Once any error has occurred, the {ONERROR} command no longer handles any errors that occur during the macro's execution. For this reason, the RESET macro branches back to a location where it can re-execute the {ONERROR} command. This technique resets the error handling protocol and ensures that subsequent errors are handled appropriately.

Other {ONERROR} commands can be found further on in the example program and in the MU_RESET program. These commands tell the program to branch back to \M, returning to the MAINMENU if an error occurs. These subsequent commands override the first {ONERROR} whenever ensuing errors occur. Notice that no message_location is specified. No message_location is needed if you do not have a use for the message.

WAITING FOR THE DATA

The next section of the BIOTEK macro illustrates another important programming concept that needs to be implemented for un-prompted instruments using the Device File Name technique. This section of the macro starts with a {READLN BUFFERA} and a call to the TRPLOOP subroutine.

Whenever a {READLN} command is issued in a program and the program has to wait until the data comes in, a series of character 153s appears in the buffer until the "real" data enters. Lotus manuals list character 153 as the "unknown character". A character 153 in your spreadsheet will appear as a filled-in box.

Lotus continues to accept character 153s until the buffer becomes filled (i.e., 240 character 153s have been received) OR until a "legitimate" data character enters the port. If enough time elapses so that 240 character 153s enter the buffer, the buffer becomes full. Lotus responds by stopping the {READLN} command just as if a carriage return/line feed had been encountered. Lotus then proceeds to the next command in the macro.

If characters other than character 153 are received, Lotus reads them in and concatenates them onto the string of character 153s until a carriage return/line feed is received or the string contains 240 characters. Lotus then proceeds to the next command in the macro.

The previous chapter mentioned that character 153s were obtained in the LS-2B fluorimeter application. In the application, the character 153s did not present a problem because the amount of time between the {WRITELN} command prompting the LS-2B for a response and the data being returned for the {READLN} was well defined and relatively short. That is, the time was never long enough to allow 240 characters 153s to be received. In that situation, very few character 153s were placed into BUFFER and string @functions removed the character 153s that had entered the system.

An un-prompted instrument, such as the Bio-Tek EL-309, presents a very different situation. When a user chooses a menu option for running a test, the program almost immediately starts screening the port for data with a {READLN BUFFERA}. Character 153s start entering BUFFERA. If the user takes minutes (or even hours) to press the "Start" key on the instrument, the 240-character limit is exceeded several times over. If a program similar to the one for the LS-2B were to be used, the program would not know to issue another {READLN} to start the process again. TRPLOOP restarts the process.

The {READLN BUFFERA} in the main program starts the reading process. It reads characters into BUFFERA until it reaches the 240-character limit OR until a carriage return is encountered.

Whenever the EL-309 transmits its data, it precedes the data with a carriage return (no line feed). This carriage return forces an end to the current {READLN BUFFERA} and thus, character 153s.

Whether the 240-character limit is reached or the string of character 153s is terminated by the leading carriage return from the EL-309, the {READLN BUFFERA} command in the main program ends. Then, the TRPLOOP subroutine is called. Once in the TRPLOOP subroutine, an {IF} command tests the first character of the string to see if it is a character 153. If it is, the program knows that either the 240-character limit was exceeded or a string of character 153s was ended by the EL-309 leading carriage return.

In either case, the program issues another {READLN BUFFERA} command to get the next string. Notice that the new {READLN} command also places its string into BUFFERA (after erasing the old string). After the next string is read in, the program branches back onto itself and the {IF} test is made again.

This process continues until the first character of the string is not a character 153. When this situation occurs, the program knows that the first string of data from the EL-309 has been received and the string is left in BUFFERA for future parsing. The program returns to the main program to read in the remaining data.

THE IMPORTANCE OF TRAP LOOPS

So that you can get a better understanding of the importance of TRPLOOP, imagine what would have happened if the program had been written without TRPLOOP.

If a user had waited several minutes before pushing the start button on the EL-309, the {READLN BUFFERA} command would have terminated after 240 character 153s were received. The program would then have proceeded to the {READLN BUFFERB} command, filling it with character 153s (or row A's data) . . . and so on.

Then, when the data had been parsed, ERRs may have appeared in all cells parsed before the start button was even pushed on the EL-309. That is, you would have missed some or all of the data. Furthermore, even if the user had pressed start instantaneously, you would have missed some data because of the preceding carriage return sent out by the EL-309 before the data.

TESTING FOR THE PRESENCE OF DATA

This section explains TRPLOOP's {IF} test in more detail and provides some alternatives to it. The syntax of the {IF} in the macro is

```
{IF @CODE(BUFFERA)<>153}{RETURN}
```

The @CODE function in this statement returns the ASCII/LICS (Lotus International Character Set) number of the first character in the BUFFERA string. Using @CODE is necessary because no keyboard equivalent exists for character 153. The < > portion of the test means "not equal to". Thus, if the first character of BUFFERA is NOT a character 153, program execution returns to the main program.

You can use other means to test whether TRPLOOP should continue. You can test the first character to see if it is a control or printable character. In this case, you could use the following {IF} test:

```
{IF @CODE(BUFFERA)<123}{RETURN}
```

The first character in the EL-309 string is a comma (character 44). This character's presence is very consistent. You could therefore use the following {IF} test to determine when to break out of a trap loop:

```
{IF @CODE(BUFFERA)=44}{RETURN}
```

Figure 8–7 shows a different variation on this scheme. The example retrieves data from a Symbol Technologies bar code reader (Symbol Technologies, Inc., 1101 Lakeland Avenue, Bohemia, New York 11716-3300). This example illustrates a more common way in which data are transmitted. That is, no preceding carriage return is transmitted to signal data receipt. Therefore, you need to look at the *end* of the string for a carriage return to determine when to break out of a trap loop and return to the calling program.

The appropriate test for this situation uses the @RIGHT function to take the last character of a string and the @CODE function to return the ASCII number of that character. For instance if the number returned was 13, then data are present in the buffer (because a carriage return, ASCII 13, was received). If a carriage return was not received, then the program would know that the buffer was filled with character 153s and, therefore, still idle. In this case, the program {BRANCH}es back onto itself to continue waiting.

The following {IF} test is taken from the program in Figure 8–7 and represents the most common format to use in this category of instruments:

```
{IF @CODE(@RIGHT(BC_BUF,1))=13}{RETURN}
```

RESETTING ERROR-HANDLING PROTOCOLS

After a program breaks out of a trap loop, the path for the {ONERROR} error trap is redefined to the main menu, via \M. This technique prevents any chance of an infinite loop back to RTRY. In addition, this redirection is important because

```
--------DP-----DQ-------DR-------DS-------DT-------DU-------DV-------DW--------
 1       /*DRIVER FOR SYMBOL TECHNOLOGIES LASER BARCODE READER*/
 2
 3           /*GET THE DATA FROM THE BARCODE READER*/
 4 GET_BC {ONERROR ERHDLR,BC_ERR}      |*OPEN COM1; SOMETIMES DOESN'T OPEN*|
 5        {OPEN "COM1",M}              |*RELIABLY; SO RETRY IF SYSTEM ERR *|
 6        {INDICATE WAND}              /*PROMPT TO WAND IN THE BARCODE*/
 7        {PANELON}{PANELOFF}          /*UPDATE THE STATUS INDICATOR*/
 8 BC_RST {READLN BC_BUF}              /*INITIALIZE THE TRAP LOOP*/
 9        {BC_LOOP}                    /*TRAP LOOP TO CLEAR OUT THE PORT*/
10        {ONERROR}                    /*DISABLE THE ERROR HANDLER*/
11        {CLOSE}                      /*CLOSE THE PORT*/
12        {RECALC BC_BUF}              /*UPDATE THE BARCODE BUFFER*/
13        {LET CURCODE,@LEFT(@RIGHT(BC_BUF,9),7)}  /*PARSE THE BUFFER*/
14        {CALC}                       /*RECALCULATE THE SPREADSHEET*/
15        {INDICATE}{PANELON}{PANELOFF}
16
17        /*LOOP FOR IDLING BARCODE READER*/
18 BC_LOOP{IF @CODE(@RIGHT(BC_BUF,1))=13}{RETURN}      /*LOOK FOR <CR>*/
19        {READLN BC_BUF}              |*BEFORE DATA ARRIVES, LOTUS   *|
20        {BRANCH BC_LOOP}             |*PUTS CHAR 153'S INTO BC_BUF  *|
21                                     |*THIS LOOP IS NECESSARY       *|
22                                     |*BECAUSE THERE MAY BE > 240 OF*|
23                                     |*THEM (WHICH WOULD ABORT THE  *|
24                                     |*{READLN})                    *|
25
26 ERHDLR {IF @EXACT(BC_ERR,"System error -- Press [HELP]")}{BRANCH BC_RST}
27
28                 /*SCRATCH PAD*/
29 BC_BUF                    /*BUFFER FOR THE INCOMING BARCODE NUMBER*/
30 CURCODE                   /*CURRENT PARSED BARCODE NUMBER*/
31 BC_ERR                    /*BARCODE ERROR MESSAGE*/
32
33 \M     {GET_BC}           |*PROGRAM CAN ALSO BE STARTED FROM MENU*|
34                           |*OR {FOR} LOOP                        *|
35
36
```

Figure 8-7

you do not want to destroy any existing data if an error occurs; nor do you want your program to lock up if a problem occurs.

The structure of the {ONERROR} command may appear a little curious. The {ONERROR} command requires its argument to be a subroutine, not a {MENU-BRANCH}. For this reason, the simple subroutine called \M was created to perform the branch to the main menu.

To re-emphasize {ONERROR} and TRPLOOP are techniques that you will almost always need to use if you are not prompting an instrument to send its data out and if you are using the Device File Name technique. If your program "idles" until data comes in, you will most assuredly need a trap loop.

SPEED CONSIDERATIONS WHILE DESIGNING DATA COLLECTION MACROS

The next series of commands in Figure 8-5 are fairly specific to the EL-309, but illustrate an important concept concerning the arrangement of your program while it is receiving data from many of the instruments in this category. You may need to arrange your program so that it can receive data very quickly, especially when an instrument sends out large amounts of data all at once, separated by carriage returns.

This technique is illustrated by the series of {READLN} commands in Figure 8–5. At first glance, these commands appear to be inefficient programming, but they are actually the fastest way to read data into a spreadsheet from a port.

Recall that your program must receive data from the EL-309 as fast as it can to eliminate the chances of a System Error. If your program reads the data into a buffer, parses the buffer, and places the data into cells, the program would not be fast enough to work reliably. Likewise, {FOR} loops could be used, but they are also inefficient and tend to be fairly slow.

Reading all the data into individual buffers and then parsing and placing the data into cells bypasses this speed problem. Another technique you can use to speed up your initial data collection is to create and erase the cells of your scratch-pad and template. The slowest process during data collection is the process of creating new cells in a spreadsheet. Once created, the process of filling the cells with data is relatively quick. Likewise, overwriting old data in cells with new data is a relatively slow process.

The fastest means of placing data in cells is to have existing, empty cells. You can establish this situation by forcing Lotus to create the cells at the same time that you construct your original template and scratch-pad. Thus, you would fill the cells with "dummy" data and then erase the data. Erasing existing cells clears previous data, but leaves the cell's memory space and attributes intact. This method is therefore faster than creating and filling new cells and it is also faster than replacing existing data in cells.

If you need a fast acquisition rate, use the /Data Fill (Lotus 1-2-3) or the /Range Fill (Symphony) command to fill your template and scratch-pad with numbers. Then, erase the numbers and use the empty cells to receive your data.

However, do not create all of the cells in your spreadsheet. Even though cells do not contain data, they use up memory. So, if you create too many cells you are likely to run out of memory and/or you may actually slow the performance of your spreadsheet. Finally, using this technique means that you will be storing bigger files on your disk. Therefore, create only the cells that you plan to use in your template and scratch-pad.

PARSING MULTIPLE DATA POINTS FROM STRINGS

The data in this example are parsed using the PARSEDATA subroutine. This subroutine illustrates how to get several data points out of the same string. In the example, each string transmitted from the EL-309 contains data from twelve wells. (See Figure 8–8 for example EL-309 strings.) This section shows you how to get individual data points from a long string of data and place the values into a template.

Start by using the concepts of Chapter 4 to determine the repetitive nature of

```
--------CL--------CM-------CN-------CO-------CP-------CQ--------CR--------CS----
13
14 ,+0000,+0331,+1021,+1063,+0734,+0269,+0289,+0741,+0737,+0356,+0407,+0365
15  ,+0101,+0282,+0854,+0874,+0947,+0273,+1631,+0884,+0312,+0430,+0381,+0434
16  ,+0206,+0321,+0403,+0400,+0449,+1619,+1650,+0778,+0746,+0867,+0910,+0829
17  ,+0299,+0266,+0328,+2081,+1679,+2272,+2336,+2055,+1600,+0530,+0571,+0786
18  ,+0416,+0276,+0747,+1716,+1727,+2251,+2319,+2101,+1943,+0372,+0351,+0635
19  ,+0505,+0319,+0755,+0462,+0472,+1688,+1680,+0844,+0346,+0425,+0368,+0704
20  ,+0600,+0301,+0333,+0577,+0832,+1725,+1634,+0996,+0613,+0392,+0665,+0448
21  ,+0698,+0275,+0410,+0864,+0784,+1757,+0409,+0917,+0802,+0423,+0595,+0344
22
23
```

Figure 8–8

the pieces of data in the output string. From this information, build a program that has loops to parse the individual data points. The next paragraph describes the algorithm used in the example program. With very little effort, you should be able to modify it to your needs.

The data are parsed using two "nested" {FOR} loops. These nested loops are illustrated in Figure 8–6. The first {FOR} loop executes the ROWPARSE subroutine for each of the eight rows. The {FOR} loop in ROWPARSE parses individual wells within an individual row. The {LET} command at the beginning of the ROWPARSE subroutine takes a row's string and places it into the cell called CURNT. Thus, the program uses the @INDEX function just once and substantially speeds up the macro's execution. (Note that BUFBLK, the buffer block, is a range that contains cells BUFFERA through BUFFERH).

CELLPARSE performs the parsing task for each well. Many methods can be used to cut an instrument's strings into components. As shown in Figure 8–8, the EL-309 separates each well from its adjacent well(s) by commas in the string and ends the string with a carriage return.

It is most convenient to count backward from the carriage return at the end of a string. The carriage return provides a reliable reference point from which to measure.

Note that the COL loop counter in the ROWPARSE subroutine increases by one each time through the loop. This incrementing serves three purposes: to count the number of times the loop runs, to serve as a column offset for the {PUT} command in CELLPARSE, and to "slide" down the string to be parsed.

The parsing function in CELLPARSE is complicated, so take a closer look at it. The equation uses the carriage return at the end of the string as a reference. Counting backward, 72 characters are between the carriage return at the end of the string and the first character of the EL-309 data. The first character of the next well's data is at a position six characters less that the current well, and so on. Based on this periodic behavior, if the COL loop counter is multiplied by 6 and subtracted from the total length of the string (72) the starting position for the current well's data can be determined. The following @function summarizes the calculation:

```
@RIGHT(CURNT,72-6*COL)
```

By using the COL loop counter in an @RIGHT function to specifically and sequentially change the starting point of a parse (as measured from the end of a string), the program can "slide" down the string.

The value of a well's absorbance is the first five characters from the starting position. Thus, an @LEFT function is needed in the equation. That is, the @LEFT function chooses the first five characters from the @RIGHT's starting position. The following is the new equation:

```
@LEFT(@RIGHT(CURNT,72-6*COL),5)
```

The chosen characters are next used by the @VALUE function to calculate the absorbance value, as follows

```
@VALUE(@LEFT(@RIGHT(CURNT,72-6*COL),5))
```

The data are next scaled by dividing @VALUE by 1000. Finally, the {PUT} command places a data point into its appropriate cell using the COL and ROW loop counters to position it correctly within the ABS range.

This combination of @RIGHT and @LEFT functions is powerful and you will find that it can be used in many applications.

EXTRACTING AND FILING JUST THE DATA PORTION OF A SPREADSHEET

After you have placed all of your data into the cells of your template, you will want to immediately protect the data by storing them away in a file. This technique is illustrated by the ARCHIVE subroutine shown in Figure 8-9.

The ARCHIVE subroutine performs an interesting and useful function. It cuts out a portion of a spreadsheet, creates a file for it, and writes the information to the file. Later, you can recall the information in the file by using the RETRIEVE subroutine (also shown in Figure 8-9). That is, you can "swap" data between your template and individual files without storing an entire spreadsheet.

This technique is very memory-efficient because it results in a much smaller file than if you were to store an entire spreadsheet. Extracting the data into another spreadsheet has another advantage. It allows you to retrieve the extracted file just as you would any other worksheet file or even combine it into another (third) worksheet file. This capability is a tremendous aid in data reduction.

ARCHIVE is similar to the subroutines used in previous chapters to save files. It is a self-modifying macro that takes the plate barcode from PLBCODE and writes it into the cell called FILESPEC. This cell is also one of the command lines of the ARCHIVE macro. After Lotus gets into the File eXtract Values function, it uses the text in the FILESPEC cell as the default for the filename prompt. The {?}

```
--------BC-------BD-------BE-------BF-------BG-------BH-------BI-------BJ----
 1                    /******FILE MANIPULATION SECTION******/
 2
 3            /*THIS SELF-MODIFYING MACRO ARCHIVES THE CURRENT*/
 4            /*PLATE INTO A FILE USING PLATE BARCODE AS ITS NAME*/
 5 ARCHIVE    {IF VERSION=1}{S}WUMAIN~    /*SYMPH: USE THE MAIN WINDOW*/
 6            {WINDOWSOFF}                /*FREEZE THE DISPLAY*/
 7            {LET FILESPEC,PLBCODE}      /*SELF-MODIFY FILESPEC INSTRUCTION*/
 8            {IF VERSION=0}/FXV          /*LOTUS 1-2-3, FILE eXtract*/
 9            {IF VERSION=1}{S}FXV        /*SYMPHONY, FILE eXtract*/
10 FILESPEC                           /*DEFAULT FNAME TO CURRENT PLATE BCODE*/
11            {?}~REPORT~             /*ALLOW CHANGES; THEN SPECIFY REPORT RANGE*/
12            {RIGHT}~{HOME}          |*THIS COMMAND IS NEEDED TO OVERWRITE*|
13            {WINDOWSON}             |*OLD FILE:  PROMPT'S DEFAULT IS NO  *|
14
15 RETRIEVE {GETLABEL "RETRIEVE WHICH PLATE'S DATA? ",FILENAME}
16            {ONERROR \M}            /*SET ERROR TRAP TO RETURN TO MAIN MENU*/
17            {IF VERSION=1}{S}WUMAIN~    /*SYMPH: USE MAIN WINDOW*/
18            {GOTO}REPORT~           /*POSITION CELL-POINTER FOR FILE-COMBINE*/
19            {IF VERSION=0}/FCCNREPORT~              |*FILE-IMPORT; TAKE ONLY*|
20            {IF VERSION=1}{S}FCCNREPORT~            |*THE "REPORT" RANGE    *|
21            IV                      /*IMPORT USING "IGNORE" AND "VALUES"*/
22 FILENAME                           /*SELF-MODIFIED PLATE NAME TO RETRIEVE*/
23            ~                       /*INITIATE THE FILE RETRIEVAL*/
24
25
```

Figure 8-9

command allows a user to change the filename. Once a user presses the carriage return, the range named REPORT is written to the file.

You can either have ARCHIVE automatically store a report in a file (as shown in the example program), or make it optional (by making it a menu choice).

RETRIEVE works the same way. It prompts for the plate barcode number, self-modifies the FILENAME cell, and retrieves the REPORT range from the corresponding file. It then overwrites the REPORT section of the template with the data from the stored file.

OTHER UTILITIES

Figures 8-10 and 8-11 show a few macros that support some other features of the program. The ENTER subroutine in Figure 8-10 prompts a user to enter concentrations for each standard. You can use a macro similar to ENTER anytime that you want a user to enter data into the cells of a template.

The {?} ~ command in the ENTLOOP macro pauses execution of the macro until a user presses return. That is, a user can move the cell-pointer around, use the function keys, enter data, edit data, etc.; the data will be entered into the highlighted cell when return is pressed. Because pressing return has no effect other than to end the {?} command, a tilde (~) is used after the {?} command to enter the data into the spreadsheet.

Also, note how the cell-pointer is managed by the ENTER subroutine to automatically move down to the next cell in the concentration table.

```
--------BH-------BI-------BJ-------BK-------BL-------BM-------BN-------BO----
 1           /***MACROS THAT MANIPULATE AND UPDATE THE DATA***/
 2
 3                     /*MACRO TO ENTER THE STANDARD CONCENTRATIONS*/
 4 ENTER     {IF VERSION=1}{S}WUSTATS~   /*SYMPHONY:GO TO STATISTICS WINDOW*/
 5           {HOME}                      /*CENTER THE FORM*/
 6           {GOTO}STDCONC~              /*MOVE CELL-PNTR TO THE TABLE*/
 7           {FOR ROW,0,7,1,ENTLOOP}     /*ENTER THE 8 CONCENTRATIONS*/
 8           {IF VERSION=1}{S}WUMAIN~    /*SYMPHONY: RETURN TO MAIN WINDOW*/
 9
10                     /*LOOP FOR ENTERING THE STANDARD CURVE*/
11 ENTLOOP   {?}~      /*GET ONE CONCENTRATION*/
12           {DOWN}    /*MOVE DOWN TO NEXT INPUT CELL*/
13
14                     /*LINEAR REGRESSION MACRO*/
15 DO_STATS  {IF VERSION=1}{S}WUSTATS~   /*SYMPH: SWITCH TO STATS WINDOW*/
16           {HOME}            /*CENTER THE SCREEN*/
17           {GOTO}STATS~      /*MOVE CELL-POINTER TO STAT REPORT*/
18           {LEFT 4}{UP 2}    /*CENTER STATS DATA ON SCREEN*/
19           {INDICATE STATS}            /*STATUS = STATISTICS PHASE*/
20           {PANELON}{PANELOFF}         /*UPDATE THE MODE INDICATOR*/
21           {WINDOWSOFF}                /*FREEZE THE DISPLAY*/
22           {IF VERSION=0}{LET STATVERS,"/DR"}  /*USE 1-2-3 MENU*/
23           {IF VERSION=1}{LET STATVERS,"/RR"}  /*USE SYMPHONY MENU*/
24 STATVERS  /RR                         /*SELF MODIFYING MACRO TARGET CELL*/
25           XSTDCONC~                   /*SET X RANGE TO STD VALUES*/
26           YSTDABS~                    /*SET Y TO ABSORBANCE FOR STDS*/
27           OSTATS~G                    /*SET OUTPUT RANGE FOR STATS*/
28           {ROWCONC}                   /*UPDATE THE CONCENTRATION TABLE*/
29           {CALC}                      /*FORCE RECALC OF SPREADSHEET*/
30           {INDICATE}                  /*SIGNAL COMPLETION*/
31           {WINDOWSON}                 /*REFRESH DISPLAY*/
32           {PANELON}                   /*TURN PANEL BACK ON*/
33           {IF VERSION=1}{S}WUMAIN~    /*SYMPHONY: RETURN TO MAIN WINDOW*/
34
35           /*UPDATE THE CONCENTRATION TABLE BASED ON ABS, SLOPE AND BIAS*/
36 ROWCONC   {FOR ROW,0,7,1,COLCONC}
37
38 COLCONC   {FOR COL,0,11,1,CELLCONC}
39
40 CELLCONC  {PUT CONC,COL,ROW,(@INDEX(ABS,COL,ROW)-BIAS)/SLOPE}
41
42
43
```

Figure 8–10

```
--------CB-------CC-------CD-------CE-------CF-------CG-------CH-------CI---
 1           /***MACRO SUBROUTINE THAT SUPPORTS THE PRINT MENU***/
 2
 3           /*PRINT FUNCS: PRINT; RETURN TO MENU1*/
 4 PRINT     {CALC}                     /*FORCE RECALCULATION OF TEMPLATE*/
 5           {WINDOWSOFF}{PANELOFF}     /*FREEZE THE DISPLAY*/
 6           {IF VERSION=0}/PPR                   /*LOTUS 1-2-3*/
 7           {IF VERSION=1}{SERVICES}PSSR         /*SYMPHONY*/
 8 REP_SPEC                             /*RANGE TO PRINT + ~  */
 9           {IF VERSION=1}Q            /*IF SYMPHONY, QUIT SUBMENU*/
10           AGPQ                       /*ALIGN; GO; PAGE ADVANCE; QUIT*/
11           {WINDOWSON}{PANELON}       /*THAW THE DISPLAY*/
12           {MENUBRANCH MAINMENU}      /*RETURN TO USER'S MENU*/
13
14
```

Figure 8–11

The DO_STATS macro in Figure 8-10 is the same macro used in Chapter 7, modified to include new range names. Like the linear regression in Chapter 7, it is simple. However, you can modify it easily to your particular needs by using @functions in the ranges (e.g., @LOG, @EXP, etc.).

Placing @functions in cells slows the speed of a spreadsheet each time it recalculates. For this reason, you may want to use a macro to place values into cells. Figure 8-10 illustrates one way to calculate the values of unknowns from a standard curve and place the results into a portion of the template.

ROWCONC, COLCONC, and CELLCONC work together to take the absorbances in the ABS range, and the SLOPE and BIAS from the linear regression section to calculate the concentrations for each well in the plate. These macros are relatively straightforward: The ROW and COL counters for the loops are used in the @INDEX function to get the absorbance of the well and also in the {PUT} command to determine the position where the concentration data are to be placed. The equation in the {PUT} command is a rearrangement of the familiar $y = mx + b$ equation. Again, you can modify this example for your particular equations.

The PRINT function of Figure 8-11 was briefly described in the section related to the menus. If you recall, a line of text is placed in the report specification (REP_SPEC) cell by the Report and Absorbance choices on the PRNMENU menu. When program execution reaches this cell, it prints the appropriate range. The remainder of the macro tests to see if it is operating under Lotus 1-2-3 or Symphony and issues the appropriate commands.

THE SCRATCH-PAD

The scratch-pad (shown in Figure 8-12) contains the following information:

- As always, VERSION keeps track of whether the program is running under Lotus 1-2-3 or Symphony.
- ERRMSG is a cell where Lotus error messages are placed by {ONERROR}s.
- BUFFERA through BUFFERH cells are buffers that receive incoming strings from {READLN} commands. These cells are collectively called BUFBLK (for buffer block). BUFBLK is used by an @INDEX during parsing.
- The cell called CURNT is a buffer that holds the current string being parsed.

ON-SCREEN PROMPTS

Figure 8-13 shows the message for the on-screen prompt that appears in Symphony. It directs the user to press the start button on the EL-309 and works like the on-screen prompt for the LS-2B described in Chapter 7.

```
--------CK-------CL-------CM-------CN-------CO-------CP-------CQ-------CR----
 1                   /******HOUSEKEEPING SCRATCHPADS******/
 2
 3 VERSION           /*0=>LOTUS 1-2-3; 1=>SYMPHONY*/
 4
 5 ERRMSG            /*THIS IS A CELL FOR ERROR MESSAGE TESTING*/
 6
 7                   /*LOOP COUNTERS*/
 8 ROW                        /*LOOP COUNTER FOR ROWS*/
 9 COL                        /*LOOP COUNTER FOR COLUMNS*/
10
11
12                   /*PLATE READER BUFFERS FOR THE ROWS*/
13 BUFFERA
14 BUFFERB
15 BUFFERC
16 BUFFERD
17 BUFFERE
18 BUFFERF
19 BUFFERG
20 BUFFERH
21
22                   /*CURRENT STRING BEING PARSED*/
23 CURNT
24
25
26
```

Figure 8-12

CREATING THE PROGRAM

Create the program in Figures 8-3 through 8-6 and 8-9 through 8-14 by entering the text into your spreadsheet. To activate the program, perform the following tasks:

1. Use the / Range Name Label Right (/RNLR) command on columns S, AD, AM, AY, BH, BC, CB, and CK to create the named ranges for each macro or named cell.

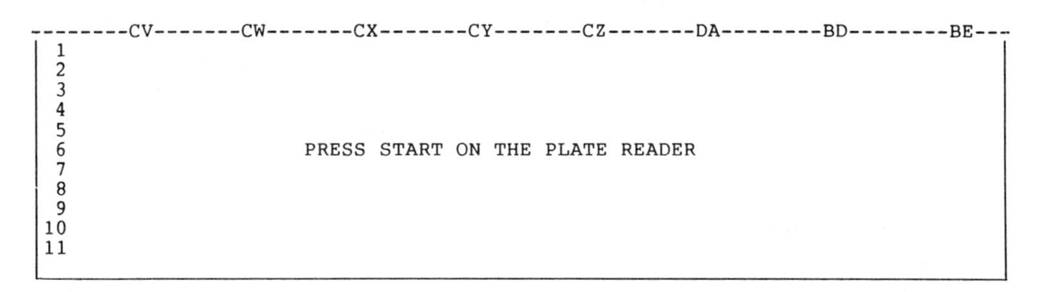

```
--------CV-------CW-------CX-------CY-------CZ-------DA--------BD--------BE---
 1
 2
 3
 4
 5
 6                  PRESS START ON THE PLATE READER
 7
 8
 9
10
11
```

Figure 8-13

2. Use the / Range Name Create (/RNC) command to create the range named BUFBLK for cells CL13..CL20.
3. Use the / Range Name Table (/RNT) command to create a table of the range names in the cells starting at DF3. (See Figure 8–14.)
4. Check the table. If any ranges are missing, add them.
5. If you are using Lotus Symphony, issue the {SERVICES} Settings Auto-execute Set (F9 SAS) command and type AUTOEXEC. Press return. AUTOEXEC will then execute each time the Symphony spreadsheet is retrieved.
6. Store the spreadsheet in a file.

As in the previous chapter, if you are using Symphony, you must complete an additional task. Create a "window" for a prompt similar to the one that appears

```
--------DE-------DF---------DG-------DH-------DI---------DJ--------DK---------
 1                          NAMED RANGES
 2                          ============
 3           ABS        B7..M14             FILESPEC    BD10
 4           ABSREP     A3..M14             MAINMENU    AE3
 5           ADD_DOS    T21                 MU_RESET    AZ16
 6           ADD_STATS  T26                 PARSEDATA   AZ34
 7           ARCHIVE    BD5                 PLBCODE     H3
 8           AUTOEXEC   T9                  PRINT       CC4
 9           BIAS       I33                 PRNMENU     AE28
10           BIOTEK     AN12                PROMPT      AZ3
11           BUFBLK     CL13..CL20          REPORT      A3..M42
12           BUFFERA    CL13                REP_SPEC    CC8
13           BUFFERB    CL14                RESET       AZ11
14           BUFFERC    CL15                RETRIEVE    BD15
15           BUFFERD    CL16                ROW         CL8
16           BUFFERE    CL17                ROWCONC     BI36
17           BUFFERF    CL18                ROWPARSE    AZ40
18           BUFFERG    CL19                RTRY        AN20
19           BUFFERH    CL20                RUNMENU     AE7
20           CELLCONC   BI40                SLOPE       H39
21           CELLPARSE  AZ44                STATMENU    AE35
22           COL        CL9                 STATS       F32
23           COLCONC    BI38                STATVERS    BI24
24           CONC       B20..M27            STDABS      D32..D39
25           CURNT      CL23                STDCOL      B7..B14
26           DATERUN    C3                  STDCONC     C32..C39
27           DO_STATS   BI15                TRPLOOP     AZ22
28           ENTER      BI4                 UTILITIES   AE13
29           ENTLOOP    BI11                VERSION     CL3
30           ERRMSG     CL5                 VIEWMENU    AE17
31           FILEMENU   AE46                \0          T4
32           FILENAME   BD22                \M          T31
33
34
35
36
37
38
```

Figure 8–14

in Figure 8–13. This window contains a help message to new users that alerts them to press the start button on the instrument.

The procedure to create the window is the same as the one described in the previous chapter. Use the following sequence:

1. Move the cell-pointer to cell CV1.
2. Issue the {SERVICES} Window Create command.
3. When prompted for the new window name, type MESSAGE and press return.
4. When prompted for the spreadsheet work environment, choose Sheet.
5. When asked for the Window area, use the up arrow key to move the highlighted area to cell DC9. Use the left arrow key to move the cell DB9. Then, press the TAB key and arrow right cell CV9. Press TAB again and arrow to CV3. Press return when you are satisfied with the location and size of the window.
6. At the next menu, choose Borders, Line, and Quit. This command sequence eliminates row and column designators for the window and makes the window look more like a message.
7. When you return to the spreadsheet environment with the cell-pointer in the MESSAGE window, enter the message text into the center of the window. If you cannot center the text because the text jumps, move the cell-pointer to the left-hand side of the window and change the window's width. (For the example, I used the / Width Set (/WS) command to set the column width to 6 and the text was perfectly centered.)
8. If you are not satisfied with the placement or size of the window, issue the {SERVICES} Window Layout command (F9 WL). This command will return you to the "Identify Window Area" prompt and allow you to repeat Step 5 to redefine the window.
9. Press the WINDOW key (F6) to remove the MESSAGE window from the display and return to the MAIN window.
10. Store the spreadsheet in a file.

Follow this procedure for another window called STATS. The STATS window should be large enough to hold standard concentrations, absorbances, and linear regression data in the statistical section of the template.

PRINTING WIDE REPORTS IN COMPRESSED MODE

Most dot matrix and laser printers can print in "compressed" mode. That is, they can decrease the width of the characters that they print and are thereby able to print more columns on a standard sheet of computer paper. The template presented in this chapter is wider than the usual 80 columns that printers can normally print. If your template is wider than 80 columns, you may want to issue the appropriate compressed mode command to your printer before printing your reports.

The easiest way to issue the compressed mode command is to configure your spreadsheet to send the control code to the printer. The control code is different for each printer, so refer to the User's Manual for your printer. Typically, control codes are non-printing characters, or strings of non-printing characters. That is, they fall within the first 26 ASCII character codes. For example, the code that places an Epson FX-80 printer into compressed mode is ASCII character 15. To return to normal mode, the code is ASCII 27 followed by an @ sign.

To specify the code for your printer in Lotus 1-2-3, issue the / Print Printer Options Setup (/PPOS) command. In Symphony, issue the {SERVICES} Print Settings Init-String (F9 PSI) command. When prompted for the initialization string, enter the string. If you are specifying ASCII characters by their number equivalents, begin the number with a backslash followed by a *three-digit* number. For example, for the Epson FX-80 you would enter \015 for compressed mode or \027@ to reset the printer to normal mode. Notice that because ASCII 15 is only two digits, it is preceded by a zero.

Once you have included the compressed mode initialization string with the print settings of your spreadsheet, set the left and right margins to 0 and 140 respectively and save the file again so that the strings and new margin settings are stored permanently with your template.

CRANK UP THE SPEED!

The program is now ready to use. If you have an EL-309, you can run the program at this point. If you have an instrument similar to the EL-309, you will find that most of this program can be used as-is for the instrument.

After you have your program up and running successfully, increase the Baud rate in a stepwise fashion until you no longer get reliable communications. Then, return to the last Baud rate that gave acceptable performance.

Keep in mind that if you change the Baud rate of your instrument, you must turn your instrument off momentarily and then turn it back on. You should also remember to reset the Baud rate of the personal computer.

WHAT'S NEXT?

The next chapter will show you how to use Lotus Measure™ as an alternative to the Device File Name method for controlling instruments from your Lotus spreadsheet. Lotus Measure should be considered if you have an instrument that sends data out quickly and continually.

The next chapter will also show you how to use Lotus Measure to obtain data from plug-in data acquisition boards and how to control instruments from a communications standard that is different from RS-232. This standard, called IEEE-488, is commonly used to control multiple instruments from the same communications port.

9

Unconventional Equipment: Lotus Measure™

The previous few chapters developed methods to control instruments and acquire their data. The programs developed accomplished this control by using DOS Device File Names and the Lotus file manipulation macro commands. The programs developed controlled input and output (I/O) directly from a Lotus spreadsheet, without using Lotus Measure.™

Lotus Measure provides some capabilities that direct control does not. However, the cost of the software package is approximately the same as Lotus 1-2-3®, if not more. Using Measure will double your software investment, and may not provide you with improved performance.

In spite of this drawback, some situations may require that you use Lotus Measure. As previously stated, some instruments output too much data, too quickly, and therefore routinely overrun the DOS buffer. You can easily recognize these instruments because they have two common characteristics:

- They send out two or more strings immediately following each other.
- They are used in applications that prevent you from slowing the Baud rate enough to avoid System Errors.

Instruments that overrun DOS buffers are not very common, but if your instrument is in this category, you will need to use Lotus Measure.

If you are using a data acquisition board (such as the MetraByte DASH-16™) that plugs directly into your personal computer, you will need Lotus Measure. Additionally, if your instrument uses the IEEE-488 interface instead of RS-232, you will need Lotus Measure.

Even though previous chapters of this book omitted discussion of Lotus Measure, the basic programming principles, steps required, structures, and techniques for instrument control described previously are essential for you to know. The differences in these topics that apply to Measure are few. The first difference is that you do not need to set communications port parameters with the DOS MODE command. The second difference is that you do not need to use {OPEN} and {CLOSE} commands. The third difference is that you must replace {WRITELN} and {READLN} commands with special Lotus Measure macro commands, such as {RSEND} and {RRECEIVE}. Thus, nearly everything that you have learned up to this point (buffers, menus, moving data, etc.) applies to Lotus Measure.

Figures 9-1 and 9-2 illustrate this point. Both programs receive data from a Mettler AE163 analytical balance with an Option 012 Data Interface. (The balance and interface are available from Mettler Instrument Corp, Box 71, Hightstown, NJ, 08520-9944.)

Figure 9-2 uses the DOS Device File Name programming techniques described throughout the previous chapters. Figure 9-1 uses Lotus Measure. Take a moment to see how similar these programs are. In fact, both programs work identically from a user's point of view.

Also try to visualize how these programs might fit into a larger program that has menus, print functions, file operations, statistical analyses, etc.

One other difference exists between the two programming techniques. It deals with the cabling between the personal computer and the instrument. The DOS Device File Name technique often requires that you disable or "fool" the handshaking (data transmission control) lines between the personal computer and instrument before the instrument will send its data. Lotus Measure, on the other hand, has the ability in many situations to control handshaking using these lines. In fact, Measure quite often requires that these lines *not* be disabled so that they are available for reliable communications.

For example, the Mettler balance requires a straight-through cable for Lotus Measure. However, it requires the following cable for DOS Device File Name programs:

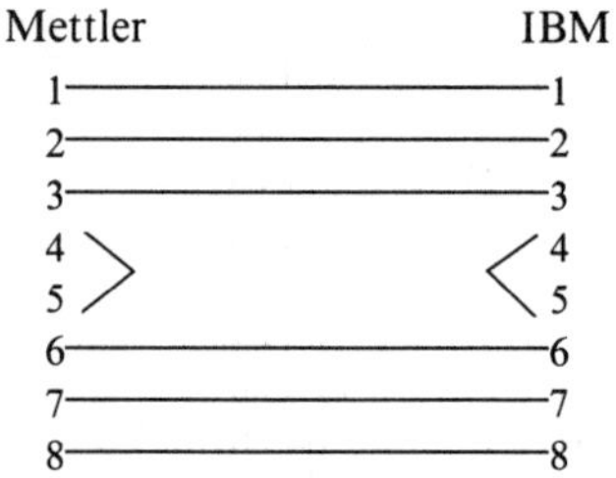

```
Mettler                 IBM
1———————————————————————1
2———————————————————————2
3———————————————————————3
4 \                   / 4
5 /                   \ 5
6———————————————————————6
7———————————————————————7
8———————————————————————8
```

This cable is the one used for the BioTek EL309 plate reader of the previous chapter. The configuration of this cable is one which is commonly required by instruments when the Device File Name technique is being used.

```
--------K-------L--------M--------N--------O--------P--------Q--------R---------
 1                       /*LOTUS MEASURE DRIVER FOR METTLER BALANCE*/
 2 \0        {LET VERSION,0}    /*RUNNING UNDER 1-2-3*/
 3           {APP3}SNRMETTLER~QQ         /*USE METTLER SETTINGS*/
 4           {METTLER}          /*START THE PROGRAM*/
 5
 6 AUTOEXEC{LET VERSION,1}      /*RUNNING UNDER SYMPHONY*/
 7           {IF @ISERR(@APP("RS232",""))}{ADD_MEAS}
 8           {S}RSNRMETTLER~QQ{ESC}      /*USE METTLER SETTINGS*/
 9           {METTLER}          /*START THE PROGRAM*/
10
11 ADD_MEAS{S}AARS232~Q         /*ATTACH LOTUS MEASURE RS232.APP*/
12
13
14 METTLER {GETNUMBER "How many Samples?  ",SAMPLES}
15           {IF VERSION=0}/REDATA~      /*ERASE DATA SECTION*/
16           {IF VERSION=1}/EDATA~       /*0=>1-2-3; 1=>SYMPHONY*/
17           {LET ROW,1}                 /*INITIALIZE ROW COUNTER*/
18 NEXT_WT {LET ROW,ROW+1}               /*INCREMENT ROW COUNTER*/
19           {WINDOWSOFF}                /*FREEZE THE DISPLAY*/
20 GET_WT  {ONERROR ER1HDLR,WT_ERROR} /*SET ERROR TRAP PROTOCOL*/
21           {INDICATE WEIGH}            /*SIGNAL WEIGHT PROMPT*/
22           {PANELON}{PANELOFF}         /*FREEZE THE PANEL*/
23 WT_RESET{RRECEIVE WT_BUF,10} /*WAIT UNTIL OBJECT PLACED ON BALANCE*/
24           {LET BUF_VAL,@VALUE(@LEFT(@RIGHT(WT_BUF,12),10))}
25           {IF BUF_VAL<0.010}{BRANCH WT_RESET} /*SECTION'S GOAL MET?*/
26 STEADY  {ONERROR ER2HDLR,WT_ERROR} /*START NEW ERROR HANDLER*/
27           {RRECEIVE WT_BUF,10}        /*WAIT FOR A STEADY STATE READING*/
28           {IF @LEFT(@RIGHT(WT_BUF,15),2)="SD"}{BRANCH STEADY}
29           {RRECEIVE WT_BUF,10}    /*GOAL WAS MET: TAKE A DATA POINT*/
30           {PUT DATA,0,ROW,@VALUE(@LEFT(@RIGHT(WT_BUF,12),10))}
31           {BEEP 5}           /*SIGNAL SUCCESSFUL READING*/
32           {WINDOWSON}{WINDOWSOFF}     /*REFRESH THE DISPLAY*/
33 WT_LOW  {ONERROR ER3HDLR,WT_ERROR} /*START NEW ERROR HANDLER*/
34           {IF ROW>SAMPLES}{QUIT}      /*LAST SAMPLE?*/
35           {RRECEIVE WT_BUF,10}  /*WAIT UNTIL OBJECT REMOVED FROM BALANCE*/
36           {LET BUF_VAL,@VALUE(@LEFT(@RIGHT(WT_BUF,12),10))}
37           {IF BUF_VAL>0.010}{BRANCH WT_LOW}   /*SECTION'S GOAL MET?*/
38           {BRANCH NEXT_WT}            /*LOOP TO GET NEXT SAMPLE*/
39
40                      /*ERROR HANDLING MACROS*/
41 ER1HDLR {IF @EXACT(WT_ERROR,"System error -- Press [HELP]")}{BRANCH GET_WT}
42
43 ER2HDLR {IF @EXACT(WT_ERROR,"System error -- Press [HELP]")}{BRANCH STEADY}
44
45 ER3HDLR {IF @EXACT(WT_ERROR,"System error -- Press [HELP]")}{BRANCH WT_LOW}
46
47 \M        {METTLER}          /*CALL METTLER BALANCE AS A SUBROUTINE*/
48           {CALC}             /*FORCE SPREADSHEET UPDATE*/
49
50           /*SCRATCH PAD*/
51 WT_BUF                       /*BUFFER FOR INCOMING WEIGHTS*/
52 BUF_VAL                      /*CELL FOR PARSING WEIGHT STRING*/
53 WT_ERROR                     /*CELL FOR ERROR MESSAGES*/
54 SAMPLES                      /*NUMBER OF SAMPLES TO BE RUN*/
55 ROW                          /*CURRENT SAMPLE NUMBER*/
56 VERSION                      /*0 => 1-2-3; 1 => SYMPHONY*/
57
58
```

Figure 9-1

```
--------V-------W--------X--------Y--------Z---------AA-------AB-------AC--------
 1          /*DOS DEVICE FILE NAME DRIVER FOR METTLER BALANCE*/
 2 \0       {LET VERSION,0}    /*RUNNING UNDER 1-2-3*/
 3          {METTLER}          /*START THE PROGRAM*/
 4
 5 AUTOEXEC{LET VERSION,1}    /*RUNNING UNDER SYMPHONY*/
 6          {IF @ISERR(@APP("DOS","")){HOOKUP}
 7          {S}AIDOS~                   /*INVOKE DOS APPLICATION*/
 8          MODE COM1:150,E,7,1~        /*SET COMMUNICATION PARAMS*/
 9          {METTLER}          /*START THE PROGRAM*/
10
11 HOOKUP   {S}AADOS~          /*ATTACH DOS APPLICATION PROGRAM*/
12
13 METTLER {GETNUMBER "How many Samples?  ",SAMPLES}
14          {IF VERSION=0}/REDATA~      /*ERASE DATA SECTION*/
15          {IF VERSION=1}/EDATA~       /*0=>1-2-3; 1=>SYMPHONY*/
16          {LET ROW,1}                 /*INITIALIZE ROW COUNTER*/
17 NEXT_WT {LET ROW,ROW+1}             /*INCREMENT ROW COUNTER*/
18          {WINDOWSOFF}                /*FREEZE THE DISPLAY*/
19 GET_WT   {ONERROR ER1HDLR,WT_ERROR} /*SET ERROR TRAP PROTOCOL*/
20          {OPEN "COM1",M}             /*OPEN COMMUNICATIONS PORT*/
21          {ONERROR ER1HDLR,WT_ERROR} /*RESET ERROR TRAP PROTOCOL*/
22          {INDICATE WEIGH}            /*SIGNAL WEIGHT PROMPT*/
23          {PANELON}{PANELOFF}         /*FREEZE THE PANEL*/
24 WT_RESET{READLN WT_BUF}    /*WAIT UNTIL OBJECT PLACED ON BALANCE*/
25          {LET BUF_VAL,@VALUE(@LEFT(@RIGHT(WT_BUF,12),10))}
26          {IF BUF_VAL<0.010}{BRANCH WT_RESET} /*SECTION'S GOAL MET?*/
27 STEADY   {ONERROR ER2HDLR,WT_ERROR} /*START NEW ERROR HANDLER*/
28          {READLN WT_BUF}    /*WAIT FOR A STEADY STATE READING*/
29          {IF @LEFT(@RIGHT(WT_BUF,15),2)="SD"}{BRANCH STEADY}
30          {READLN WT_BUF}    /*GOAL WAS MET: TAKE A DATA POINT*/
31          {PUT DATA,0,ROW,@VALUE(@LEFT(@RIGHT(WT_BUF,12),10))}
32          {BEEP 5}           /*SIGNAL SUCCESSFUL READING*/
33          {WINDOWSON}{WINDOWSOFF}     /*REFRESH THE DISPLAY*/
34 WT_LOW   {ONERROR ER3HDLR,WT_ERROR} /*START NEW ERROR HANDLER*/
35          {IF ROW>SAMPLES}{QUIT}               /*LAST SAMPLE?*/
36          {READLN WT_BUF}    /*WAIT UNTIL OBJECT REMOVED FROM BALANCE*/
37          {LET BUF_VAL,@VALUE(@LEFT(@RIGHT(WT_BUF,12),10))}
38          {IF BUF_VAL>0.010}{BRANCH WT_LOW}  /*SECTION'S GOAL MET?*/
39          {BRANCH NEXT_WT}            /*LOOP TO GET FOR NEXT SAMPLE*/
40
41                   /*ERROR HANDLING MACROS*/
42 ER1HDLR {IF @EXACT(WT_ERROR,"System error -- Press [HELP]")}{BRANCH GET_WT}
43
44 ER2HDLR {IF @EXACT(WT_ERROR,"System error -- Press [HELP]")}{BRANCH STEADY}
45
46 ER3HDLR {IF @EXACT(WT_ERROR,"System error -- Press [HELP]")}{BRANCH WT_LOW}
47
48 \M       {METTLER}          /*CALL METTLER BALANCE AS A SUBROUTINE*/
49          {CALC}             /*FORCE SPREADSHEET UPDATE*/
50          {CLOSE}            /*CLOSE THE PORT*/
51
52          /*SCRATCH PAD*/
53 WT_BUF                      /*BUFFER FOR INCOMING WEIGHTS*/
54 BUF_VAL                     /*CELL FOR PARSING WEIGHT STRING*/
55 WT_ERROR                    /*CELL FOR ERROR MESSAGES*/
56 SAMPLES                     /*NUMBER OF SAMPLES TO BE RUN*/
57 ROW                         /*CURRENT SAMPLE NUMBER*/
58 VERSION                     /*0 => 1-2-3; 1 => SYMPHONY*/
59
60
```

Figure 9-2

In short, you will need to follow the same steps to interface an instrument to Lotus Measure as the ones that you used for the Device File Name technique. However, you will need to follow a few additional steps, which will be discussed in detail as they become relevant.

BRIEF DESCRIPTION OF LOTUS MEASURE

Lotus Measure has two versions. One version is for use with Lotus 1-2-3 and the other is for use with Symphony. When ordering Measure, you must specify which version you want.

Lotus claims that Measure can control the collection of data from over 8,000 types of instruments. It has five different "drivers" (modules) that allow you to control not only instruments that use the RS-232 communications standard, but other protocols as well.

The following is a brief summary of the protocols:

IBMGCA module. This Measure module was designed to be used with the IBM Game Control Adapter (GCA). By using this module and the GCA, you can build a crude A/D converter system. That is, the IBM GCA has four analog inputs and four digital inputs. If you have a GCA, you could theoretically use the adapter as a variable resistor or a digital switch to input data into a spreadsheet. However, data from this module are not of very good quality and can be collected only at very slow speeds. For these reasons, this system is not recommended.

RS-232 module. This module adds seven macro commands. These commands allow you to communicate with instruments that have RS-232 interfaces. This module also contains some other features that allow you to use instruments that cannot be interfaced using the programming techniques described in the previous few chapters. One of these features is support for both hardware and software handshaking. For example, you can enable XON/XOFF software handshaking. XOFF is a communications control character (ASCII 19) that instructs a personal computer or instrument to suspend transmission. XON is another control character (ASCII 17) that instructs the personal computer or instrument to resume transmission. Thus, XON/XOFF can regulate the flow of data between devices and thereby allow them to communicate even if they process the data at different rates. For example, an instrument will send some of its data and then wait for the personal computer to signal that it is ready to receive more data. This feature, then, avoids over-running the DOS buffer in cases where the personal computer cannot process the data as fast as the instrument can send it. This feature also allows you to use faster Baud rates with instruments that send out long data strings. The Bio-Tek EL-309 in Chapter 8 is an example of an instrument that could benefit from an XON/XOFF.

Another feature of this module is a "time-out". A "time-out" is essentially a timer that starts when a communications command is issued. The personal computer waits for input or a response from the instrument. If the instrument takes longer than the "time-out" setting to respond, a Lotus error message is displayed and the macro is aborted.

MBC16 module. This Measure module is designed to work with the DASH-16, a data acquisition plug-in board. This board is available from MetraByte™, 440 Myles Standish Blvd., Tauton, MA. The module will also support a number of other data acquisition plug-in boards from the manufacturers listed in Chapter 6. Before you order a board, consult the manufacturer to make certain that the board is supported by Measure.

NAT488/GPIB module. The IEEE-488 interface is a data transmission system comprised of a group of specialized cables that are controlled by a plug-in board in your personal computer. These cables carry information between your personal computer and one or more instruments. Unlike RS-232 (which can be connected to only one instrument at a time), the IEEE-488 interface can connect up to fifteen IEEE-488 compatible instruments simultaneously. IEEE-488 provides a uniform and organized method of sending information between your personal computer and a series of instruments or from one instrument to another. Thus, you can control more than one instrument from the same port. This capability solves the problem of having only two communications ports (COM1 and COM2) and many instruments to interface.

Also, unlike RS-232, IEEE-488 is *extremely fast.* Baud rates in RS-232 rarely exceed 19,200 bps; while Baud rates for IEEE-488 are routinely 1 million to 8 million bps! This rate is usually quick enough for even the fastest experiments.

The NAT488 module is the Lotus 1-2-3 Measure module required to control the IBM GPIB Adapter and the National Instruments (12109 Technology Blvd., Austin, TX) GPIB-PCII™ and GPIB-PCIIA™ boards and their compatibles. The module is called GPIB in the Symphony version of Measure. With one of these boards and Measure you have the ability to support up to fifteen IEEE-488 instruments simultaneously. These Measure modules add 21 new macro commands to the spreadsheet. These commands allow you to control sending, receiving, and gathering of information along a chain of instruments.

HP488/HP-IB module. The HP488 (Lotus 1-2-3) and HP-IB (Symphony) Measure modules support all HP-IB (Hewlett Packard®-Interface Bus) IEEE-488 compatible plug-in boards. They work identically to the NAT488/GPIB modules.

INSTALLING AND STARTING LOTUS MEASURE

With the Measure version for Lotus 1-2-3, one or more of the modules are added to the ".SET" driver file for your spreadsheet. Once added, the Measure module(s) automatically load each time the spreadsheet is started. When you want to use a module's menu, you must press a special key combination (such as [ALT] F9) to activate the menu. The {APP} command is a Lotus 1-2-3 Measure macro command that automatically displays a module's main menu. You can use the {APP} command to invoke one of Measure's menus from within your macro. {APP3} for example is equivalent to pressing [ALT] F9.

With the Measure version for Symphony, each Measure module acts like other Symphony add-in application programs. That is, akin to the DOS.app and STAT.app programs described in previous chapters, you must attach the application program to Symphony using the {SERVICES} menu (F9 Application Attach). This command adds the name of the module to the {SERVICES} menu. Thereafter, you just choose the module's name from the {SERVICES} menu and the application program will be invoked. Likewise, you can issue an {S} macro command in a program and invoke an add-in directly to have it perform specific tasks.

To summarize: With the Lotus 1-2-3 version of Measure, you can use the module's macro commands directly in your program as soon as the spreadsheet has been started. With the Symphony version of Measure, you can use the module's macro commands only after it has been attached.

INSTALLING LOTUS MEASURE FOR LOTUS 1-2-3

Each Measure data acquisition module that you plan to use needs to be added to your 1-2-3 driver set. A Lotus driver set is a file that brings together all the software programs that drive the pieces of hardware used by Lotus. For example, a driver set includes software drivers for your monitor, printer, and hard disk. After you complete this section of the book, your 1-2-3 driver set will also hold the drivers for your data acquisition board, RS-232 interface, and/or IEEE-488 interface.

Begin by making a copy of the Measure files in the subdirectory containing your 1-2-3 program: Issue the DOS CHANGE DIRECTORY command (CD\subdirectory, where subdirectory contains 1-2-3). Place one Measure diskette into drive A, type copy a:*.* and press return. Using *.* tells DOS to copy all of the files on the disk. Repeat for all remaining Measure diskettes.

Next, copy your existing Lotus driver set to one that has the name of the module that you want to use. That is, issue one of the following copy commands for the particular Measure module(s) that you plan to use:

```
copy 123.set ibmgca.set
copy 123.set mbc16.set
copy 123.set nat488.set
copy 123.set hp488.set
copy 123.set rs232.set
```

Copying the Lotus driver set starts the process of creating a new 1-2-3 driver set with the name of the Measure module that you plan to use. By doing so, you retain the usual functionality of the spreadsheet if you start it with the usual command (123); while adding Measure functionality if you specify the alternate .set name on the DOS prompt line when you start 1-2-3. For example, to use the RS-232.set, you would type:

```
123 rs232.set
```

With this command, you specify the driver set that contains the Measure RS232 drivers. Lotus then adds the appropriate Measure functionality.

As noted earlier, Measure slows the speed of Lotus 1-2-3. It also consumes memory. However, if you will always be using Measure drivers, use the procedures given below to add the drivers directly to 123.set. By adding the drivers directly, you will not be required to enter the name of the alternate set each time you invoke Lotus. If you add the drivers to 123.set, it is recommended that you make a backup of the 123.set file under the name orig.set, so that you will always have a copy of your original (faster) driver set that you can go back to if need be.

The next step is to add the appropriate Measure driver to the new Lotus driver. This process uses a program called NEWLIB (on the Measure distribution disk) to create the new driver set. Issue the appropriate command for the module that you want to use:

```
newlib ibmgca.set ibmgca.drv -u
newlib mbc16.set mbc16.drv -u
newlib nat488.set nat488.drv -u
newlib hp488.set hp488.drv -u
newlib rs232.set rs232.drv -u
```

Press return. A list of the contents of the driver set file will be displayed on the screen. The Measure driver you selected will have been added to the list. The -u at the end of the line instructs NEWLIB to *u*pdate the driver set, adding the new Measure driver. Other options that you can use are

- -d: removes a driver from the driver set
- -l: lists the drivers in the driver set
- -h: gives a list of these four options (-u, -d, -l, -h)

INSTALLING LOTUS MEASURE FOR LOTUS SYMPHONY

Because the Measure data acquisition modules for Symphony are add-in application programs, you just need to copy each module that you plan to use into the subdirectory containing your Symphony program. Issue the DOS CHANGE DIRECTORY command (CD\subdirectory, where subdirectory contains Symphony). Place one of the Measure diskettes into drive A, type copy a: * . * and press return. Using * . * tells DOS to copy all of the files on the disk. Repeat for the remaining diskette.

STARTING LOTUS 1-2-3 WITH MEASURE

Once you have installed the Measure driver into the driver set, you will not need to repeat the procedure. Thereafter, when you want to use one of the Measure modules, you would enter the name of the appropriate driver set on the command line. By doing so, the Measure module is added automatically. For example, if you want to use the RS232.set, you would type

```
123 rs232.set
   <return>
```

After a few seconds, a title screen for Lotus Measure appears; after a few more seconds, the spreadsheet appears. To access the RS-232 Measure menu system, press [ALT] F9.

STARTING LOTUS SYMPHONY WITH MEASURE

Because Lotus Measure for Symphony is a collection of add-in application programs, you start Symphony as you normally do. That is, you type

```
 access
<return>
<return>
```

To attach the appropriate module, issue the {SERVICES} Application Attach (F9 AA) command. When presented with the list of application programs, choose the module and press return. Then press {ESC}ape until you get to the main SERVICES menu. The module's name will appear on the menu. To access the module, type the first letter of the module's name. To quit the SERVICES menu, press {ESC}ape.

LOTUS MEASURE RS232 MODULE

The most common method of communications for scientific instrumentation is RS-232. This section will show you how to configure and use the RS232 module of Lotus Measure.

Analytical balances are relatively ubiquitous in laboratories. The example used in this section is the Mettler AE163 balance mentioned earlier in this chapter.

The following are the usual steps to create a Measure RS232 application. Try to notice the similarities to those used in all of the previous Device File Name applications.

1. Determine the configuration of the cable/obtain the cable. Either using information from your User's Manual or Appendix A of this book, determine the type of cable that you need for your instrument. You should either buy or make the cable well in advance of when you want to begin developing the program. The example balance is listed in its User's Manual as a DCE (transmits on pin 3). You can therefore use a straight-through cable. After you connect the cable from your instrument to the personal computer, secure the cable at both ends with screws.
2. Set the communications parameters on the instrument. As is the case for the Device File Name method, you must set the DIP switches or jumper blocks on the instrument to their appropriate positions for the Baud rate, number of character bits, parity, and number of stop bits.

   In this example, the switches are located inside the Option 012 Data Interface. To set the switches on this instrument, remove the interface from the back of the balance by GENTLY pulling backwards at the top. This action disconnects the plug. Next, move the interface backward at an angle. You will see a ribbon cable connector. GENTLY pull the connector out. Then, pull the interface tabs from the back of the balance. Remove the four corner screws from the cover plate and remove the plate. You will see a circuit board. Remove the four screws that hold the board in position and pull it up slightly to expose a DIP switch block. Move the switches to the following positions:

Parity	OFF
Odd	OFF
Out Only	OFF
Cont	ON

   These settings will allow continuous monitoring with no parity.

   Next, locate jumper pins. They will appear as a set of double row "porcupine" looking needles facing upward. On one of the sets of pins is a small block of plastic. Remove it. Printed on the circuit board is a table of Baud rates for each set of pins. Place the plastic block on the set that equates to 4800 Baud.

   Gently replace the board and screw it back in. Do not over-tighten. Screw the cover back on CAREFULLY! Replace the ribbon cable onto its connector at the back of the balance and press the plug adapter of the interface into the balance plug. Plug the balance in and turn it on.

   Remember to turn your instrument off momentarily and then back on when changing communications settings. Because most instruments check the positions of DIP switches and jumper pins only at startup, the old values will be used unless the instrument is restarted again.
3. Start the Lotus spreadsheet by typing 123 RS232.set for Lotus 1-2-3 or AC-

CESS for Symphony. In Lotus 1-2-3 bring up the Measure RS-232 module by pressing [ALT] F9. In Symphony, press {SERVICES} Application Attach RS232 and return; then press {ESC}ape and choose RS232 ({ESC}R).

4. The main menu for the RS-232 module is shown in Figure 9–3. Choose Interface from the menu by pressing return. Another menu will appear. Select each of the choices on that menu to set up the following parameters (modified for your instrument). The parameters listed below are pertinent to the Mettler balance:

Comm-Port:	COM1
Baud:	4800
Stop-Bits:	1
Parity:	NONE
Length:	7

5. When data are received from an instrument, Measure automatically parses them. The default method uses two consecutive spaces (\032\032) to separate one string from the next. That is, two consecutive spaces appearing in a string of characters signals the end of one string and the beginning of the next. Measure parses the string by ending its current read command and placing only the first string in the buffer. To use this default parsing process would require that you also use some advanced programming techniques.

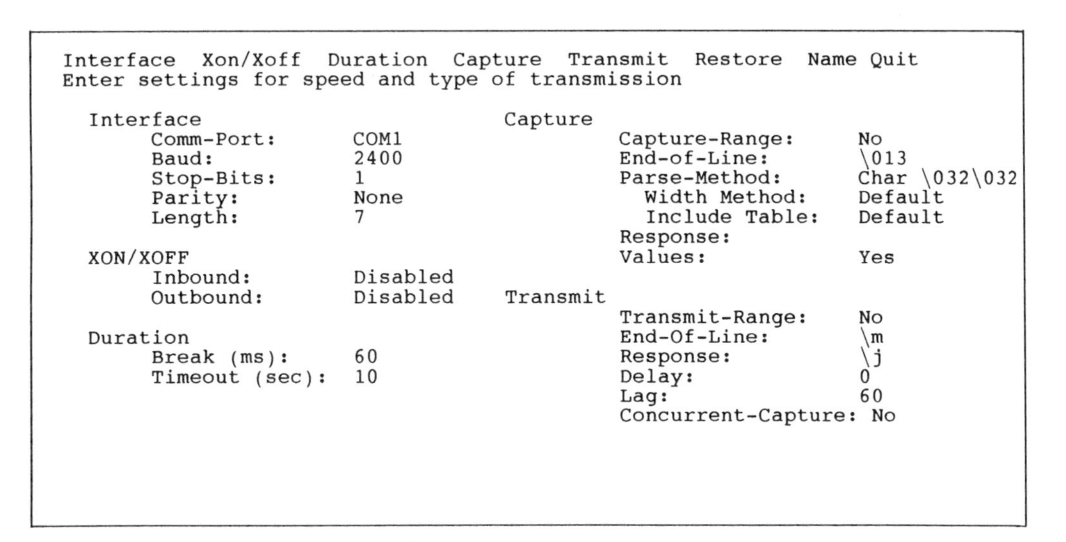

```
Interface  Xon/Xoff  Duration  Capture  Transmit  Restore  Name Quit
Enter settings for speed and type of transmission

  Interface                         Capture
       Comm-Port:       COM1                 Capture-Range:      No
       Baud:            2400                 End-of-Line:        \013
       Stop-Bits:       1                    Parse-Method:       Char \032\032
       Parity:          None                   Width Method:     Default
       Length:          7                      Include Table:    Default
                                             Response:
  XON/XOFF                                   Values:             Yes
       Inbound:         Disabled
       Outbound:        Disabled    Transmit
                                             Transmit-Range:     No
  Duration                                   End-Of-Line:        \m
       Break (ms):      60                   Response:           \j
       Timeout (sec):   10                   Delay:              0
                                             Lag:                60
                                             Concurrent-Capture: No
```

Figure 9–3

A string from the Mettler balance example will illustrate why these advanced techniques would be needed. The following example string from the Mettler balance contains at least two spaces between the status portion and the data portion of the string.

```
DS  0.0000 g
```

Thus, the first time a Measure read command is issued you would get the status character(s) in the buffer. The second time you would get the data string. This alternating of strings would continue, but the order depends on when you started your program and is therefore unpredictable, thus leading to parsing and/or @VALUE errors.

To make your programming as simple as possible, always work with an entire string for all instrument applications. You can usually keep Measure from parsing a string by using the familiar carriage return/line feed combination as the Parse-Method. This method is explained below.

First select Capture from the Settings menu. From the Capture menu, select Parse-Method and then Character. You will then be prompted for the separator. The \032\032 default will be displayed, so press ESCape to clear it. Enter \M\J or \013\010 and press return. Either combination can be used; they both mean ''carriage return/line feed''.

If this parse method will not work for your instrument, specify a character that would never be found in your data (e.g., STX, ASCII 002, or ETX, ASCII 003).

6. Use the Quit option to return to the main menu.
7. Next, select Name from the main menu. Name allows you to save the settings.

   When you start a Lotus 1-2-3 session, Measure activates and sets the communications parameters to their default values. When you start a Symphony session, an auto-executing macro attaches and activates Measure. Attaching Measure to Symphony also sets the communications parameters to their default values.

   In either case, the default values of the communications parameters will probably not be the ones that you want to use for your instrument. Saving the settings for your instrument is important because it will allow you to quickly reset the default values to the appropriate values from your macro program whenever you recall the spreadsheet.

   After selecting the Name option, you are presented with a second menu. Choose the Save option. When prompted for a name, enter an appropriate, descriptive title (e.g., METTLER) and press return.
8. Select Quit to return to the spreadsheet.
9. Design and enter your template and a ''bare bones'' macro program. Follow the principles and rules that you learned in previous chapters to design your template and program. An example Measure program is shown in Figure 9–1.

```
Interface  Xon/Xoff  Duration  Capture  Transmit  Restore  Name Quit
Enter settings for speed and type of transmission

  Interface                          Capture
       Comm-Port:       COM1                   Capture-Range:        No
       Baud:            4800                   End-of-Line:          \013
       Stop-Bits:       1                      Parse-Method:         Char \M\J
       Parity:          Even                     Width Method:       Default
       Length:          7                        Include Table:      Default
                                               Response:
  XON/XOFF                                     Values:               Yes
       Inbound:         Disabled
       Outbound:        Disabled         Transmit
                                               Transmit-Range:       No
  Duration                                     End-Of-Line:          \m
       Break (ms):      60                     Response:             \j
       Timeout (sec):   10                     Delay:                0
                                               Lag:                  60
                                               Concurrent-Capture: No
```

Figure 9–4

Any new programming techniques in the example are explained in the next section.

10. "Activate" your template by issuing the / Range Name Create command and specifying the names for the ranges of cells that will contain your data (e.g., DATA for C5..D1000 in this example). Test the template.
11. "Activate" the macro program by issuing the / Range Name Label Right command for the column(s) containing your macro and scratch-pad names.
12. If you are using Symphony, specify the name of the auto-executing macro. Issue the {SERVICES} Settings Auto-execute Set command. The name in this example is AUTOEXEC.
13. Save the file under the name of the instrument (e.g., METTLER).

```
--------A--------B--------C--------D--------E--------F--------G--------H-----
 1                    METTLER BALANCE SAMPLE WEIGHINGS
 2
 3
 4                    TARE WT  FINAL WT  NET WT
 5                  1
 6                  2
 7                  3
 8                  4
 9                  5
10
11
12
13
14
```

Figure 9–5

```
--------AF-------AG-------AH-------AI---- ----AJ-------AK-------AL------AM---
 1                              NAMED RANGES
 2                              ==============
 3                    ADD_MEAS L11          ROW       L55
 4                    AUTOEXEC L6           SAMPLES   L54
 5                    BUF_VAL  L52          STEADY    L26
 6                    DATA     C5..D1000    VERSION   L56
 7                    ER1HDLR  L41          WT_BUF    L51
 8                    ER2HDLR  L43          WT_ERROR  L53
 9                    ER3HDLR  L45          WT_LOW    L33
10                    GET_WT   L20          WT_RESET  L23
11                    METTLER  L14          \0        L2
12                    NEXT_WT  L18          \M        L47
13
14
15
16
17
```

Figure 9-6

14. Start the program by retrieving the file. Later, you can press [ALT] M to restart it, or add a menu system like those described in previous chapters.

15. Test the program and establish communications.

16. After communications have been established, add the amenities and menus to the program.

17. Final test the template and the macro program.

18. Save the spreadsheet in a file. Save several backup copies.

Note that these steps are remarkably similar to those required to interface an instrument using the Device File Name technique shown in Figure 2-1.

A MEASURE RS-232 EXAMPLE: THE METTLER PROGRAM

Both the Device File Name program (Figure 9-2) and the Measure program (Figure 9-1) work in essentially the same way. This section examines some of the differences between the two programs and also looks at some of the techniques that are typical of a Measure program.

The auto-executing macros in Figure 9-1 no longer use the DOS MODE command to set communications parameters. Instead they issue commands that set up the RS-232 module. This module, then, takes responsibility for setting up the communications.

The '\0 auto-executing macro is for Lotus 1-2-3. It issues the {APP3} command. The {APP} command is a Lotus 1-2-3 Measure macro command that automatically displays a Measure main menu. It can also execute a module menu choice from within a macro. It is equivalent to pressing [ALT] F9.

The other Measure modules also use the {APP} command, but have different numbers. They are

Module	Macro Command	Key Equivalent
IBMGCA	{APP4}	[ALT] F10
MBC16	{APP2}	[ALT] F8
NAT488	{APP1}	[ALT] F7
HP488	{APP1}	[ALT] F7
RS-232	{APP3}	[ALT] F9

The AUTOEXEC macro is the auto-executing macro for Symphony. It checks to see if the RS-232 module has already been attached. If it has not, ADD_MEAS is called to attach it.

The next line in both auto-executing macros ({APP3}SNRMETTLER~QQ and {S}RSNRMETTLER~QQ{ESC}) demonstrate how to issue the Name Retrieve command. Recall that in Step 7 above, Measure settings were stored in a file called METTLER. Therefore, these commands will retrieve the METTLER settings for use by your macro. They eliminate the need for your users to set up Measure each time they use it. The two Q commands quit the Measure menu system. Symphony requires an {ESC}ape to quit the SERVICES menu.

Whenever Measure is used, screen "flicker" is prominent. Therefore, you should "freeze" the display and panel for as much of the time as possible. The {WINDOWSOFF} {PANELOFF} set of commands is necessary to stop screen flicker while the METTLER setup is being retrieved. It also speeds the retrieval process by eliminating the need for Lotus to refresh the screen and panel.

The {RRECEIVE} in the example is a macro command specific to the RS-232 module of Measure. It receives data from the balance and stores them in a worksheet. The syntax of this command is

```
{RRECEIVE receive_range,wait_time}
```

The receive_range is the name of a cell or group of cells into which the data are stored. In this context, it is similar to the {READLN BUFFER} command used previously. However, unlike the {READLN} command, if a range of cells is specified in the {RRECEIVE} command, it will continue to read data until the entire range is filled, or until wait-time has been exceeded.

The wait_time argument is the amount of time {RRECEIVE} will wait for a response from the instrument. If an incoming string is received within the wait_time, it is placed in the receive_range specified cell of the spreadsheet and the macro continues with the *next cell* in the program. If the wait_time elapses before a string is received, the {RRECEIVE} aborts and the macro continues in the *same cell.* In this instance, {RRECEIVE} acts like an {IF} command. This feature can be a handy method of handling errors.

For example, you could modify the {RRECEIVE}s in the METTLER program to {MENUBRANCH} back to a menu if a response is not received within 30 seconds:

```
{RRECEIVE WT_BUF,30}{MENUBRANCH MAINMENU}
```

OTHER MEASURE RS-232 COMMANDS

As stated earlier, Measure adds seven new macro commands. {APP} and {RRECEIVE} have already been described. The third most commonly used Measure command is {RSEND}. This command is the Measure equivalent of the Lotus {WRITELN} command that outputs strings to instruments. That is, you could replace the {WRITELN} command with {RSEND} in all of the programs in this book and the programs would run the same. The syntax of this command is

```
{RSEND transmit_range}
```

The transmit-range can be either a cell address or a range. If a range is specified, all strings in the range are sent to the instrument. The {RSEND} command depends on the settings in Measure. With the Settings Transmit menu choice in Measure, you can set parameters that customize the way that {RSEND} transmits data to an instrument. The following defines each of the parameters:

Transmit_Range. This parameter specifies a range of cells to be sent if a range is not specified in an {RSEND} command. The transmit_range is generally used infrequently because you can specify it in the macro command.

End-of-Line. An End-of-Line character signals the end of a command string to your instrument. The instrument uses this character to determine when it has received an entire command. When an instrument reaches an End-of-Line, it executes the command. The End-of-Line character is usually a carriage return (character 13), but can be any of the first 26 ASCII character codes. Check the User's Manual for your instrument to see which End-of-Line character it requires.

Response. After a string is sent to some instruments, they respond by returning a character. This parameter allows you to specify the character that {RSEND} waits for before transmitting the next line of data in the transmit_range.

Delay. The delay time is the delay between the output of one string to the instrument and the next string to be sent. Recall that timing is critical when you are working with communications. In Chapter 7, {WAIT} commands allowed the Perkin Elmer LS-2B time to both process commands that were sent to it and to return an acknowledgment. The delay time specification is an equivalent method of accomplishing the same goal. However, it is not as flexible, because it applies the same delay to all commands being sent.

Lag. The lag time is the delay between each character within a string that is sent out to the communications port. It allows you to "fine-tune" transmission for "finicky" instruments; but its use is rare.

Concurrent-capture. Some instruments start transmitting their data as soon as they begin to receive commands. The concurrent-capture parameter allows you to capture this transmitted data while {RSEND} continues to send data.

The remaining Measure RS232 macro commands are used only rarely. The following are brief descriptions of these other commands:

{**RBREAK**}. This command sends a string of NULL characters (ASCII 0s) to an instrument. Some instruments (but very few) recognize these NULL characters, stop processing and respond with a character or a string. Consult your User's Manual to see if your instrument recognizes NULL characters. If you try to send a string of NULLs to an instrument that does not recognize them, you will often get peculiar responses.

The {RBREAK} command is most often used in error-handling routines. For example, as previously explained, {RRECEIVE} can branch to an error-handling routine if the wait_time is exceeded. {RBREAK} is one command that can be used to immediately stop an instrument so that your program can clear up this problem.

{**RREQUEST**}. This command essentially combines {RSEND} and {RRECEIVE}. That is, it sends out a command to an instrument and receives a response back. This command would have been useful for the ACROSYSTEMS® ACRO-931. If you recall, strings were sent to prompt it to return temperatures. Then {READLN}s were used to read in data.

The syntax of the {RREQUEST} is also a combination of the {RSEND} and {RRECEIVE} commands:

```
{RREQUEST out_buf,trigger,wait,in_buf}
```

where,

- out_buf is the cell or range of cells holding the output string(s).
- trigger is a string that some instruments can transmit to signal the start of their data strings. Measure will wait until it receives this string before accepting data. (See your instrument's User's Manual for details on the trigger's format.)
- wait is the length of time to wait for a response to return.
- in_buf is the cell or range of cells to place the strings returned from the instrument.

{PARITYON}. This command looks for parity errors that may occur when data are received using {RRECEIVE} or {RREQUEST}. The syntax is

```
{PARITYON branch_macro}
```

The branch_macro argument represents the macro that the {PARITYON} will branch to if an error is found. This macro is usually an error-handler.

{PARITYOFF}. This command disables {PARITYON}, thereby stopping the check for parity errors. This command is equivalent to setting the parity to NONE in the Settings Interface menu.

TAKING A SNAPSHOT OF METTLER

Chapter 6 discussed several ways to design programs for instruments that need to have their data collection "triggered". The METTLER macro is an example of this category of instrument. The METTLER example provides an additional method to the ones presented in Chapter 6 for triggering data collection.

The METTLER macro can be broken into five main sections. Each of these sections corresponds to a sequence of events that takes place when repetitive weighings are made. To get a feel for what each section does, think about how the balance's display reacts during a weighing session.

At the beginning, a user tares the balance, setting the display to 0.00000. The display will remain at this setting until the user places the first weight on the pan. Then the display will move through a series of intermediate weights until it displays a steady state. This steady state is the weight that the user records. Next, the user removes the weight and the display returns to 0.00000 again. All subsequent samples are handled in the same fashion.

Each section of the program corresponds to a division of the weighing process; each program section tests for the completion of the corresponding division in the cycle before the program advances to the next section. This program is similar to the programs shown in Figures 6-17 through 6-20.

If you will notice, {ONERROR} commands have been added to each of the sections to handle errors. They are all similar. The Lotus error messages are tested to see whether they were due to System Errors. If they were, program control {BRANCH}s to the beginning of the particular section. With custom {ONERROR} commands for each section, you eliminate the possibility of losing your place in the program.

The WT_RESET, STEADY, and WT_LOW sections each wait until a specific goal has been achieved. The WT_RESET section waits until the weight on the balance is greater than 0.010 grams. This "program trip" weight is set above zero to provide a "barrier zone" to account for small drifts in the balance tare or air currents moving across the pan. Thus, the "barrier zone" ensures that the program does not erroneously proceed into the next section.

The STEADY section tests to see if a steady state has been achieved. The Mettler balance precedes the data string with an "SD" if the weight is in transition and an "S" if it is at steady state. Therefore, testing for an "SD" will show if another weight from the balance needs to be screened or if the current weight is appropriate. If the steady state has been achieved, the program reads in another weight, parses it, and places it into a cell of the spreadsheet.

The test in the WT_LOW section determines when a weight has been removed from the balance. When a weight is removed, the weight transmitted from the balance falls to less than 0.010 grams. The {IF} test detects this decrease and a test is made to see if enough samples have been weighed. If not, the program {BRANCH}s back to NEXT_WT to start the process over again.

EXPANDING THE METTLER PROGRAM TO FIT YOUR NEEDS

You can fit this program, with a few modifications, into a larger one that has menus, print functions, file operations, statistical analyses, etc. For example, suppose you are performing a "tare" on a series of containers, adding samples to the containers, and re-weighing them. Or, suppose you are weighing samples, drying them, and then re-weighing to find the percent moisture.

To automate either of these scenarios, you would need two columns of data in your template and a menu system to switch between the columns, statistics, etc. You would also need to modify the {PUT} command to allow you to "switch" between the columns, while still allowing you to use the same program for the two columns. That is, you would want to create a COL cell in the scratch-pad, insert {LET} commands in your menu choices to place either a 0 or a 1 in the COL cell and modify the {PUT} command in the WAIT_STDY section to read as follows

```
{PUT DATA,COL,ROW,@VALUE(@LEFT(@RIGHT(WT_BUF,12),10))}
```

THE MBC16 MODULE: PLUG-IN BOARDS

Chapter 6 discussed plug-in boards and stand-alone A/D converters. You may want to take a moment now to review the concepts of that chapter and the list of board manufacturers. Although the focus of that chapter was on stand-alone A/D converters, most of the information applied directly to plug-in A/D converters as well. Stand-alone converters and plug-in boards differ in two major ways. The first is that you MUST use Lotus Measure to access plug-in boards. You cannot use the Device File Name technique for plug-in boards. The second is that although some plug-in A/D converters have the ability to run in "background" mode, Lotus Measure cannot access this mode.

(Courtesy of MetraByte Corp.)

Other than those two differences, choosing and using a plug-in board follow the same rules as stand-alone interfaces. Figures 9–12 and 9–14 illustrate this similarity. These figures use the Measure MBC16 module and a plug-in board to perform the same tasks as the programs that run the ACRO-912 stand-alone interface in Figures 6–15 and 6–20, respectively. Note that each pair of programs is nearly identical. A subsequent section is this chapter will review the programs in these figures in more detail. The following discussion, however, explains how to set up a typical board and how to configure the Measure MBC16 module.

CONFIGURING A/D PLUG-IN BOARDS

With equipment that communicates via RS-232 connectors, you need to set the communications parameters and attach the equipment to the personal computer with cables. Plug-in boards are quite different. They do not use Baud rate, parity, number of character bits, number of stop bits or cables. A plug-in board uses parameters called "base address", "channel configuration", "voltage range", and "bipolar/unipolar input range". These parameters are set by switches on the plug-in boards; not by software string commands. Each of these parameters are explained below. They require configuration on both the board and in the Measure MBC16 module.

Base Addresses

The base address is similar in theory to the addresses discussed in Chapter 2 for the ACRO-900™. Just as the ACRO-900 uses a number to keep track of the memory location of each module, your personal computer keeps track of the data on plug-in boards by *I/O* (input/output) memory addresses.

This set of addresses is different from the set of addresses used for RAM (Random Access Memory, the memory that stores your programs and data). That is, each piece of data contained on the plug-in board's memory has a unique *I/O* memory location.

The address of the first piece of data is called the "base address". All other data on the board can be found as an offset from the base address. For example, the MetraByte DASH-16™ has 16 analog inputs and uses 16 consecutive address locations starting at a base address. Because the addressing for *R*andom *A*ccess *M*emory (which you can use for your programs) is completely separate from I/O memory addressing, the two cannot conflict.

I/O memory locations in personal computers are usually specified in hexadecimal (base 16). IBM reserves certain I/O addresses for standard devices. These addresses are

Address	Device	Address	Device
000-1FF	Internal System	378-37F	LPT1:
200-20F	Game Adapter	380-38C	SDLC Comm.
210-217	Expansion Unit	380-389	Binary Comm. 2
220-24F	Reserved	3A0-3A9	Binary Comm. 1
278-27F	Reserved	3B0-3BF	Mono Dsp/LPT1:
2F0-2F7	LPT2:	3C0-3CF	Reserved
2F8-2FF	COM2:	3D0-3DF	Color Graphics
300-31F	Prototype Card	3E0-3E7	Reserved
320-32F	Hard Disk	3F0-3F7	Floppy Disk
		3F8-3FF	COM1:

This table contains the standard IBM options, but if you have other plug-in cards, they may use addresses not listed in the table. Thus, you may have I/O memory addressing conflicts with other devices in your computer. Because of possible conflicts, most plug-in board manufacturers have DIP switches (similar to the ones used for setting RS-232 Baud rates) on their boards that allow you to set the base address for the board. Usually, a good choice is to use a base address of hexadecimal 300 (decimal 768) or 310 (decimal 784). These addresses are used by Prototype Cards, which are rarely used. However, it is always a good idea to consult the User's Manuals for each of the other boards in your personal computer to see if their base addresses are the same as the one that you plan to use.

If you have a plug-in card, set the DIP switches to a base address. Use the User's Manual that accompanied the board as a guide. You will need to know the base address for the Measure MBC16 module later, so write it down.

Channel Configuration

The input channels on most plug-in boards can be configured to one of two different formats. The configuration that you use will depend on how your instruments and/or sensors (i.e., devices) supply output voltage signals to the board.

If the input configuration of a board is ''differential'', the board's input channels are *paired* and the voltage data received from an experiment or instrument is determined as the *difference* in the pair of signals.

If an input configuration is ''single-ended'', one wire is treated as a ground (zero volts) and the other wire is treated as a voltage referenced to ground. In this configuration, *each* channel of the plug-in board accepts a single input signal from one of the devices and there is one common ground for all of the devices.

To determine the differential or single-ended input requirements for your plug-in board, you will need to know two pieces of information about your data collection equipment. First, determine if your equipment is grounded by looking at the power plug. If the power plug is three-pronged, it's grounded. If the plug is two-pronged, it is not (unless a wire has been specifically attached from the device to a ground).

Next, determine whether the output signal from the equipment is differential or single-ended. Telling the format of the output of your equipment may not be straight-forward because every manufacturer seems to use slightly different terminology.

If you are using a commercially available instrument or sensor, look in the Specifications section of the User's Manual. If the device is differential the signal output will be listed as ''double ended'', ''difference signal'', ''differential output'', ''referenced to each other'', or the manual will list ''HI'' and ''LO'' voltage outputs. If the device is single-ended, the signal output will be listed as ''single-ended'', ''relative to ground'', or the manual will list a ground lead.

Alternatively, you can have an electronics professional look at the circuit diagram of the output. If the output is single-ended it has a single output amplifier whose signal is relative to ground. Figure 9–7 shows the output circuit for a single-ended signal. If the output is differential, it has two complementary outputs from one or two amplifiers and the signals will be relative to each other. Figures 9–8 and 9–9 show two possible output circuits for differential signals. Based on the grounding method and the output signal of your device, determine which board configuration to use.

- If your instrument is not grounded (two prong plug) *AND* your device has single-ended output, use a single-ended configuration for your board.

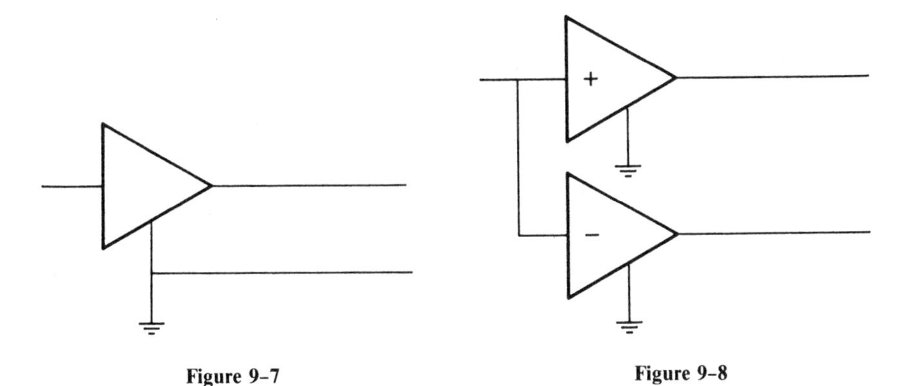

Figure 9–7

Figure 9–8

- If your device is grounded (three prong plug) *AND* has single-ended output, use a differential configuration for your board.
- If your device has differential output, use a differential configuration, whether it is grounded or not.

After you have determined which configuration to use, set the "Channel Configuration" switch on the board. Most of the plug-in boards require that the *entire* board be configured to either a differential or single-ended configuration. If you plan on using your plug-in board for various voltage signal configurations, reconfiguring the entire board may be inconvenient because you will have to tear apart your computer and remove the board each time you want to change the configuration. For this reason, if you have both single-ended and differential signals, you may want to configure the board to differential configuration and connect the wires accordingly.

If you cannot decide on a configuration, try the single-ended configuration and see if the format gives accurate data. If the data are inaccurate or if you have lengthy distances between your devices and your personal computer, use the differential configuration. With differential configuration, you can input single-ended signals into an A/D converter that has been configured to differential, but you cannot perform the reverse input. The only drawback to using differential input is that instead of having 16 inputs, you will have only 8.

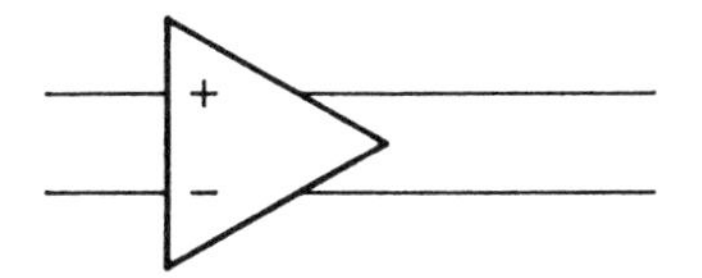

Figure 9–9

Voltage Ranges

Another parameter to set is the voltage range. Some board manufacturers call this range "gain". These switch settings are important because Measure cannot send string commands to plug-in boards to change their voltage ranges. The voltage range that you set depends on the input mode to which the plug-in board is configured. (See next section.)

Bipolar and Unipolar Input Modes

Most plug-in boards can be configured to one of two different methods for handling input voltage signals. These methods are referred to as "modes", called "bipolar mode" and "unipolar mode". These two modes pertain to the format of the analog signal values being processed.

In bipolar mode, voltages that are transmitted from an experiment or instrument take on both positive and negative values. In bipolar mode, outputs range from equal negative to positive full scale limits. For example, an experiment or instrument may have output ranges of −10 to +10, −5 to +5, −2.5 to +2.5, −1 to +1, or −0.5 to +0.5 volts full scale.

In unipolar mode, the values will lie between zero volts and some voltage. For example, an experiment or instrument may have output ranges of 0 to 10, 0 to 5, 0 to 2, or 0 to 1 volts full scale.

To configure your plug-in board, first determine which mode your experiment or instrument uses. If you are using a commercially available instrument or sensor, look in the Specifications section of the User's Manual. If the device is unipolar, the output range will usually (but not always) be listed as "variable 0-##" (where ## is some voltage, usually positive). If the device is bipolar, the signal output will normally be listed as "0 +/− ## full scale" (where ## is a voltage) or "+/− ##".

These nomenclatures are generally indicative of the polarity mode of the signals. However, when using these terminologies to determine the mode, be very cautious and be acutely aware that they may not be indicating what they appear to be.

If signal output information is not available, use your intuition. Analog voltages tend to parallel the processes they monitor. If you have a process that can have both positive and negative values, then the output will most likely be bipolar. Examples of these processes are pressure/vacuum, pH, electrodes, temperature, and most transducers that measure +/− processes. If your process can have only positive values, the output will most likely be unipolar. Examples of these processes are light intensity, resistance, and strain.

If you cannot determine the correct mode, try the unipolar format and see if the format agrees with the data. If the format and data do not agree, use bipolar mode. That is, you can input unipolar signals into an A/D converter that has been configured to bipolar mode, but you cannot input the reverse. One drawback to using bipolar input is that you will sacrifice one-half the resolution in the conversion.

After you have determined which mode to use, set the mode and full scale switches on the board. Most plug-in boards require that the *entire* board be configured to either bipolar or unipolar mode. If you plan on using your plug-in board for various voltage signal modes, re-configuring the entire board may be inconvenient because you will have to tear apart your computer and remove the board each time you want to change the polarity mode. For this reason, if you have both unipolar and bipolar signals, you may want to configure the board to bipolar mode.

At this point, take the time to write down the mode and the full scale ranges to which you have configured your plug-in board. Also determine an equation that could be used, if needed, to convert the output of the board into voltages. This formula is important and you will most likely need to use it when you configure the Measure MBC16 module.

VOLTAGE CONFIGURATION FORMULAS

Most plug-in boards do not report their results in volts. They divide the voltage range into "steps" and report voltages as the number of "steps" between the lower and upper limits of a range. A 12-bit converter divides the voltage range into 4096 steps. For example, if the A/D converter is in unipolar mode, the steps range from 0 to 4095. Thus, an input range of 0 to 10 volts translates to 2.44 mV per step (10 V/4906). A 16-bit converter divides the range into 65536 steps. Thus, a 16-bit converter has 0.152 mV per step (10 V/65536).

In bipolar mode, steps range from −2048 steps to +2047 steps for a 12-bit converter. That is, the number of steps are evenly distributed about zero volts.

It is very important that you identify whether your plug-in board uses this step reporting method or reports converted voltages directly. If the former is the case, you will need to either make the conversion from steps to voltage manually, or allow Measure to make the conversion before the value is placed in the buffer of your spreadsheet.

To determine which reporting method is used, develop a program like the one shown in Figure 9–12 and examine the data that are received from your plug-in board. You can also configure the Measure MBC16 module without a conversion formula and use the module's Observe function to test the values returned by the board directly. If the numbers received are decimal voltages, a conversion formula is not needed. However, if the numbers are integers within the full scale step ranges, you must determine a formula to convert the steps to voltages.

Determining a conversion formula is not complicated. The formula you use will depend on the resolution of your A/D converter, whether it is in bipolar or unipolar mode, and the full scale voltage range.

To develop a sound formula, first determine the total number of steps between the upper and lower voltage limit. The number of steps depends on your board's resolution and is 2 to the power of the resolution. For example, a 12-bit resolution has a total of 4096 steps (2 to the power of 12); a 16-bit resolution has a total of 65536 steps (2 to the power of 16); and a 17-bit resolution has a total of 131072 steps (2 to the power of 17).

Next, incorporate the resolution into a formula that also depends on the voltage range and whether the mode is bipolar or unipolar. For *bipolar* mode, the general equation is

([]-steps)*(voltage/steps)

where, [] tells Measure to use the data coming in from the board, steps is *one half* the resolution, and voltage is the highest voltage in the range.

It is very important that you use *one half* the total resolution of the converter in this formula. Remember, in bipolar mode half the readings are above zero volts, the other half are below. Subtracting half the resolution from the [] value normalizes it to zero and dividing by half of the resolution determines the voltage within the range above (or below) zero.

For example, for an A/D converter with a 12-bit resolution, a full scale voltage of +/− 1 volt and bipolar mode, the equation would be

([]-2048)*(1/2048)

For *unipolar* mode, the situation is quite different. In this mode, the general equation is

[]*(voltage/steps)

For example, for an A/D converter with a 16-bit resolution, full scale voltage of 0 − 5 volts, and unipolar mode, the equation would be

[]*(5/65546)

Thus, in unipolar mode, it is not necessary to normalize to zero and you must use the *total* resolution in the equation.

INSTALLING AN A/D PLUG-IN BOARD

Once you have configured the plug-in board, written down the settings, and identified the formula to use, install the board into your personal computer. Follow the instructions given in your User's Manual. You may also need to follow the recommendations in the User's Manual to calibrate the board.

Complete the installation by connecting your analog equipment to board using shielded wire. Before you make the connection, review the grounding and differential/single-ended output information that you determined above (in the "Channel Configuration" section).

This information will place your device into one of four categories. The category will determine how you form the connection:

- If your device has a single-ended output *AND* a three-pronged (grounding) plug, use your board in differential mode. In this case, connect the plus (or "HI") signal output from your device to the "HI" connector of one of the input pairs of your board. Connect your device's other output to the "LO" input of the board's pair. (Make sure that the two input channels correspond.)
- If your device has a single-ended output *AND* a two-pronged (non-grounding) plug, use your board in single-ended mode. In this case, the signal ground leads from your devices must be connected to the Low Level Ground pin of the plug-in board and the plus (or "HI") leads must be connected to the input pins for the channels that you plan to use.
- If your device has a differential output *AND* a three-pronged (grounding) plug, use your board in differential mode. In this case, connect the "HI" signal output from your device to the "HI" connector of one of the input pairs of your board. Connect your device's "LO" output to the "LO" input of the board's pair. (Ensure that the two input channels correspond.)
- If your device has a differential output *AND* a two-pronged (non-grounding) plug, use your board in differential mode. In this case, connect the "HI" signal output from your device to the "HI" connector of one of the input pairs of your board. Connect your device's "LO" output to the "LO" input of the board's pair. (Ensure that the two input channels correspond.) Then, connect the board's "LO" input to the board's Low Level Ground with a jumper wire.

Shielded wires can be fastened directly to a type "D" connector, similar to those used for RS-232 connections (but with more pins—usually 37). This connector may then be pushed onto the connector of the board.

Carefully study the diagram of the connector in your plug-in board's User's Manual to ensure that you are using the correct pins. The pins are small and in a compact array, thereby making it easy for you to make a mistake.

Also note that one of the end pins (pin 1) is usually a "+5" volt power line and the other end pin (pin 19) is usually a ground; the end pins are not usually inputs for signals.

Alternatively, you can purchase a "Screw Terminal Board". This stand-alone, intermediate module is a circuit board that has screw terminals that make it easy for you to connect the wires from your instrument or experiment to the module. The terminal board module is connected to your plug-in board via a ribbon cable.

CONFIGURING MEASURE'S MBC16 MODULE

The next step is to configure the Measure MBC16 module to the same values to which you set your plug-in board. If you are using Lotus 1-2-3, start the spreadsheet by typing 123 MBC16.set and pressing return. Bring up the Measure MBC16 module by pressing [ALT] F8. If you are using Symphony, start the spreadsheet by typing ACCESS and then pressing return twice. Attach the MBC16 add-in by pressing {SERVICES} Application Attach (F9 AA) and choose MBC16. Then, press {ESC}ape until you get to the main SERVICES menu and choose MBC16.

The main menu for the MBC16 module is displayed. Your first task is to place the settings for each channel of your plug-in board into individual ID-Settings sheets of the module.

An "ID" is the combination of a channel name, its type, which board the channel is on, a spreadsheet range for its data, and a conversion formula.

An ID therefore, contains all of the information that the MBC16 module needs to know about a channel to get its data and place it into the spreadsheet. To access the ID-Settings menu, press I and then press I again for ID. When prompted for a name for the ID, use any name up to eight characters in length, but do not use a number as a name. (The example uses the name CHAN0.) Next, you will see a display like the one shown in Figure 9–10.

Select each choice on the ID-Settings menu in turn to set the parameters for the appropriate channel of your plug-in board. The pertinent settings for the example programs in Figures 9–7 and 9–8, using channel 0 on a DASH-16 board in bipolar mode, are shown below:

```
Type: Analog
Board Number: 0
Channel Number: 0
Range: VOLT_BUF
```
(The cell in your template that you want your data to be placed into; i.e., your buffer)

If you are using channels to monitor digital or binary data, use the Type menu choice to specify the appropriate data form. (Binary means that the data are treated as a combination of four subsequent digital channels.)

```
Type  Board  Channel  Range  Formula  Gain  View  Duplicate  Next-ID  Quit
Enter channel type: analog, digital, binary, counter

   ID-Name:   CHAN0

              Type:   Analog                    Range:   W49..W49
      Board Number:   0                         Size:    1
    Channel Number:   0
              Gain:   None
              View:   Yes

   Formula:   ([]-2048)*(1/2048)
                                                A/D ID-Settings C:\123\
```

Figure 9–10

For now, enter the following formula. It will not apply a correction to the data received from the board and will allow you to decide whether your board returns steps or voltages.

[ ]*1.0

Note that Measure allows you to have up to four plug-in boards. Each board is given a number from 0 to 3. The number that you supply will correspond to a base address defined elsewhere in the module's menu (see below). In this case, board number zero has been assigned to address 300 hexadecimal.

Setting a Range in Measure is different than setting a range in Lotus 1-2-3 or Symphony. When you move a named cell or range in Lotus 1-2-3 or Symphony, the spreadsheet automatically makes the appropriate corrections. Measure does not make these types of corrections. *Measure keeps a range designation the same regardless of what happens in the spreadsheet.* Therefore, if you add or delete rows in your program, or move your buffer, you MUST go in and manually change Measure's range designation or your data will be placed into the wrong cells.

If you plan on using more than one input channel on the card, you must return to the ID-settings menu and repeat this procedure for each channel that will collect data (assigning different names for each channel).

Next, choose the Hardware option from the ID-settings menu. This option presents you with the addresses that the MBC16 module uses for each board (0–3). If you have configured a board to an address that is different from the one listed, change it now. Then, select the Quit option to return to the ID-settings menu.

From the ID-settings menu, choose S for Stage-Settings. A stage is an assemblage (collection) of channel IDs and a procedure for collecting data from them. A stage is like a subroutine macro that gets a piece of data from each of the listed IDs using the step-by-step procedure specified in the Stage-Settings sheet.

If you look in the Measure manual (under Stage-Settings), you will find that the MBC16 module can perform some of the same functions that you could write a macro program to perform. It can trigger readings, apply formulae, build in delays, etc. However, controlling these functions in a macro program achieves greater flexibility and is easier to troubleshoot.

Because of these advantages to macro programming, keep the Measure setup as simple as possible and disable as much of the module as possible. You can then attain close control of the data collection from the macro program. Perhaps even more importantly, if a problem arises you will not have to worry about trying to identify the cause of the problem in two different places; all you will need to do is look at the program.

Keeping the Measure interface as simple as possible makes it "transparent" That is, except for when the module provides data to the macro program, the module will not be noticeable to the program.

However, to collect any data from Measure, you must at least place the ID names of the channels to use, the rate of data collection, and the number of data points to collect into the Stage-Settings sheet. From the Stage Settings menu (Figure

9–11) choose 1, 1, and then enter the name of the ID. The example has only one ID, CHAN0.

Next you are prompted for a "rate divisor". The rate divisor is a method that Measure uses to apply different data acquisition rates to each ID in a stage. That is, the rate divisor determines the relative acquisition rate for each ID by dividing the Rate (see below) by the value of the rate divisor. In this example, use a divisor of one.

Next, choose Rate from the menu. When prompted for time units, type S (for seconds) and press return. When prompted for the rate per second, type 2000 and press return. Because you will want to build time delays into the macro program, enter a value of up to 4000 (which is almost equivalent to disabling the delay between data collection points within Measure). Sometimes this rate of 4000 samples per second will prove too fast for your board/spreadsheet combination and you will get a "RATE TOO FAST" error message. For this reason, use a maximum rate of 2000 for DASH-16™ applications.

If you have other IDs, repeat this procedure to add each ID to the stage. Alternatively, you can use the procedures given in the "Using Multiple Input Channels with the MBC16 Module", below.

Finally, choose the Number option and enter 1. Choosing 1 will retrieve a single point from the channel(s) each time the stage is run and place it into your buffer(s).

Complete the process by choosing the Name option to store the settings in a file. The example uses the name DASH. A subsequent section of this chapter will show you how to retrieve this file with your auto-executing macro. By retrieving the file au-

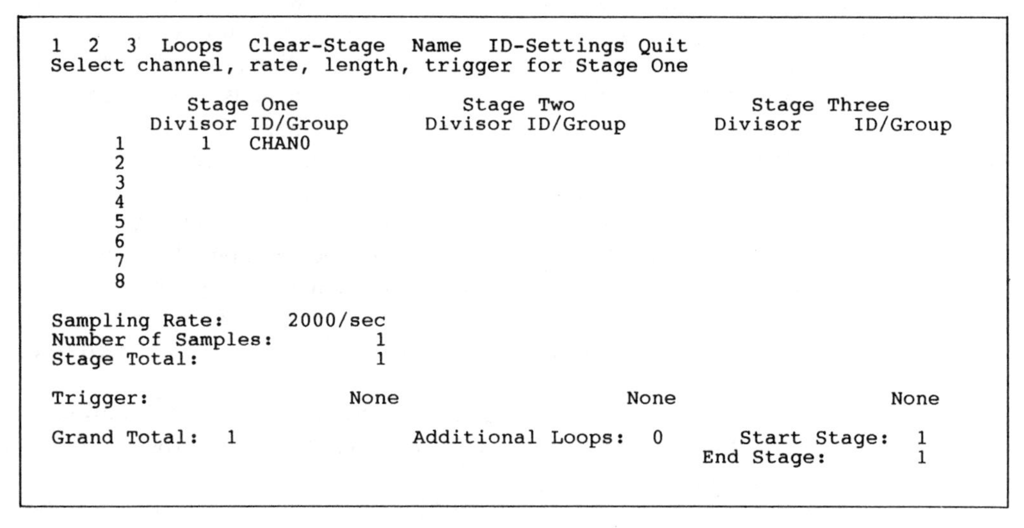

```
1  2  3  Loops  Clear-Stage  Name  ID-Settings Quit
Select channel, rate, length, trigger for Stage One

          Stage One              Stage Two              Stage Three
       Divisor ID/Group       Divisor ID/Group       Divisor    ID/Group
    1     1    CHAN0
    2
    3
    4
    5
    6
    7
    8

Sampling Rate:      2000/sec
Number of Samples:         1
Stage Total:               1

Trigger:                 None                  None                  None

Grand Total:  1             Additional Loops:  0       Start Stage:  1
                                                     End Stage:      1
```

Figure 9–11

tomatically, you eliminate setting up the IDs and stage every time you retrieve your spreadsheet.

TESTING MBC16 SETTINGS

Quit the Stage-settings menu to return to Measure's main menu. Your next task is to test the system. The first test confirms that you have correctly entered all necessary settings and that no information is missing. Measure automates this procedure with the Verify option from the main menu. Select this option. One of two displays appears. If the verify did not find any errors, the screen will state "Verify complete-No errors". If information is missing, the settings conflict, *OR* if the board is not detected by the software, a screen will appear listing the settings that are missing or a message stating the board's absence. It will also provide you with information on how many errors were detected. If you are presented with the second screen, you will need to rectify the error(s) before proceeding.

Once you have achieved internal consistency in module settings using Verify, you can start the testing that ensures that the data are coming in from all plug-in board channels and that the data are accurate. The Go menu choice of the main menu will help you perform the test.

With a known voltage source or your instrument connected to the input channel(s), select the Go option a couple of times. View the data that were placed into the cell(s) of the spreadsheet specified in the ID-settings sheet(s). At this point, you should be able to tell whether your board is reporting results in voltages or steps.

If your board is reporting results in steps, you will see *integers* that are representative of the step ranges for your converter and its configuration settings. For example, if your board has a 12-bit resolution, integers in the range of −2048 to +2047 or 0 to 4095 (depending on the full scale voltage and whether the board was configured to be bipolar or unipolar, respectively) will be reported. At this point, you should return to each ID-settings sheet and enter the formula that you determined while you were configuring the board in the last section of this chapter. Test the new formula using the Go menu choice again. When it is verified to be correct, save the new settings (overwriting the old settings file) by issuing Stage-settings Name Save (SNS). Then, press Quit and/or {ESC}ape until you return to the spreadsheet.

PROGRAMMING FOR MBC16

Once you have accurate data in your buffer(s), you are ready to begin programming. From this point on, use the programming techniques that have been presented throughout this book. The only differences that you will encounter are shown in Figure 9–12. The program in this figure performs the same tasks as the program

```
--------V--------W--------X--------Y--------Z--------AA-------AB-------AC-------
 1 /*LOTUS MEASURE TEST PROGRAM TO MONITOR VOLTAGE OR CHART RECORDER*/
 2 /*OUTPUTS: TAKES READINGS AND DETERMINES THE PERIODIC SEQUENCE*/
 3 /*OF EVENTS DURING THE SAMPLING PROCESS.*/
 4 \0         {LET VERSION,0}                /*RUNNING UNDER LOTUS 1-2-3*/
 5            {APP2}INRDASH~QQ               /*USE DASH-16 SETTINGS*/
 6            {LET GET_VOLT,"{APP2}GQ"}      /*SELF-MODIFY "GO" MACRO LINE*/
 7            {VOLT_TEST}                    /*START THE PROGRAM*/
 8
 9 AUTOEXEC   {LET VERSION,1}                /*RUNNING UNDER SYMPHONY*/
10            {IF @ISERR(@APP("MBC16",""))}{ADD_MEAS}
11            {S}MINRDASH~QQ{ESC}            /*USE DASH-16 SETTINGS*/
12            {LET GET_VOLT,"{S}MGQ{ESC}"}       /*MODIFY "GO" MACRO LINE*/
13            {VOLT_TEST}                    /*START THE PROGRAM*/
14
15 ADD_MEAS   {S}AAMBC16~Q       /*ADD IN LOTUS MEASURE MBC16.APP*/
16
17 VOLT_TEST{LET DATE,@DATEVALUE(@NOW)}           /*DATE STAMP THE EXPERIMENT*/
18            {WINDOWSOFF}{PANELOFF}              /*FREEZE THE DISPLAY*/
19            {GETNUMBER "How Many Time Points?",POINTS}
20            {GETNUMBER "How Long Is The Interval Between Measurements?",DELAY}
21            {INDICATE I/O}{PANELOFF}    /*FREEZE PANEL*/
22            {GOTO}START~      /*MOVE CELL-POINTER TO STARTING POSITION*/
23            {LET BEGIN,@TIMEVALUE(@NOW):VALUE}  /*GET THE STARTING TIME*/
24            {IF DELAY>0}{FOR ROW,0,POINTS,1,BIGLOOP}   |*THESE GIVE MAXIMAL *|
25            {IF DELAY=0}{FOR ROW,0,POINTS,1,SMALLOOP}  |*PERF FOR ZERO DELAY*|
26            {CALC}            /*FORCE RECALCULATION OF THE SPREADSHEET*/
27            {GOTO}ELAPSCOL~                /*RESET CELL-POINTER TO START*/
28            {IF VERSION=0}/RFDT3.{DOWN POINTS}~ /*TIME FORMAT, 1-2-3*/
29            {IF VERSION=1}/FT3.{DOWN POINTS}~   /*TIME FORMAT, SYMPHONY*/
30            {WINDOWSON}{PANELON}           /*THAW THE DISPLAY*/
31            {INDICATE DONE}   /*CHANGE INDICATOR TO "DONE"*/
32
33 BIGLOOP    {INDICATE WAIT}                     /*SHOW DELAY PHASE*/
34            {WAIT @NOW+@TIME(0,0,DELAY)}        /*WAIT FOR SPECIFIED TIME*/
35            {INDICATE}                          /*FREE INDICATOR BOX*/
36 SMALLOOP   {LET CURRENT,@TIMEVALUE(@NOW)}      /*GET CURRENT TIME*/
37            {LET DELTA,CURRENT-BEGIN}           /*FIND ELAPSED TIME*/
38            {PUT ELAPSCOL,0,ROW,DELTA}          /*PUT ELAPSED TIME IN SHEET*/
39            {PUT PTCOL,0,ROW,ROW}               /*PUT POINT NUMBER IN SHEET*/
40            {WINDOWSON}{WINDOWSOFF}             /*ADD DATA TO SCREEN*/
41            {DOWN}                              /*MOVE DOWN TO NEXT ROW*/
42            {GET_VOLT}                          /*GET A VOLTAGE READING*/
43
44                      /*MACRO WHICH GETS ONE VOLTAGE*/
45 GET_VOLT   {S}MGQ{ESC}                    /*ISSUE THE "GO" COMMAND FROM MBC-16*/
46            {PUT ABSDATA,0,ROW,@ABS(@VALUE(VOLT_BUF)):VALUE}
47
48 \M         {VOLT_TEST}       /*CALL UP VOLTAGE TEST AS A SUBROUTINE*/
49                      /*SCRATCH PAD*/
50 VERSION              /*0=>LOTUS 1-2-3; 1=>SYMPHONY */
51 ROW                  /*COUNTER FOR LOOP; ROW OFFSET FOR {PUT}*/
52 POINTS               /*NUMBER OF DATA POINTS TO COLLECT*/
53 BEGIN                /*TIME EXPERIMENT STARTED*/
54 DELAY                /*USER DEFINED DELAY BETWEEN DATA POINTS*/
55 CURRENT              /*CURRENT TIME*/
56 DELTA                /*ACTUAL DELAY, FROM TIME ZERO*/
57 VOLT_BUF             /*A BUFFER FOR INCOMING DATA*/
58
59
```

Figure 9-12

that runs a stand-alone (RS-232 controlled) A/D converter. Take a moment to compare Figure 9–12 to Figure 6–15. The differences are outlined below.

The auto-executing macros in Figure 9–12 no longer use the DOS MODE command to set communications parameters. Instead, they access the MBC16 menu system to retrieve the CHAN0 settings stored in the file named DASH. The '\0 auto-executing macro for Lotus 1-2-3 issues the {APP2} command; the {APP2} command is a Lotus Measure command that automatically accesses the MBC16 main menu. The AUTOEXEC macro for Symphony first checks to see if the application has been attached; if not, the macro attaches it. Next, it accesses the MBC16 main menu from the SERVICES menu.

The remainder of the macro command issues the ID-settings Name Retrieve sequence (INR) followed by the name of the file to retrieve (DASH). The Q's quit the Measure macro system. An {ESC}ape command is required to quit the SERVICES menu in Symphony.

The next difference is the method of obtaining data for the buffer. If you recall, the program in Figure 6–15 opened the RS-232 communications port, issued a {WRITELN} command, read the data in with a {READLN} command, and closed the port. With plug-in boards, you access the MBC16 menu system, issue the Go command and then return to the spreadsheet. Measure automatically uses the settings of the MBC16 stage to retrieve the data and place it into your buffer.

The command that uses the Measure menu to get data for Lotus 1-2-3 is

```
{APP2}GQ
```

For Symphony, the command is

```
{S}MGQ{ESC}
```

To customize the program for the particular Lotus Measure version being used, the two auto-executing macros have {LET} commands that place the appropriate command into the GET_VOLT cell. The {S}MGQ{ESC} in the GET_VOLT cell of Figure 9–12 originated from this action, meaning the program was started under Symphony.

Placing the appropriate command into the cells of macro programs at spreadsheet retrieval (startup) is faster and more efficient than using two back-to-back {IF} commands. For instance, the following set of commands could have been used in GET_VOLT:

```
GET_VOLT{IF VERSION=0}{APP2}GQ
        {IF VERSION=1}{S}MGQ{ESC}
```

However, this method is much slower than customizing a cell at startup and using a command directly as the program executes.

Another way that the programs differ is in the parsing functions of the {PUT} commands in the GET_VOLT macros of Figures 6–15 and 9–12. The RS-232 program in Figure 6–15 uses an @CLEAN function to remove a carriage return at the end of a data string. Plug-in boards typically do not terminate their data with a carriage return. Thus, if an @CLEAN had been placed in the {PUT} command of Figure 9–12, it would have placed an ERR into the VOLT_VAL cell. Removing the @CLEAN from the {PUT} commands in Figure 9–12 eliminates this problem.

COMPONENT ORIENTED DIFFERENCES

The plug-in and RS-232 programs have two other differences. The most important of these differences is concerned with timing. The 12-bit successive approximation converter of the DASH-16 is much faster than the 17-bit dual slope integration converter of the ACRO-912. Furthermore, because the DASH-16 is a plug-in board it is not required to use the slower RS-232 port. This attribute makes data acquisition for the DASH-16 very fast. However, when data acquisition is very fast, tests for steady states and increasing/decreasing signals can be unreliable (because the small differences that are obtained when fast sampling rates are used can be affected by noise and resolution inaccuracies). If you experience unreliability in data acquisition "triggers", you may want to slow the acquisition rate by using {WAIT} commands in your program.

Figures 9–14 and 9–15 illustrate the addition. Figure 9–14 is the plug-in equivalent to the RS-232 program in Figure 6–20. (Figure 9–13 contains the auto-executing macros that customize data collection commands for the Measure version.)

The program in Figure 9–14 is unreliable because the speed of the DASH-16

```
--------V--------W--------X--------Y--------Z--------AA-------AB-------AC-------
 1           /*AUTOEXECUTING MACROS FOR MBC16 MODULE OF LOTUS MEASURE*/
 2
 3 \0        {LET VERSION,0}    /*RUNNING UNDER 1-2-3*/
 4           {WINDOWSOFF}{PANELOFF}      /*FREEZE THE DISPLAY*/
 5           {APP2}INRDASH~QQ            /*CALL UP STORED DASH-16 SETTINGS*/
 6           {LET GO_LO,"{APP2}GQ"}      |*                             *|
 7           {LET WAIT_HI,"{APP2}GQ"}    |*SELF-MODIFY "GO" MACRO LINES*|
 8           {LET GO_STDY,"{APP2}GQ"}    |*                             *|
 9           {VOLT_TRIP}                 /*START THE PROGRAM*/
10
11 AUTOEXEC  {LET VERSION,1}    /*RUNNING UNDER SYMPHONY*/
12           {WINDOWSOFF}{PANELOFF}      /*FREEZE THE DISPLAY*/
13           {IF @ISERR(@APP("MBC16","")))}{ADD_MEAS}
14           {S}MINRDASH~QQ{ESC}         /*CALL UP STORED DASH-16 SETTINGS*/
15           {LET GO_LO,"{S}MGQ{ESC}"}   |*                             *|
16           {LET WAIT_HI,"{S}MGQ{ESC}"} |*SELF-MODIFY "GO" MACRO LINES*|
17           {LET GO_STDY,"{S}MGQ{ESC}"} |*                             *|
18           {VOLT_TRIP}                 /*START THE PROGRAM*/
19
20 ADD_MEAS  {S}AAMBC16~Q       /*ATTACH LOTUS MEASURE MBC16.APP*/
21
22
```

Figure 9–13

```
--------V--------W--------X--------Y--------Z--------AA-------AB-------AC-------
25      /*PROGRAM TO MONITOR CHART RECORDER OUTPUTS TO TAKE READINGS*/
26        /*IE, VOLTAGE MONITORED; WHEN DOWNWARD TREND IS DETECTED*/
27   /*THE LARGEST OF THE LAST 4 READINGS IS PLACED INTO THE TEMPLATE*/
28                     /*LOTUS MEASURE: MBC16 MODULE*/
29
30 VOLT_TRIP{GETNUMBER "How many Samples?  ",SAMPLES}
31          {WINDOWSOFF}{PANELOFF}      /*FREEZE THE DISPLAY*/
32          {IF VERSION=0}/REDATA~      /*ERASE DATA SECTION*/
33          {IF VERSION=1}/EDATA~       /*0=>1-2-3; 1=>SYMPHONY*/
34          {LET ROW,0}                 /*INITIALIZE ROW COUNTER*/
35 GET_VOLT {LET ROW,ROW+1}             /*INCREMENT ROW COUNTER*/
36          {LET REF1,0}       |*                                  *|
37          {LET REF2,0}       |*INITIALIZE THE DATA ARCHIVE SYSTEM*|
38          {LET REF3,0}       |*                                  *|
39 WAIT_LO  {IF ROW>SAMPLES}{QUIT}                /*LAST SAMPLE?*/
40 GO_LO    {S}MGQ{ESC}         /*USE MBC16 MENU; CHOOSE "GO" TO GET A VOLTAGE*/
41          {LET VOLT_VAL,@VALUE(VOLT_BUF)}     |*WAIT UNTIL VOLTAGE    *|
42          {IF VOLT_VAL>0.02}{BRANCH WAIT_LO}  |*RETURNS TO A LOW VALUE*|
43 WAIT_HI  {S}MGQ{ESC}         /*USE MBC16 MENU; CHOOSE "GO" TO GET A VOLTAGE*/
44          {LET VOLT_VAL,@VALUE(VOLT_BUF)}
45          {IF VOLT_VAL<0.02}{BRANCH WAIT_HI}  /*WAIT UNTIL VOLTAGE CLIMBS*/
46 WAIT_STDY{LET REF1,REF2}                |*SHUFFLE PREVIOUS READINGS*|
47          {LET REF2,REF3}                |*UP INTO ARCHIVE SYSTEM   *|
48          {LET REF3,VOLT_VAL}
49          {WAIT @NOW+@TIME(0,0,1)}    /*PAUSE FOR 1 SECOND*/
50 GO_STDY  {S}MGQ{ESC}         /*USE MBC16 MENU; CHOOSE "GO" TO GET A VOLTAGE*/
51          {LET VOLT_VAL,@VALUE(VOLT_BUF)}
52          {IF REF3<VOLT_VAL}{BRANCH WAIT_STDY}/*DETECT DOWNWARD TREND*/
53          {PUT DATA,0,ROW,@MAX(REF1,REF2,REF3,VOLT_VAL)}
54          {WINDOWSON}{WINDOWSOFF}     /*REFRESH THE DISPLAY*/
55          {BRANCH GET_VOLT}              |*LOOP BACK TO GET_VOLT*|
56                                         |*FOR NEXT SAMPLE      *|
57 \M       {VOLT_TRIP}        /*CALL VOLTAGE TRIP AS A SUBROUTINE*/
58          {CALC}             /*FORCE SPREADSHEET UPDATE*/
59
60                     /*SCRATCH PAD*/
61 VOLT_BUF                    /*BUFFER FOR INCOMING VOLTAGES*/
62 VOLT_VAL                    /*CELL FOR PARSING VOLTAGE STRING*/
63 SAMPLES                     /*NUMBER OF SAMPLES TO BE RUN*/
64 ROW                         /*CURRENT SAMPLE NUMBER*/
65 VERSION                     /*0 => 1-2-3; 1 => SYMPHONY*/
66 REF1                        |*ARCHIVE SYSTEM FOR      *|
67 REF2                        |*PREVIOUS VOLTAGE VALUES*|
68 REF3                        |*                        *|
```

Figure 9–14

gives false positive tests for the {IF} commands. Therefore, a {WAIT} command is issued just prior to each data acquisition command, as shown in Figure 9–15. These {WAIT} commands slow the reading process and help ensure that rejection limits are not exceeded. If you use {WAIT} commands, examine each section of your program to ascertain where pauses are needed.

The final difference can be found in the {IF} commands. In the RS-232 program, the limits for {IF} commands were 0.001 volt. In this program, limits of 0.02 volt were needed to get reliable testing. That is, when 0.001 was used for the DASH-16, extraneous data were placed in the example template because the section tests tripped prematurely. The 0.001/0.02 difference is due to the difference in sensitivity and resolution of the ACRO-912 and the DASH-16. The tolerance of the two sys-

```
--------V--------W--------X--------Y--------Z--------AA-------AB-------AC-------
25     /*PROGRAM TO MONITOR CHART RECORDER OUTPUTS TO TAKE READINGS*/
26      /*IE, VOLTAGE MONITORED; WHEN DOWNWARD TREND IS DETECTED*/
27 /*THE LARGEST OF THE LAST 4 READINGS IS PLACED INTO THE TEMPLATE*/
28          /*LOTUS MEASURE: MBC16 MODULE; WITH PAUSES*/
29
30 VOLT_TRIP{GETNUMBER "How many Samples?  ",SAMPLES}
31          {WINDOWSOFF}{PANELOFF}     /*FREEZE THE DISPLAY*/
32          {IF VERSION=0}/REDATA~     /*ERASE DATA SECTION*/
33          {IF VERSION=1}/EDATA~      /*0=>1-2-3; 1=>SYMPHONY*/
34          {LET ROW,0}                /*INITIALIZE ROW COUNTER*/
35 GET_VOLT {LET ROW,ROW+1}            /*INCREMENT ROW COUNTER*/
36          {LET REF1,0}       |*                                 *|
37          {LET REF2,0}       |*INITIALIZE THE DATA ARCHIVE SYSTEM*|
38          {LET REF3,0}       |*                                 *|
39 WAIT_LO  {IF ROW>SAMPLES}{QUIT}              /*LAST SAMPLE?*/
40          {WAIT @NOW+@TIME(0,0,1)}            /*PAUSE FOR 1 SECOND*/
41 GO_LO    {S}MGQ{ESC}        /*USE MBC16 MENU; CHOOSE "GO" TO GET A VOLTAGE*/
42          {LET VOLT_VAL,@VALUE(VOLT_BUF)}     |*WAIT UNTIL VOLTAGE    *|
43          {IF VOLT_VAL>0.02}{BRANCH WAIT_LO}  |*RETURNS TO A LOW VALUE*|
44 PAUSE_HI {WAIT @NOW+@TIME(0,0,1)}    /*PAUSE FOR 1 SECOND*/
45 WAIT_HI  {S}MGQ{ESC}        /*USE MBC16 MENU; CHOOSE "GO" TO GET A VOLTAGE*/
46          {LET VOLT_VAL,@VALUE(VOLT_BUF)}
47          {IF VOLT_VAL<0.02}{BRANCH PAUSE_HI} /*WAIT UNTIL VOLTAGE CLIMBS*/
48 WAIT_STDY{LET REF1,REF2}            |*SHUFFLE PREVIOUS READINGS *|
49          {LET REF2,REF3}            |*UP INTO THE ARCHIVE SYSTEM*|
50          {LET REF3,VOLT_VAL}        |*                          *|
51          {WAIT @NOW+@TIME(0,0,1)}    /*PAUSE FOR 1 SECOND*/
52 GO_STDY  {S}MGQ{ESC}        /*USE MBC16 MENU; CHOOSE "GO" TO GET A VOLTAGE*/
53          {LET VOLT_VAL,@VALUE(VOLT_BUF)}
54          {IF REF3<VOLT_VAL}{BRANCH WAIT_STDY}/*DETECT DOWNWARD TREND*/
55          {PUT DATA,0,ROW,@MAX(REF1,REF2,REF3,VOLT_VAL)}
56          {WINDOWSON}{WINDOWSOFF}     /*REFRESH THE DISPLAY*/
57          {BRANCH GET_VOLT}          |*LOOP BACK TO GET_VOLT*|
58                                     |*FOR NEXT SAMPLE      *|
59 \M       {VOLT_TRIP}        /*CALL VOLTAGE TRIP AS A SUBROUTINE*/
60          {CALC}             /*FORCE SPREADSHEET UPDATE*/
61
62                   /*SCRATCH PAD*/
63 VOLT_BUF                    /*BUFFER FOR INCOMING VOLTAGES*/
64 VOLT_VAL                    /*CELL FOR PARSING VOLTAGE STRING*/
65 SAMPLES                     /*NUMBER OF SAMPLES TO BE RUN*/
66 ROW                         /*CURRENT SAMPLE NUMBER*/
67 VERSION                     /*0 => 1-2-3; 1 => SYMPHONY*/
68 REF1                        |*ARCHIVE SYSTEM FOR     *|
69 REF2                        |*PREVIOUS VOLTAGE VALUES*|
70 REF3                        |*                       *|
71
```

Figure 9–15

tems to electrical noise also differs. The ACRO-912 uses a dual slope integration converter; the DASH-16 uses a successive approximation converter. The former technique has excellent noise immunity at the expense of speed. (See Chapter 6.)

These differences point out the important fact that you will need to optimize {WAIT}s and {IF} test conditions in *your* program for *your* data acquisition board and *your* instrument. The only way that you can do so is by testing the system with a series of samplings which includes both the high and low values that you expect to detect. You can then "fine tune" the numbers in the {WAIT} and {IF} tests to your exact application.

USING MULTIPLE INPUT CHANNELS WITH THE MBC16 MODULE

In all previous examples, only one of the DASH-16 channels was used. However, the DASH-16 has 16 single-ended or 8 differential analog input channels. Lotus Measure allows you to get data from as many of these channels as you want using a single Go menu command. You can also use a single Go menu command to obtain concurrent data capture from up to four boards (64 channels).

The best method of obtaining concurrent data capture is to combine the IDs for each channel that you want to use into a "group". You can combine up to 8 IDs into a "group" and you can have up to 8 groups in each of three stages (a maximum of 192 IDs).

Begin by modifying your program so that it has a buffer for each channel. Figure 9–16 illustrates the changes that are required. The scratch-pad has been modified to include named cells for each of 8 channels (CHAN_0 through CHAN_7). These cells will receive data. Cells W61 through W68 have been given the range name BUF_BLK by issuing the /RNC command. Also, the {PUT} command has been moved into a loop that transfers the contents of each channel into the template using the @INDEX function to choose the appropriate cell within BUF_BLK.

Next, you must build an ID for each channel. Use the same method you used for CHAN0, Figure 9–10. Repeat the procedure for each channel and give a unique name to each ID. When prompted for a range, enter the name of the cell in the scratch-pad corresponding to the channel number.

```
--------V--------W--------X--------Y--------Z--------AA-------AB-------AC-------
44              /*LOOP TO GET ONE VOLTAGE FROM EACH DASH-16 CHANNEL*/
45 GET_VOLT {S}MGQ{ESC}                /*ISSUE THE "GO" COMMAND FROM MBC-16*/
46          {FOR INC,0,7,1,PUT_CHAN}   /*PUT DATA INTO TEMPLATE*/
47
48          /*PUT VALUE FOR EACH MEMBER OF DASH GROUP INTO TEMPLATE*/
49 PUT_CHAN {PUT ABSDATA,INC,ROW,@ABS(@VALUE(@INDEX(BUF_BLK,0,INC))):VALUE}
50
51 \M       {VOLT_TEST}        /*CALL UP VOLTAGE TEST AS A SUBROUTINE*/
52                  /*SCRATCH PAD*/
53 VERSION          /*0=>LOTUS 1-2-3; 1=>SYMPHONY*/
54 ROW              /*COUNTER FOR LOOP; ROW OFFSET FOR {PUT}*/
55 POINTS           /*NUMBER OF DATA POINTS TO COLLECT*/
56 BEGIN            /*TIME EXPERIMENT STARTED*/
57 DELAY            /*USER DEFINED DELAY BETWEEN DATA POINTS*/
58 CURRENT          /*CURRENT TIME*/
59 DELTA            /*ACTUAL DELAY, FROM TIME ZERO*/
60 INC              /*LOOP COUNTER FOR PLACING VOLTAGES INTO TEMPLATE*/
61 CHAN_0           |*                                 *|
62 CHAN_1           |*                                 *|
63 CHAN_2           |*                                 *|
64 CHAN_3           |*BUFFERS FOR INCOMING DATA        *|
65 CHAN_4           |*FOR EACH OF THE DASH-16 CHANNELS*|
66 CHAN_5           |*                                 *|
67 CHAN_6           |*                                 *|
68 CHAN_7           |*                                 *|
69
```

Figure 9–16

Always remember that Measure keeps the range designation the same regardless of what happens in the spreadsheet. Therefore, if you add or delete rows in your program or move your buffers, you MUST go in and manually change Measure's range designation or your data will be placed into the wrong cell(s).

Then, choose Group from the ID-settings menu. You will be presented with a display like the one shown in Figure 9–17; but, it will not contain any IDs. Repeatedly choose the Add-ID and add each ID to the group. When you are finished, choose the Quit menu choice.

Go to the Stage-Settings menu. A display like the one shown in Figure 9–11 appears. Choose 1 and then choose 1 again. Specify the Group and Divisor in the same way that you specified the CHAN0 ID. Finish the process by saving the settings using the Name menu choice.

You can verify and test the Group in the same way that you did for an ID. However, this test will produce eight readings; one from each channel.

Now, each time you issue the Go command from your macro, you will get a reading from each channel placed into your scratch-pad and the values will be transferred to your template by the {PUT} command in the {FOR} loop.

IEEE-488/GPIB/HP-IB

The *G*eneral-*P*urpose *I*nterface *B*us (GPIB) is often referred to by its IEEE (Institute of Electrical and Electronic Engineers) standard designation "IEEE–488". This standard is a formal specification of the HP-IB interface developed by Hewlett-Packard® Corporation (with some modifications). The interface can monitor and control up to 15 GPIB compatible instruments simultaneously from a single GPIB plug-in interface board. With this control, you can produce a large increase in the data acquisition productivity of your laboratory.

```
Add one ID to this group
Add-ID  Delete-ID  Quit

    Group Name:  DASH
 IDs:     Chan0 Chan1 Chan2 Chan3 Chan4 Chan5 Chan6 Chan7

                          A/D Group Settings C:\Symphony\DASH.ACF
```

Figure 9–17

Actually, two different configurations can be used for interconnecting your instruments. They are shown in Figures 9–18 and 9–19. These two configurations are referred to as the "star" and the "daisy-chain," respectively. A combination of these two configurations is also possible.

Think of these configurations as a kind of "party-line" in which each instrument talks to and listens to any of the other members of the party-line. This party-line is usually referred to as a "bus" in computer books. These two terms are used interchangeably in this chapter.

The GPIB connector, shown in Figure 9–20, has both male and female connectors at each end of the cable. This design allows you to interconnect instruments in both the star and the daisy-chain configurations without a decline in performance. The length of your cable cannot exceed 2 meters (6.7 feet) times the number of instruments. The combined length of all cabling for all of your instruments cannot exceed 20 meters (67 feet), unless you use a booster (extender). If you need a bus extender, check with your GPIB board manufacturer. If unavailable, contact National Instruments™ (12109 Technology Blvd., Austin, TX, 78727)

Because of all the data being received from multiple sources, at any time, and traveling across the same party-line, control is extremely important to the entire system. The traffic cop in the system is a GPIB board plugged into your personal

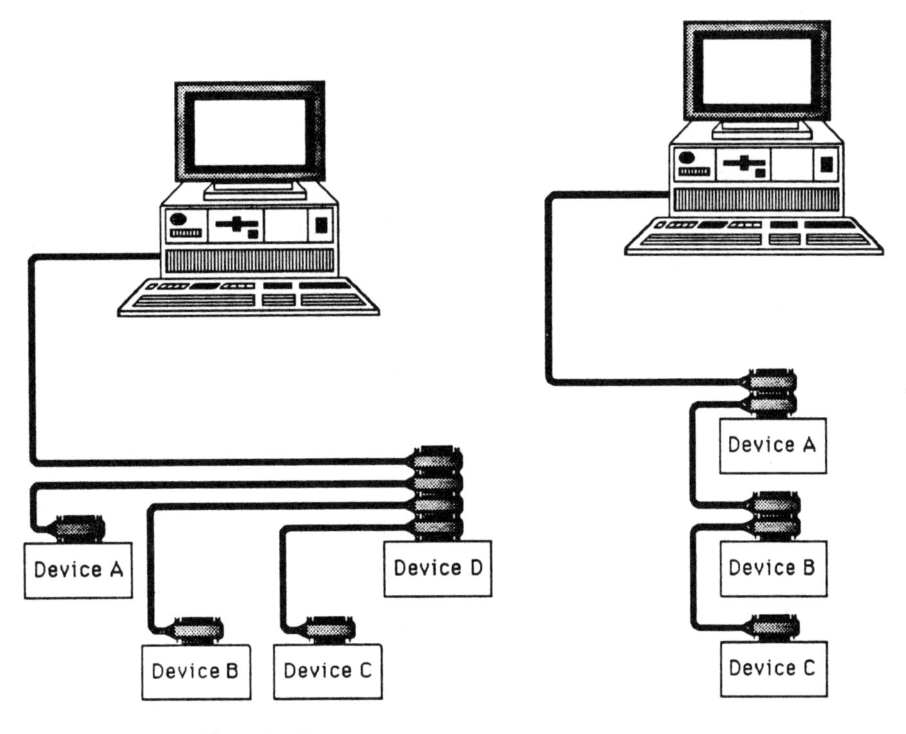

Figure 9–18

Figure 9–19

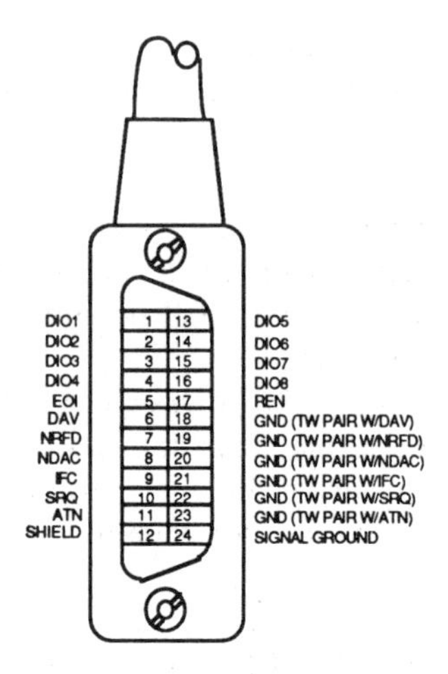

Figure 9–20

computer. It is named the "controller" and is somewhat like a telephone company switching center. In this capacity, the controller monitors the telephone line to see when someone wants to use the line and, when someone does, it decides who is going to talk to whom, who is going to listen, and when. When the controller notices that an instrument wants to make a call, it connects the caller to the receiver or to the computer. This analogy is somewhat different from a real party-line because even when all instruments use the same line, only privileged parties can listen in at any given time.

The controller oversees the flow of all data and the commands that are communicated between the computer and the instruments on the party-line. It also regulates their roles. Instruments on the party-line can be "talkers", "listeners", or both. A talker transmits data to the personal computer or to one or more listeners. The Mettler balance described in the Measure RS-232 module illustrates the RS-232 analogy of a talker; it only sends data. A listener is an instrument that receives data. Printers and plotters are listeners. Instruments that can both receive data and commands and send data back are talker/listeners. The Perkin-Elmer LS-2B Fluorimeter described in Chapter 7 illustrates the RS-232 analogy of a talker/listener.

One other category of instruments exists. This category is called "extended talker/listeners". Extended talker/listeners are usually instruments that are modular, with each module being able to be accessed separately by passing a second address to the controlling unit of the instrument. The ACRO-900™ (described in Chapters 2 through 6) illustrates the RS-232 analogy of this category. If you recall, the ACRO-900 had modules that could be accessed for data. These modules were accessed via a coordinating unit within the ACRO-900 control module. GPIB extended talker/listener modules can be similarly accessed. They are supplied with the address of a module to access. The mechanism of transferring this address information is however, somewhat different.

GPIB ADDRESSES

The GPIB system has three different types of addresses. The first type of address is the computer I/O memory location of the plug-in board. This type of address was discussed in the Measure MBC16 module section of this chapter.

The second type of address is called a "primary" address. A unique GPIB address is assigned to the plug-in board and each instrument attached to the bus. Thus, the plug-in board and each instrument have unique primary addresses (from zero to thirty) that the controller uses to send messages to the devices.

The third type of address is called a "secondary" address. These GPIB addresses specify which extended talker/listener module at a primary address is being accessed. Secondary addresses range from 00 to 31.

Although some rare exceptions exist, each instrument on a bus must have a unique address so that direct communications with the instrument can be made without interference from other instruments on the party-line (bus). For conventional applications, ensure that no two instruments are assigned the same address. Also, an instrument will typically have only a primary address unless it is modular. If you suspect that it also uses secondary addresses, consult your User's Manual.

How these addresses are used by the controller to govern communications is somewhat complicated, but the following discussion will help you to understand how the macro commands in your program can target the correct instrument. Primary addresses range from 0 to 30. (Zero is usually assigned to the GPIB plug-in board.) However, the GPIB system actually works in hexadecimal.

Hexadecimal numbers are described in Chapter 10; for now, just be aware that hexadecimal (HEX) is a numbering system based on 16 (as opposed to 10 for decimal). In contrast to ten decimal digits (0–9), hexadecimal numbers range from 0 through 9 and A through F. These digits represent the numbers 0 through 15. This system is used extensively in personal computers because two HEX digits fit conveniently into one byte (8 bits) of binary data; computers like to work with bytes.

Primary addresses range from HEX 00 to HEX 1E (decimal 0 to 30). Listener addresses are formed by adding HEX 20 (decimal 32) to primary addresses; talker addresses are formed by adding HEX 40 (decimal 64) to primary addresses. For example, if an instrument has a primary address of HEX 10, it will have a listen address of HEX 30 and a talk address of HEX 50.

Secondary addresses range from 00 to 31 decimal. These addresses begin after the last primary talker address. Thus, "real" secondary addresses are in the range of HEX 60 to HEX 7E.

Instruments are commanded to listen or talk by issuing listen addresses and talk addresses down the bus. That is, when you execute one of the read or output commands from your macro program, the controller will respond by issuing two encoded addresses down the bus. These two addresses disable *all* instruments from their talk and listen modes. These addresses are actually global (universal) commands and are called the "untalk" and "unlisten" commands.

Next, the controller issues an encoded address to reactivate the communications of the instrument with which it wants to communicate. This encoding scheme contains two pieces of information, the instrument's address and whether it should listen or talk.

The following table lists the address codes and the corresponding encoded address commands for talking and listening:

Primary Address	Listen Commands	Talk Commands	Secondary Address
0	space	@	`
1	!	A	a
2	"	B	b
3	#	C	c
4	$	D	d
5	%	E	e
6	&	F	f
7	'	G	g
8	(	H	h
9	)	I	i
10	*	J	j
11	+	K	k
12	,	L	l
13	-	M	m
14	.	N	n
15	/	O	o
16	0	P	p
17	1	Q	q
18	2	R	r
19	3	S	s
20	4	T	t
21	5	U	u
22	6	V	v
23	7	W	w
24	8	X	x
25	9	Y	y
26	:	Z	z
27	;	[	{
28	<	\	|
29	=	]	}
30	>	^	~
31	?	_	

The unlisten and untalk commands are universal, acting on every device on the bus. The universal command for untalk is "_" and the universal command for unlisten is "?". All other commands act only on the instrument being addressed, the address determined by the primary address and, if appropriate, the secondary address.

THE ROLE OF THE CABLE IN GPIB CONTROL

Before you can fully understand how control works in the IEEE-488 system and how Measure macro commands interact with the system, you need some background information about the cable. IEEE-488 cables and RS-232 cables are very different. In a simple RS-232 cable, data are transmitted one bit at a time on pin 2. Data are received in a similar fashion on pin 3. Thus, RS-232 transmission is referred to as serial transmission and requires that 8 one-bit pulses be sent for each character. Pins 4, 5, 6, and 20 are usually used to control the flow of data (i.e., handshake) between an instrument and a personal computer. These pins merely pause data transmission if a buffer becomes full.

The IEEE-488 cable, however, has eight data lines. (See Figure 9–20.) Each line transmits one bit of a character. Simultaneous transmission of all eight bits of a character is referred to as "parallel" transmission.

These lines have a dual purpose. They are also used to issue GPIB interface commands to instruments. GPIB interface commands are different from the string commands that you use to control the function of an instrument and from the strings of measurement data that you receive from it. They are commands that activate or deactivate whether an instrument talks or listens to the GPIB bus.

That is, when a controller wants an instrument to talk or listen, it issues the encoded addresses shown in the above table and these addresses act as the command. Because these lines have the dual purpose described above, another line (called the attention line, or ATN) is used to specify whether the eight lines are carrying data or GPIB interface command messages. Thus, when the ATN line is true (i.e., 1), the lines are carrying GPIB interface command messages; when the ATN line is false (i.e., 0), the lines are carrying data.

IEEE-488 cables have other important lines. In addition to the data/GPIB address command lines, IEEE-488 cables have three handshake lines. These lines control the flow of data in much the same way as the RS-232 handshake lines do. However, they are more efficient at ensuring that message bytes are sent and received without transmission errors.

IEEE-488 cables also have five lines called the "management lines". The management lines include the ATN line described above and four other lines that are used to send messages throughout the system. For example, the SRQ line is the *S*ervice *R*e*Q*uest line. One common use of the SRQ line is when an instrument has completed its data collection and is ready to transmit data to the computer. The instrument uses the SRQ line to signal the personal computer that it needs the computer to do something (i.e., it requests "service").

This line is important to a programmer. For example, if a program was set up to monitor the SRQ line from Measure, when a line indicated that an instrument needed service, the program could ask each instrument connected to the system if it was the one requesting service. This process is known as "polling" and, depending on the instruments used, you could poll instruments all at once (parallel) or one at a time (serial). Then, after you have determined which instrument needed service, you could use a subroutine in your program to issue commands to read in data, transmit data, change the instrument's settings, etc.

Another important line is the End-Or-Identify (EOI) line. The most common use of this line is to indicate the end of a data string. Its function is therefore similar to a carriage return/line feed. Most instruments that send out data are able to set this line to true at the end of a data string. However, some instruments use a special pre-defined character, analogous to RS-232.

As you can see, GPIB is more complicated than RS-232 because you not only control the actions of your instruments, but you also control *which* instrument is being controlled. However, most of the details of control are carried out by Measure and your plug-in board.

The next sections outline necessary steps to perform when implementing a GPIB system in your laboratory.

CONFIGURING GPIB PLUG-IN BOARDS

First, configure your GPIB board and plug it into your computer. GPIB plug-in boards require three parameters to be set. These parameters are: the base address, the DMA channel, and the interrupt line. These parameters are set by switches or jumpers on the plug-in board. Consult your User's Manual for instructions.

The base address is the computer I/O memory location of the plug-in board. This type of address was discussed in the previous Measure MBC16 module section

(Courtesy of National Instruments, Inc.)

(Courtesy of National Instruments, Inc.)

of this chapter. Follow the guidelines given in that section and your User's Manual, but be aware that the usual base address is HEX 2E1 for the IBM® GPIB Adapter and the National Instruments GPIB IBM and GPIB-PCIIA™ board. Other allowable addresses are

22E1	HEX
42E1	HEX
62E1	HEX

The usual base address for the GPIB-PCII, the IOtech GP488A, and the Capital Equipment PC-488™ boards is HEX 2B8. However, these boards can be set to any address in the following ranges, provided they do not interfere with another board in your personal computer:

100–1E8	HEX
200–318	HEX
330–3E8	HEX
3F8	HEX

If your computer is an IBM PC,™ an IBM PC/XT,™ or a PC/AT,™ the base address that you use must be higher than HEX 200 or the computer will not recognize your GPIB board.

The *D*irect *M*emory *A*ccess (DMA) channel transfers data directly between your instruments and the personal computer's main memory (RAM). By using this method, you will bypass the central processing unit (CPU) of the personal computer when processing the data. The data will be placed directly into memory at a very high speed. Bypassing the CPU means that you will not tie up the CPU and will substantially increase the speed of your program.

Three DMA channels can be used. They are 1, 2, and 3. The appropriate channel for GPIB is usually 1 because 2 and 3 are often used by other devices in the computer. If DMA channel 1 is used by another plug-in board or device in your computer, you may need to disable the channel on one or the other. Only certain plug-in cards allow you to share DMA channels. (e.g., Plug-in cards with hard drives on them DO NOT allow DMA sharing.)

The Interrupt line is a line used by the GPIB controller to get the attention of the personal computer's CPU. When the line is asserted, the computer stops its current action and "services" the request. Several interrupt lines (2, 3, 4, 5, 6, and 7) can be used by plug-in boards, disk drives, etc. Because the computer may receive more than one interrupt at the same approximate time, a priority is established for servicing. Once an interrupt has been serviced, the computer returns to the tasks that it was performing before the interrupt was asserted.

The most common interrupt line used for GPIB boards is 7, but it may need to be changed if another plug-in board or device in your computer also uses the line.

CONFIGURING INSTRUMENTS FOR GPIB

RS-232 requires that you set communications parameters (such as Baud rate, parity, number of character bits, and stop bits). RS-232 also requires that you determine the wiring of the cables for each of your instruments. With GPIB, neither of these tasks are necessary.

The only parameter required to be set for GPIB is the primary address. If your instrument is an extended talker/listener, you must also set secondary addresses.

These addresses are usually set by switches on the instrument. For conventional applications, each instrument must have its own unique address. If more than one instrument has the same address, communications on the bus may become confused.

Consult your User's Manual to determine how to set the primary (and, if necessary, secondary) address on the instrument. Record the addresses for each instrument because you will need to know them when you configure Measure.

Next, connect the instruments into one of the configurations shown in Figures 9–18 and 9–19. You may also use a combination of the two configurations. Remember, your cable should not exceed a length of 2 meters times the number of instruments or a total cable length of 20 meters without a booster.

GATHERING THE PERTINENT INFORMATION

Read the following list adhering to the instructions, before configuring Measure:

1. Decide on an illustrative name for your board and each instrument that you want to interface. For example, if you are interfacing a spectrophotometer, you might call it SPEC.
2. Record the I/O address that you set your board to.
3. Decide on time limits to be imposed when executing macro command(s). These time limits are called ''time-outs'' and are safety features that ensure that Measure does not wait indefinitely for an instrument to respond. An example of a situation where an instrument would not respond is one in which the instrument is turned off. If an instrument does not respond within the specified time-out period, the macro command terminates. This feature prevents your program from unnecessarily locking up, thus requiring the user to press [CONTROL] [BREAK].
4. Look in your instrument's User's Manual(s) to determine how I/O transmissions to and from the board terminate when executing board calls. That is, know how each instrument determines the end of a string of characters. In RS-232, this method is usually a carriage return/line feed combination. In GPIB, it can be an end-of-string character, an END message (i.e., EOI line asserted), and/or a byte count.
5. Know your board's interrupt level.
6. Know your board's DMA channel.
7. Set and know the address of each instrument.
8. Know the polling protocol that each of your instruments requires (i.e., parallel or serial, see below).

CONFIGURING THE MEASURE NAT488/GPIB AND HP488/HP-IB MODULES

Your next step is to set the Measure NAT488/GPIB or the HP488/HP-IB module to the values that correspond to the configuration settings that you made on your instruments and plug-in board.

The NAT488 module is the Lotus 1-2-3 Measure module required for National Instruments GPIB boards and GPIB interface boards from most other manufacturers. The module is called GPIB in the Symphony version of Measure.

The HP488 (Lotus 1-2-3) and HP-IB (Symphony) modules are for interface boards manufactured by Hewlett-Packard® (3000 Hanover St., Palo Alto, CA 94304).

The remainder of this chapter discusses the National Instruments GPIB-PC board. The HP488/HP-IB modules should work similarly.

If you are using Lotus 1-2-3, start the spreadsheet by typing 123 NAT488.set and pressing return. Bring up the Measure GPIB module by pressing [ALT] F7.

If you are using Symphony, start the spreadsheet by typing ACCESS and pressing return twice. Attach the GPIB add-in by pressing {SERVICES} Application Attach (F9 AA) and choose GPIB. Then, press {ESC}ape until the main SERVICES menu is displayed, and choose GPIB.

When the main menu for the GPIB module is displayed, your first task is to place the settings of your first instrument into the setup section of the menu. Select Setup. You will be presented with a menu of default names for fifteen devices (instruments) and the GPIB controller board. Select DEVICE1. You will be presented with a display like the one in Figure 9–21.

Select each of the menu choices in turn on the settings menu to set the parameters for your first instrument. These settings will be referred to whenever the instrument's name appears in a Measure macro command. The following are explanations of the settings as they apply to the example programs.

The programming examples below use a Nelson A/D interface. (This interface was described in Chapter 6.) DEVICE 1 was renamed to "NELSON". You can use any name that consists of only alphabetic characters, numbers, and/or the underscore character.

The primary address (1, in this example) corresponds to the address set on the back of the Nelson interface. The primary address for each instrument must be unique to the instrument.

The Input-EOS is an ASCII character that signals the end of a read operation. That is, it signals the *E*nd *O*f a *S*tring of characters coming from the instrument. The Input-EOS is usually a line-feed (ASCII 10). To specify a different EOS for a character that falls within the first 26 ASCII codes, use a three-digit number preceded by a backslash. To specify an alphanumeric character, enter the character.

The Output-EOS is an ASCII character that is added to the end of a write ({OUTPUT}) operation by the GPIB board. The Output-EOS is used by the instru-

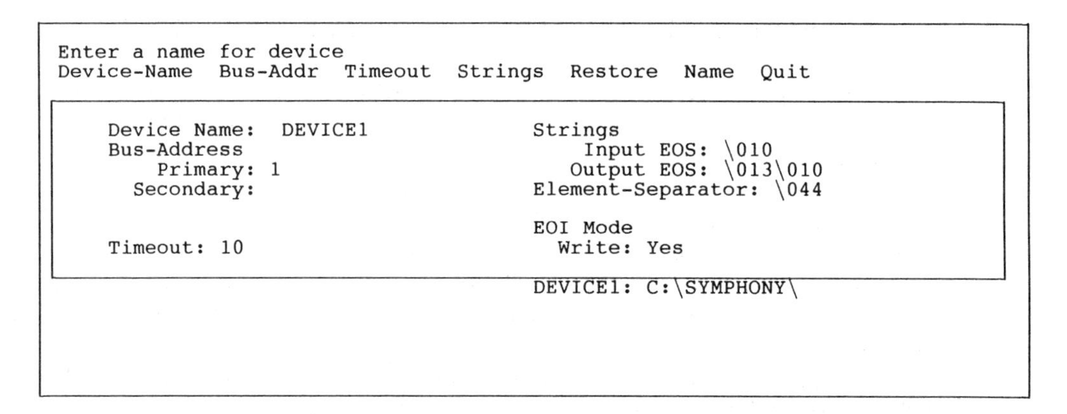

Figure 9–21

ment to determine the end of a command string or data. This EOS can be either one or two characters. It is usually a carriage return/line feed combination. Changes can be made to the Output-EOS using the same rules for the Input-EOS.

An EOI of "yes" means that the EOI line in the connector will be asserted at the end of each string of characters sent to, or received from, an instrument. You should set the EOI to "yes" for most, if not all, of your instruments.

An asserted EOI line is an *electrical* method of signaling the end of a string. It is a common method used by GPIB instruments. Nearly all talkers are able to set this line to true at the end of a data string, but some may use a special pre-defined character (Output-EOS). This lack of standardization can present a problem. For this reason, pay particular attention to the User's Manual of each instrument to see the convention being used for this line.

The Element Separator \044 meaus that each number in a string is separated from its neighbors by commas. The most common method of separating data points in a string is by commas. When an Element Separator is encountered during a read macro command, the data prior to the separator are placed in a cell. The data that follow until the next element separator are placed into the next cell, and so on. For example, in the following string,

```
16075,16079,16056,16022
```

16075 would be placed into the first cell of a range, 16079 would be placed into the second cell, 16056 into the third cell, and 16022 into the fourth.

Setting the time-out can be tricky. When you use a Measure macro read command to input data from an instrument, you must allow enough time for Measure to read in all of the data *and* store the data in spreadsheet cells. This extra time can be especially important if you are using the "count" feature of the various read commands (see below). A "count" that reads in 100 readings can take up to 2 seconds. Thus, if your time-out is too short, you will miss some data. If you enter a time-out of 0, you disable the time-out and the macro command does not terminate until you press [CONTROL][BREAK]. So, you must decide which time-out is best for your situation.

When you complete setting these parameters, Quit this menu and choose Setup again. This time, select DEVICE2. Select each of the choices in turn on the Settings menu to set the parameters for your second instrument.

Repeat this procedure for each instrument. Select sequential DEVICE names from the Setup menu for each instrument that you plan to add to the system. When presented with the settings sheet, change the DEVICE name to one that is more descriptive of the instrument. Select each of the appropriate choices in turn on the menu to set the parameters for the instrument. Make sure that the address for each instrument is unique to the instrument and that it matches the address that was used when you configured the instrument.

Next, select BOARD from the settings menu. BOARD allows you to assign values to the system that will be used when no instruments are named in a macro command. Read commands and output commands expect instrument names to be specified. If no instrument names apear in these macro commands, the BOARD settings are used. Likewise, other macro commands (such as {WAITSRQ}) are universal commands and do not take instrument names as arguments. These commands also use the BOARD settings.

Select each of the appropriate choices in turn on the menu to set the parameters for your plug-in board. The values that you use should be a compromise best-fit for all instruments in your system.

BOARD was renamed to "GPIB" for this example. The default values for most of the remaining parameters were used because they were appropriate. In the examples given below, the board waits a substantial length of time for an instrument to request a service. In these situations, disabling the time-out is desirable. Remember, the time-out can be disabled by entering a zero.

It is important that the board have its own Primary Address in a GPIB system. That is, the board must be treated as a talker/listener just like any other instrument in the system. The primary board address is typically set to 0 (30 in the HP-IB module). Therefore, each instrument in the system can be set sequentially from 1 through 15.

When you are finished, return to the main Measure menu and choose the Name option. This Name option is different from the one on the Device Settings menus. It allows you to store the settings for the entire bus in a single file. The example was given the name GPIB. A subsequent section will show you how to retrieve this file with your auto-executing macro. By using the GPIB file, you will not be required to set up the board and instrument settings each time you retrieve your spreadsheet.

After the settings are stored in a file, quit the Measure menu and return to the spreadsheet.

A SIMPLE GPIB EXAMPLE PROGRAM

Once you have configured Measure for your GPIB system, you are ready to begin programming. From this point on, use the programming techniques that were presented throughout the earlier chapters of this book.

Figures 9–22 and 9–23 show an elementary GPIB program. Figure 9–22 illustrates the auto-executing macros that initialize a GPIB system, whether it contains a single instrument or several instruments.

Figure 9–23 illustrates the fundamental commands needed to communicate with a single GPIB instrument. Admittedly, this example is more like an RS-232 application than a GPIB application. However, it is a foundation only and will be built upon in the next sections. By the end of this chapter, you will know how to set up a highly interactive network/conglomerate of instruments, printers, A/D converters, etc.

The example program uses a Nelson 900 A/D converter to monitor absorb-

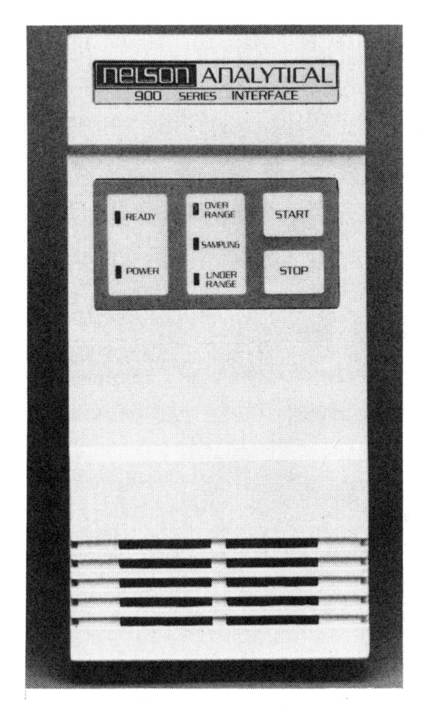

(Courtesy of Nelson Analytical, Inc.)

```
--------V--------W--------X--------Y--------Z--------AA-------AB-------AC-------
 1          /*AUTOEXECUTING MACROS FOR NAT488/GPIB MODULES OF LOTUS MEASURE*/
 2
 3 \0        {LET VERSION,0}            /*RUNNING UNDER 1-2-3*/
 4           {WINDOWSOFF}{PANELOFF}     /*FREEZE THE DISPLAY*/
 5           {APP1}NRGPIB~QQ            /*CALL UP STORED GPIB SETTINGS*/
 6           {INIT}                     /*INITIALIZE THE GPIB BOARD*/
 7           {DMAON 1}                  /*TURN ON DMA CHANNEL NUMBER 1*/
 8           {STATUSON TEST_BUF}        /*KEEP TRACK OF COMMUNICATIONS STATUS*/
 9           {CLEAR}                    /*RETURN ALL INST'S TO POWER-UP STATE*/
10           {BRANCH NELSON}            /*START THE PROGRAM*/
11
12 AUTOEXEC {LET VERSION,1}             /*RUNNING UNDER SYMPHONY*/
13           {WINDOWSOFF}{PANELOFF}     /*FREEZE THE DISPLAY*/
14           {IF @ISERR(@APP("GPIB","")}{ADD_MEAS}
15           {S}GNRGPIB~Q{ESC}          /*CALL UP STORED GPIB SETTINGS*/
16           {INIT}                     /*INITIALIZE THE GPIB BOARD*/
17           {DMAON 1}                  /*TURN ON DMA CHANNEL NUMBER 1*/
18           {STATUSON TEST_BUF}        /*KEEP TRACK OF COMMUNICATIONS STATUS*/
19           {CLEAR}                    /*RETURN ALL INST'S TO POWER-UP STATE*/
20           {BRANCH NELSON}            /*START THE PROGRAM*/
21
22 ADD_MEAS {S}AAGPIB~Q       /*ATTACH LOTUS MEASURE GPIB.APP*/
23
24
```

Figure 9-22

```
--------V--------W--------X--------Y--------Z--------AA-------AB-------AC-------
 1          /*SIMPLE GPIB PROGRAM TO GATHER HPLC DATA IN BACKGROUND MODE*/
 2
 3 NELSON    {WINDOWSOFF}{PANELOFF}      /*FREEZE THE DISPLAY*/
 4           {CLEAR NELSON}              /*CLEAR THE CONVERTER*/
 5           {BLANK TEMPLATE}            /*ERASE DATA AREA OF TEMPLATE*/
 6           {OUTPUT NELSON,"I"}         /*RESET THE INTERFACE*/
 7           {OUTPUT NELSON,"OE"}        /*SELECT EMULATION MODE*/
 8           {OUTPUT NELSON,"C 1"}       /*SELECT CHANNEL ONE*/
 9           {OUTPUT NELSON,"R 0"}       /*SET VOLTAGE RANGE*/
10           {OUTPUT NELSON,"S 10"}      /*SET SAMPLING RATE TO 1 / SECOND*/
11           {OUTPUT NELSON,"Q 001000 23"}        /*SET TIMER FOR 100 SECONDS*/
12           {OUTPUT NELSON,"B"}         /*BEGIN CONTINUOUS SAMPLING*/
13
14 GET_DATA  {OUTPUT NELSON,"E"}         /*END CONTINUOUS SAMPLING*/
15           {OUTPUT NELSON,"HU"}        /*PROMPT NELSON TO RETURN THE DATA*/
16           {NREAD NELSON,TEMPLATE,101}          /*READ DATA IN AS NUMBERS*/
17           {CALC}         /*FORCE RECALCULATION OF THE SHEET*/
18           {WINDOWSON}{PANELON}        /*THAW THE DISPLAY*/
19
20 \M        {BRANCH GET_DATA} /*GET THE DATA FROM BACKGROUND MODE*/
21
22 \R        {BRANCH NELSON}   /*BEGIN BACKGROUND MODE FOR DATA COLLECTION*/
23
24                   /*SCRATCH PAD*/
25 VERSION           /* 0=>LOTUS 1-2-3; 1=>SYMPHONY */
26 DELAY             /*USER DEFINED DELAY BETWEEN DATA POINTS*/
27 TEST_BUF          |*BUFFER FOR {STATUSON} RESULTS*|
28 STATUS            |*                             *|
29
30
31
```

Figure 9–23

ances from an HPLC (High Performance Liquid Chromatograph) column. A description of the Nelson 900 was given in Chapter 6. The example program places the Nelson 900 into background mode and takes 1000 one-second readings. A second part of the program prompts the Nelson 900 for its data and places the data into the template.

GPIB AUTO-EXECUTING MACROS

The auto-executing macros in Figure 9–22 illustrate some of the important steps required to initialize your program and the GPIB board. The new Measure macro commands included are {INIT}, {DMAON 1}, {STATUSON}, and {CLEAR}. The function of each of these new commands is described in the next four sections.

INITIALIZING THE GPIB SYSTEM

You MUST execute the {INIT} command before using any other Measure GPIB command. The {INIT} command initializes the GPIB plug-in board, specifies the board as the system controller, and ensures that certain key variables are set up correctly. You need only to execute the {INIT} command once in a session, but you must execute it *first*. If you try to execute another Measure macro command before {INIT}, the macro will abort and you will get an error message.

TURNING ON THE DMA CHANNEL

The {DMAON} command turns on the direct memory access channel. If you recall, DMA provides a direct "pipeline" between your instrument and the computer's memory. Using DMA, each time you issue a Measure read command you transfer information directly from your instrument to the memory that holds the data for the cells of your spreadsheet. This method is much faster than reading the data into a buffer and then "hopscotching" it into your template because it does not tie up the computer's CPU. DMA channel 1 matches the channel that was set up on the board when it was configured.

SETTING UP A MECHANISM FOR SYSTEM STATUS CHECKS

Each time a Measure macro command executes, it returns a status code relating to its success and to the condition of the bus. The {STATUSON} command stores this information in two vertical cells of a spreadsheet. The upper cell contains the number of bytes sent or received during the last send or read command. The lower cell contains an encoded number that indicates the status of the system. By testing the number in the *lower* cell, you can evaluate the current state-of-affairs of your GPIB system.

The status code consists of a number from 0 to 15. The number is formed by Measure from the addition of numbers that specify certain conditions. The following table shows the various conditions that are tested and the number that is added if a condition is "true":

Condition	Number Added
OK	0
SRQ Line Asserted	1
ERROR On The Bus	2
TIME-OUT Occurred During Execution Of The Command	4
END Detected On A READ Command (EOI Line Asserted)	8

For example, if a TIME-OUT was detected and SRQ was asserted, the number would be: 1 + 4 = 5. A zero means that the system status is acceptable.

Though not obvious, the number that you get actually contains a "bit map" of information. This technique of encoding information via a number is common.

Decimal numbers in a computer are actually stored as a sequence of binary digits. Each binary digit is able to take on a value of zero or one and each successive digit of the sequence represents a multiplier for successively increasing powers of two. For example, the bit sequence of 00000101 can be converted into a decimal five by multiplying each bit in the string by the appropriate power of 2 (128, 64, 32, 16, 8, 4, 2, and 1). That is

$$5 = (0*128)+(0*64)+(0*32)+(0*16)+(0*8)+(1*4)+(0*2)+(1*1)$$

In this case, each status condition test (i.e., the attributes) is either a zero or a one and forms the first (right-most) 4 bits in a binary number. In other words, they are multipliers for successive powers of two. The following table gives the position of the bits:

Power of 2:	3	2	1	0
END	1	0	0	0
Time-out	0	1	0	0
ERROR	0	0	1	0
SRQ	0	0	0	1
OK	0	0	0	0

For the TIME-OUT/SRQ example:

$$\text{Status} = (0 * 2\char`^3) + (1 * 2\char`^2) + (0 * 2\char`^1) + (1 * 2\char`^0) = 5$$

You can therefore look at the status as if it were made up of a "true" or a "false" (one or zero) for each attribute and you can use the Measure {TEST} command to pinpoint which of the attribute(s) to test. The {TEST} command takes care of the details of bitmapping for you. So, the only thing that you need to know is the bit value of the attribute that you want to test for. Use the following table to determine the bit value for each attribute that can be tested for by the {TEST} command:

Status Test	Bit Value
END	8
Time-out	4
ERROR	2
SRQ	1

The use of the {TEST} command will be explained in more detail; but for now you must understand that you are setting up a mechanism in the auto-executing macros for capturing status data that can be used to keep track of the performance of the system as the program progresses.

The following are some examples of common uses for this tracking system:

1. If you have one instrument that transmits data frequently and another that transmits infrequently, you can write a program to control or read the data from the first instrument, while periodically testing the status of the system to see if the second instrument needs servicing (attribute 1). When the {TEST} command gives a true for attribute 1, you can temporarily sidetrack to a macro subroutine to service the second instrument and then return to the first instrument.
2. You could test attribute 2 to see if an error has occurred. If it has, then you could temporarily call an error-handling subroutine to deal with the error.
3. You could test for attribute 4 to see if a time-out has occurred and call a subroutine to issue a message that prompts the user to turn an instrument on.
4. You could test for attribute 8 to see if the EOI line had been asserted. If it was not, then you would know that not all of the data had been received for a read command and you could call a subroutine to get the remainder of the data.

You should place the cells for your {STATUSON} in your scratch-pad. Heed one word of caution, however. If you move cells during program development and {STATUSON} is in your auto-executing macro, you should re-issue the command. That is, although under normal circumstances you need only to execute the command once, you need to re-execute it to specify a new location if cells have been moved.

RESETTING INSTRUMENTS TO THEIR POWER-ON STATES

The last new command in the auto-executing macros, the {CLEAR} command, returns all instruments to their power-on states. That is, they "reset" to the same state that they would have been in had you just turned them on. If you want to target specific instruments to be cleared, name them in the command. For example, to clear and reset the instruments named NELSON and DVM; you can issue the following command:

```
{CLEAR NELSON,DVM}
```

The {CLEAR} is a commonly used command in error-handling subroutines that are called when a {TEST} command detects an error using attribute 1.

Another common use of the {CLEAR} command is shown in Figure 9–23. In this situation, it is used to "wipe the slate clean" prior to using an instrument. Resetting an instrument to its initial power-on state is a wise procedure to follow because it will decrease the number of surprises that you will encounter. By resetting all parameters to their power-on default values and then specifically setting the parameters pertinent to your program, you can ensure that the instrument is set up consistently.

WRITING TO GPIB INSTRUMENTS

Figure 9–23 shows some other new Measure commands. The first one is the {OUTPUT} command. This command is the GPIB equivalent of the RS-232 {WRITELN} and {RSEND} commands and works similarly. However, it has some additional features that enhance its functionality in the GPIB process. For example, a string that is sent out can be terminated in one of three ways, depending on the instrument's settings sheet. A string can be terminated by an end of string (EOS) character, an electric signal on the EOI line, or by a time-out.

The {OUTPUT} command uses an instrument's name as one of its arguments to determine where a string is to be transmitted. When an {OUTPUT} command executes, Measure asserts the ATN line (i.e., places it in GPIB command mode) and issues a universal unlisten/untalk ("?_") command. Next, Measure sends out the listen address of the instrument that is specified in the {OUTPUT} command. For example, if NELSON were at address 1, a "!" would be transmitted. The ATN line would then be placed in data mode and the data string would be transmitted. If an Output-EOS were specified in the instrument's settings sheet, it would be appended. If the Write-EOI mode were specified in the settings sheet, the EOI line of the connector would be asserted. All of these details are automatically taken care of by Measure.

As you know, GPIB is oriented toward merging many instruments into a single unit. To facilitate this objective the {OUTPUT} command has been designed to allow you to send out the same string to an entire list of instruments. For example, to send the same string of characters from a cell called OUT_BUF to the instruments named NELSON, DVM, and SPEC, you would use the following command:

```
{OUTPUT NELSON,DVM,SPEC,OUT_BUF}
```

However, if you use a list of instruments, make sure that the time-out of one of the instruments in the list is long enough to output the data to all of the instruments. If a time-out occurs before all data has been sent, the data will not be completely sent. The default time-out that is used is the one for the instrument with the longest time-out on the list. All other settings are taken from the first instrument in the list. So, be careful.

If you do not specify an instrument, the settings are taken from the board settings sheet and all of the instruments in the system that have been addressed to listen (by a previous output command) receive the information.

One other difference remains between {OUTPUT} and the RS-232 {RSEND} and {WRITELN} commands. With the RS-232 commands, you could only transmit strings of characters. If you had numbers in a cell, you were forced to use an @STRING function to convert the number into a string before transmission. The {OUTPUT} command also sends alphabetic characters and numbers as strings of ASCII characters. However, if the data in a cell were a number, the {OUTPUT} command would automatically convert the number to a string. Furthermore, the

{OUTPUT} command would output the number in the format of the cell. For example, if a cell containing 10.35 is formatted as currency with three digits, it would be sent out as $10.350.

READING DATA FROM GPIB INSTRUMENTS

The next new command in Figure 9–23 is the {NREAD} command. It shows similar enhancements.

Although all data that comes into the GPIB board are strings of binary zeros and ones, the data may be interpreted in three different ways. The data placed into the buffer of your scratch-pad or into your template depend on whether you use an {LREAD}, an {NREAD}, or a {BREAD} macro command.

The {LREAD} command interprets data as a string of characters. This command is the read command most similar to the {READLN} command used in previous chapters and the {RRECEIVE} command used in the Measure RS-232 module previously described in this chapter. Using this command will require that you parse a data string and use the @VALUE function to turn it into numbers.

The {NREAD} command works similarly. It also reads data as a string of characters. However, it converts the string into a number before it places the value into a cell(s). This command is like using a {READLN} or {RRECEIVE} followed by a conversion with an @VALUE function. This function is convenient if your instrument sends out only numerical data with no alphabetic characters.

The {BREAD} command reads data in as straight binary data (zeros and ones), converts the bytes into a number and stores the number in the spreadsheet. {BREAD} treats every byte as a component of a specific number format; the format is specified in the {BREAD} command. The available formats allow you to convert one, two, or three bytes (8, 16, or 24 bits) into signed or unsigned integers. Signed integers are whole numbers that can be positive, negative, or zero. Unsigned integers are whole numbers that can only be positive or zero.

The following table shows the different format choices:

Format	# Bytes	Signed/Unsigned	Range
1S	1	SIGNED	−127 to +127
1U	1	UNSIGNED	0 to +255
2S	2	SIGNED	−32767 to +32767
2U	2	UNSIGNED	0 to +65535
3S	3	SIGNED	−32767 to +32767

The 3S format reads three bytes and converts them into a signed integer. The first byte is used as the high byte (highest value), the second byte as the middle byte, and the third byte as the low byte (lowest value). This format allows you to interface instruments that have seven-bit character transmissions. Consult your instrument's

User's Manual to determine if your instrument transmits its data in binary format and how the bytes are to be decoded. From this information you can choose which of the five listed formats to use.

Instruments that transmit their data in straight binary format require advanced programming knowledge and are fairly uncommon. For these reasons, {BREAD} is not discussed in further detail. However, the {LREAD} and {NREAD} commands will be used often. Both of these commands have the same syntax:

```
{LREAD device,range,cycle}
{NREAD device,range,cycle}
```

Like the {OUTPUT} command, the GPIB read commands also have additional features that enhance their functionality in the GPIB process. For example, a string that is received can be terminated in one of five ways, depending on the instrument's settings sheet. A string can be terminated by an end of string (EOS) character, an electric signal on the EOI line, a filled destination range, an error on the bus, or by a time-out.

The read commands use **a single** instrument's name as one of its arguments. This name determines which instrument should transmit a string. When a read command executes, Measure asserts the ATN line (i.e., places it in GPIB command mode) and issues a universal unlisten/untalk ("?_") command. Next, Measure sends out the talk address of the instrument specified in the read command. For example, if NELSON were at address 1, an "A" would be transmitted. The ATN line would then be placed in data mode and the data string would be received. If an Input-EOS were specified in the settings sheet of the instrument, the read would terminate when the EOS was detected. If the EOI line were specified in the settings sheet of the instrument, the read would terminate when the EOI line was asserted.

When an Element Separator is specified in the instrument's settings, it is used as a separation boundary for adjacent data points. For example, when the Element Separator is a comma, the following string

```
12234,14566,16678
```

would automatically be parsed and a 12234 placed into the first cell of the range; 14566 would be placed into the second cell of the range; and 16678 would be placed into the third cell. Measure takes care of these parsing and placing functions automatically.

Another convenient feature of this command is the CYCLE argument. This argument is like placing the read command in a {FOR} loop, only it executes faster. That is, CYCLE specifies the number of reads that are to be executed. This feature requires that your instrument be able to transmit its data either continuously or after it receives a single prompt from an {OUTPUT} command. You cannot prompt for individual data points while using CYCLE. The number of cycles can range from 1 to 32767 and each subsequent read is placed into the next lower cell in the specified range. In other words, it fills the range from top to bottom.

The CYCLE argument has a couple of idiosyncrasies of which you need to be aware. The first is concerned with time-out. The time-out that you use for an instrument must be long enough to cover *all* reads specified by CYCLE. If it is not, you will not recover all data. This data recovery problem will occur if you specify a large number of cycles and the instrument has a short time-out. In this case, all of the cycles may not be completed before the time-out occurs and you would lose some data.

The second quirk is concerned with Element Separators and the number of data points that you receive. Each cycle ends with an EOI or EOS, not an Element Separator. Therefore, if each string contains several data points separated by Element Separators, each cycle will actually return more than one data point. For example, if an instrument transmitted strings with four data points separated by commas, a CYCLE count of five would lead to 20 cells of the range being filled, not 5. Because of this functionality, you must plan ahead when specifying the CYCLE count and the size of the receiving range.

The third peculiarity is concerned with the size of a cell. If a string longer than 240 characters is received, the string will be split and placed into successive rows of a range. For example, if 300 characters are received, the first 240 characters are placed into the first cell of the range and the remaining 60 characters are placed into the second cell.

Figure 9–23 has an {NREAD} command example. In the example, the instrument settings sheet that is used is the one called NELSON. {NREAD} is executed POINTS times and places each subsequent data point into the range named TEMPLATE. This range was defined as C3..C3000 when the template was prepared.

TESTING {STATUSON} CODES

The auto-executing macros in Figure 9–22 set up a mechanism to check on the status of the communications being carried out by the GPIB controller. This mechanism was the {STATUSON} command. If you recall, each time a Measure macro command executes, it returns a code relating both to its success and the condition of the bus. If a {STATUSON} command had been issued earlier in the program, the status code is placed in two vertical cells. The lower cell of this range contains an encoded number relating to the status of the system. By testing this number, you can evaluate the current state-of-affairs of your GPIB system.

Figure 9–24 shows you how to implement this evaluation. The {TEST} command in the REDO macro subroutine decodes the status number and tests it. If the test result is true, the macro commands following the {TEST} command *in the same cell* are executed. If the test is false, the macro commands in the cell are ignored and the program continues by executing the cell directly below it. The {TEST} command works similarly to the {IF} commands used in previous chapters.

The syntax of the {TEST} command is

```
{TEST cell_address,data_arguments}{MACRO}
```

The cell_address is the *second* cell of the {STATUSON}'s range. Therefore, you must create a name for that cell and specify it in the {TEST} command. You cannot use the name specified in the {STATUSON} command. For this reason, the name STATUS was given to the second cell in Figure 9–23 and used in the {TEST} command.

The data arguments refer to the bit value(s) of the attribute(s) for which you want to test. This system was explained in the auto-executing macro discussion. Use the following table to determine the bit value for each {TEST} command's attributes:

Status Test	Bit Value
END	8
Time-out	4
ERROR	2
SRQ	1

In the example in Figure 9–24, a test is made to see if attribute 4 is true. That is, if a time-out occurred, the {TEST} command would execute the {QUIT} command and abort the macro.

This example illustrates a simple response to {TEST}. You can just as easily call a subroutine to deal with the situation. For example, you could have called a macro that would have issued a prompt to the user to turn an instrument on, issued initialization and start strings, and returned to the REDO macro to continue with the program.

If you test for more than one attribute, all attributes must lead to true results before the {TEST} command executes a true response. For example, if you were testing attributes 1 and 4 (SRQ and time-out) and both were not true, the macro commands following the {TEST} command in the cell would be ignored.

One word of caution: The {TEST} command, being a Measure command, causes an update to the STATUS cell after it executes. This update is unfortunate because it disallows sequential {TEST}s on the STATUS cell. For this reason, you cannot test for one attribute and then another, thus making it impossible to see if one OR another condition exists and then acting accordingly.

If you need to test for two or more different conditions independently by using sequential {TEST} commands, use a {LET} command just prior to the first {TEST} command. This sequence makes a copy of the status number in another cell. Test the cell which contains the copy. By doing so, the status information does not change after testing and the second {TEST} command reflects the actual conditions of the system.

```
--------V--------W--------X--------Y--------Z--------AA-------AB-------AC-------
 1 /*GET DATA FROM ONE INSTRUMENT,WHILE MONITORING ANOTHER FOR SERVICE REQST*/
 2
 3 START_RUN{WINDOWSON}{PANELOFF}        /*FREEZE THE DISPLAY*/
 4           {CLEAR NELSON}               /*CLEAR THE CONVERTER*/
 5           {BLANK BUFFER}               /*ERASE DATA AREA OF TEMPLATE*/
 6           {OUTPUT NELSON,"I"}          /*RESET THE INTERFACE*/
 7           {OUTPUT NELSON,"OE"}         /*SELECT EMULATION MODE*/
 8           {OUTPUT NELSON,"C 1"}        /*SELECT CHANNEL ONE*/
 9           {OUTPUT NELSON,"R 0"}        /*SET VOLTAGE RANGE*/
10           {OUTPUT NELSON,"S 10"}       /*SET SAMPLING RATE TO 1 / SECOND*/
11           {BGN_SCAN}                   /*START COLLECTING DATA WITH SCANNER*/
12           {SPOLL SCANNER,SPOLL_BUF}    /*CLEAR OUT SRQ FROM POWER-UP*/
13           {FOR INC,0,10000,1,REDO}     /*GET 10000 DATA POINTS*/
14
15 REDO      {OUTPUT NELSON,"T"}          /*PROMPT FOR A READING*/
16           {NREAD NELSON,TEMPLATE}      /*READ IN A VALUE*/
17           {TEST STATUS,4}{QUIT}        /*TEST FOR TIME-OUTS*/
18           {SPOLL SCANNER,SPOLL_BUF}    /*POLL SCANNER FOR SERVICE REQUEST*/
19           {TEST STATUS,4}{QUIT}        /*TEST FOR TIME-OUTS*/
20           {IF SPOLL_BUF>=64}{SERV_SCAN}          /*SERVICE IT IF SRQ ASSERTED*/
21
22 BGN_SCAN                        /*SUBROUTINE TO START SCANNER COLLECTING DATA*/
23                .
24                .
25                .
26
27 SERV_SCAN                       /*SUBROUTINE TO GET DATA FROM THE SCANNER*/
28                .
29                .
30                .
31
32
33 \R        {BRANCH START_RUN}          /*START THE PROGRAM*/
34
35                        /*SCRATCH PAD*/
36 VERSION                /* 0=>LOTUS 1-2-3; 1=>SYMPHONY */
37 INC                    /*INCREMENT COUNTER FOR LOOP*/
38 TEST_BUF               |*BUFFER FOR {STATUSON} RESULTS*|
39 STATUS                 |*                              *|
40 SPOLL_BUF              /*BUFFER FOR SERIAL POLL DATA*/
41
42
```

Figure 9-24

The following is an example of this concept. It transfers the STATUS information to a cell named STATTEST. This process freezes the value of the status information for sequential (independent) testing for both SRQ and TIMEOUT.

```
          :
{LET STATTEST,STATUS}
{TEST STATTEST,1}{QUIT}
{TEST STATTEST,4}{QUIT}
          :
```

POLLING AND SERVICING INSTRUMENTS

Two of the most powerful features of GPIB are its ability to control one instrument while simultaneously monitoring several others and its ability to wait until an instrument needs attention and then service it. These abilities are useful in situations in which one instrument in a system requires frequent attention and the other instruments need attention only infrequently.

For example, consider a system that has an A/D converter being continually prompted to collect data and another instrument that collects its data in background mode. (That is, it was initialized and started by the program and thereafter worked independently.) When the second instrument finishes collecting data, it would signal the GPIB board that it had data to transmit. (It would signal by asserting the SRQ line in the connector.)

Figure 9–24 illustrates how to test and respond to service requests. In this example, the program actively controls a Nelson 900 A/D converter while monitoring a second instrument (hypothetically called a ''SCANNER'') that is working in background mode.

Whether you use a program like the one in Figure 9–24 to control just two instruments or a program that controls a system with many instruments, you must be able to determine *if* any of the instruments in the system need servicing. If you are using a program that either checks the status cell of {STATUSON} for an SRQ or waits for an SRQ with a {WAITSRQ} command, you must also be able to determine *which* of the instruments need servicing. This capability is needed because the SRQ may be asserted by any of the instruments in the system. It may also be asserted by multiple instruments simultaneously.

Each of these situations require that you carry out a ''poll'' of the instruments to determine which instrument(s) on the bus need service.

Two methods are used for polling instruments. One method is called a ''parallel poll'' and the other is called a ''serial poll''. Consult the User's Manuals for your instruments to determine which system of polling they support. You should also see which instruments assert the SRQ line to get the controller's attention when they need service.

Parallel Polls

A parallel poll is much faster than a serial poll, but less common. Not all devices are capable of responding to a parallel poll. In this type of poll, all instruments on the bus are asked to transmit their status bits on different lines of the bus. Thus, all instruments report their statuses in ''parallel'' and in a single cycle.

The Measure command that conducts parallel polls is the {PPOLL} command. The syntax of this command is

```
{PPOLL cell_address}
```

When this command is issued, a number is placed into the specified cell of the spreadsheet. Each of the eight bits of this number correspond to each of the eight data lines in the connector and specify which of the instrument(s) require servicing. For example, if 64 were placed in a cell (a binary 01000000), it would indicate that the instrument configured to respond on data line 6 returned a 1 and therefore asserted a service request; while all other instruments returned zeros. Similarly, a response of 3 (binary 00000011) would indicate that the instruments configured to respond on data lines 1 and 2 returned 1's, thereby requiring service.

The Measure command used to configure instruments and their responses to parallel polls is the {CPPOLL} command. Use this command early in your program to specify on which of the eight data lines each instrument should respond and the way in which each should respond. The syntax of this command is

```
{CPPOLL INSTRUMENT,CONFIGURATION,STATE}
```

where, CONFIGURATION is the data line on which to respond and STATE is a status bit that the instrument is to issue if it needs service. You can set the STATE to either zero or one; however, zero is usually used to denote no SRQ and one is usually used to denote an asserted SRQ.

Serial Polls

Serial polls are more common, but not ubiquitous. A serial poll is a slower method of determining which instrument asked for service. This type of poll checks each instrument individually. That is, to use the Measure {SPOLL} command you specify the instrument that you want to poll in the command; the {SPOLL} command responds by placing a number from 0 to 255 in a cell.

The number placed in the cell is dependent on the instrument, so consult your User's Manual for its meaning. The general format of the number is a number whose bits encode a message.

The usual convention is such that if an instrument has asserted its SRQ line, bit 7 is 1 (e.g., 01000000). The bits to the right of bit seven are usually a message from the instrument that you can use to determine the course of your service. For example, bit pattern 01010010 means that an instrument has asserted a service request; the 010010 portion may mean that it wants to send data.

In this pattern, the number that corresponds to 1 in bit position seven is 64. You can use this information to create a test for the number placed in a cell and use the test to determine if an instrument needs to be serviced. If the number is greater than 64, you will know that the instrument asserted the SRQ line and requested service.

The syntax of the {SPOLL} is

```
{SPOLL INSTRUMENT,CELL_ADDRESS}
```

where CELL_ADDRESS is the address of the cell into which the status byte is to be placed.

Some older versions of the Measure manual state that you can use the {TEST} command to decide the returned number. Although this statement is not incorrect, it is misleading because it actually refers to the status byte returned by {STATUS-

ON}, not {SPOLL}. These two commands return different information. The status byte from the {STATUSON} command indicates whether *any* of the instruments have asserted the SRQ line. The data returned from {SPOLL} is *specific to the instrument being polled.* The data from {STATUSON} can be tested with the {TEST} command; the data from {SPOLL} cannot, since the data are instrument dependent. Therefore, you must develop your own method to decode each instrument's status bytes.

In the example program of Figure 9-24, each time the REDO macro executes and a value is read in from NELSON, a serial poll is made on SCANNER. The result is tested to see if the SRQ bit (bit 7) was asserted. If the value returned is greater than or equal to 64, SCANNER is serviced by the hypothetical macro subroutine SERV_SCAN. This type of subroutine typically prompts the instrument for its data. The program then continues with the task of controlling NELSON.

Serial polling has a couple of quirks of its own. Some instruments need an {SPOLL} to be performed on them before you can obtain valid results. This requirement is set because an SRQ is normally asserted at power-on. This is the function of the {SPOLL} in the START_RUN macro of Figure 9-24. The {SPOLL} clears the power-on SRQ. Then, when the instrument needs service, it asserts a valid SRQ.

Also, some instruments assert an SRQ after they collect *each* data point and/or whenever they place text in their output buffers. In these cases, you will want to poll the instrument and check the message to the right of bit seven. This message will usually specify what type of service, if any, is being requested. Consult the User's Manual of your instrument for the specifications of each status bit and how to decode them.

WAITING FOR SRQS/MONITORING SEVERAL INSTRUMENTS

The {WAITSRQ} command is a convenient way to wait until an instrument needs servicing. It pauses the macro program until any of the instruments in the system have asserted the SRQ line, or until the controller board times out. Because no instruments are named in the {WAITSRQ} command, the controller board's settings are used. If you use {WAITSRQ}, you will probably want to disable the time-out of the board by placing a zero in the controller's settings sheet.

In Figure 9-25, the program waits until one of the instruments on the list requires servicing. The Measure command that performs this function is the {WAITSRQ} command in the RUN macro. When an SRQ is detected, a serial poll is conducted on each of the instruments to determine which ones need servicing. The {FOR} command in the AGAIN macro and the ASK_STAT macro conduct this survey.

The macro technique that is used to sequentially poll the instruments is the self-modification technique described throughout previous chapters of this book. Each time through the loop, the two {LET} commands in ASK_STAT create two

```
--------V--------W--------X--------Y--------Z--------AA-------AB-------AC-------
 1            /*GPIB PROGRAM TO MONITOR SEVERAL INSTRUMENTS*/
 2           /*IE, THIS IS A DEMONSTRATION OF SERIAL POLLING*/
 3
 4 RUN        {CLEAR}  /*RESET ALL INSTRUMENTS TO POWER-ON STATE*/
 5            {SETUP}  /*INITIALIZE ALL INSTRUMENTS TO PARAMETERS FOR EXPT*/
 6 AGAIN      {WAITSRQ}            /*WAIT FOR A SERVICE REQUEST*/
 7            {FOR INC,0,3,1,ASK_STAT}   /*WHICH INST(S) NEED SERVICE?*/
 8            {BRANCH AGAIN}    7*CONTINUE MONITORING SRQ LINE*/
 9
10                /*SELF-MODIFY INSTRUMENT SERIAL POLLS AND TESTS*/
11 ASK_STAT {LET SPOL_SPEC,+"{SPOLL "&@INDEX(TABLE,0,INC)&",SPOLL_BUF}"}
12          {LET SPOL_TST,+"{IF SPOLL_BUF>=64}{"&@INDEX(TABLE,0,INC)&"}"}
13 SPOL_SPEC{SPOLL MOTOR,SPOLL_BUF}     /*THIS CHANGES FOR EACH LOOP*/
14 SPOL_TST {IF SPOLL_BUF>=64}{MOTOR}   /*THIS CHANGES FOR EACH LOOP*/
15
16
17 SETUP                  /*SUBROUTINE THAT INITIALIZES ALL OF THE INSTRUMENTS*/
18                 .      /*AND STARTS THEM RUNNING*/
19                 .
20                 .
21
22 MOTOR                  /*SUBROUTINE TO DETERMINE MOTOR'S POSITION*/
23                 .      /*AND MOVE MOTOR TO ANOTHER POSITION*/
24                 .
25                 .
26
27 DVM                    /*SUBROUTINE TO GET DATA FROM THE DIGITAL VOLT METER*/
28                 .      /*AND THEN RESTART THE DVM*/
29                 .
30                 .
31
32 SCOPE                  /*SUBROUTINE TO GET DATA FROM THE OSCILLOSCOPE*/
33                 .      /*AND THEN RESTART THE SCOPE*/
34                 .
35                 .
36
37 WAVE                   /*SUBROUTINE TO SERVICE THE WAVEFORM GENERATOR*/
38                 .      /*AND THEN RESTART THE WAVEFORM GENERATOR*/
39                 .
40                 .
41
42                            /*TABLE OF SUBROUTINE NAMES*/
43                        MOTOR     /*MOTOR THAT CONTROLS POSITION OF SENSOR*/
44                        DVM       /*DIGITAL VOLT METER*/
45                        SCOPE     /*OSCILLOSCOPE*/
46                        WAVE      /*WAVEFORM GENERATOR*/
47            /*SCRATCH PAD*/
48 VERSION                /* 0=>LOTUS 1-2-3; 1=>SYMPHONY */
49 POS                    /*MOTOR POSITION COUNTER*/
50 INC                    /*COUNTER FOR {NREAD}s*/
51 DELAY                  /*USER DEFINED DELAY BETWEEN DATA POINTS*/
52 TEST_BUF               |*BUFFER FOR {STATUSON} RESULTS*|
53                        |*                             *|
54 SPOLL_BUF              /*BUFFER FOR SPOLL DATA; WILL BE TESTED BY SPOL_TST*/
55
56
```

Figure 9–25

lines of the program based on the INC counter. That is, the origin of SPOL_SPEC and SPOL_TST is a string concatenation of the instrument's name and the remainder of the text for the commands. The instrument's name is retrieved from the table at the bottom of the program by the @INDEX commands in the two {LET}s. TABLE, in this example, is a named range of cells X43 through X46.

The self-modifying macro in Figure 9–25 is a quick and efficient way to poll several instruments. An alternative method is to create a macro and enter an {SPOLL} and {IF} test for each instrument. However, the {FOR} loop is much easier.

As in the example of Figure 9–26, service macros have not been specified. This omission should allow you to focus on the polling technique.

OTHER GPIB MACRO COMMANDS

The following are the remaining GPIB macro commands and their uses. Because they are self-explanatory, examples of their usage are not presented.

The {TRIGGER list} command signals a list of instruments to begin collecting data. This method of starting instruments ensures that all instruments start simultaneously. It is faster than issuing individual character strings to each instrument using a series of {OUTPUT} commands.

The {LOCAL list} command places an instrument or list of instruments into local mode. A user can then control an instrument from its front panel. If you do not specify an instrument, all instruments on the bus are placed in local mode.

The {REMOTE list} command places an instrument or list of instruments in computer control mode. Most instruments do not require this command to be issued because they automatically convert to remote mode as soon as they are addressed to listen or talk.

The {LOCKOUT} command disables the front panel controls of all instruments in the system. This command prevents a user from making changes to an instrument's settings that would adversely affect its performance while it is being controlled from a GPIB program.

The {SEND data_argument} command sends the address-specific encoded talk, listen, untalk, and unlisten commands (see table, above). Because {OUTPUT}, {LREAD}, {NREAD}, {BREAD}, {LOCAL}, etc., automatically coordinate this function for you, you will probably not need to use this command often (unless you want to add to the list of "activated" instruments).

WHAT'S NEXT?

The next chapter shows you a couple of ways to obtain data directly from the digital display of an instrument. This method is particularly useful for older instruments that do not have RS-232 connectors. It is also a good alternative to using analog outputs or chart recorder outputs for data.

The chapter also shows you how to handle digital signals with stand-alone interfaces.

10

Getting Data from an Instrument's Digital Display

If your instrument has the ability to display numbers but does not have an RS-232 connector, you can use the information in this chapter to learn how to capture the data from the instrument's display and place the data into your template. This alternative is presented in opposition to using chart recorder output, as described in Chapter 6. Although this alternative can require more preparation, it will provide you with a perfect correlation between the data on the instrument's display and the data in your spreadsheet.

Prior to this chapter, you worked with instruments or interfaces that transmitted data sequentially as individual characters. The characters were transmitted using the well-defined transmission protocols of the RS-232 standard. Each character was transmitted as a sequential series of individual binary digits, the binary digits were reassembled and then the digits were converted back into characters using the ASCII standard. Transmitting data one bit at a time is known as "serial" transmission.

In this chapter, you will learn an alternative method of data transmission. This method represents data within instruments. In this method, all binary digits of a character exist simultaneously in separate wires. This method is similar to the parallel transmission that is commonly used for parallel printers.

BINARY CODED DECIMAL

One of the most common methods that scientific instruments use to display numbers is to have each digit of the display controlled by four wires. Each of the four wires for a digit represent a number, most commonly 8, 4, 2, and 1. The number displayed is the *sum* of the numbers represented by the wires that are switched on. That is, the binary value of these four groups is from 0000 to 1001, which code for decimal digits from zero through nine. This method of representing decimal numbers is called *B*inary *C*oded *D*ecimal (BCD).

The objective of BCD is to represent numbers in a form that transistors, switches, and integrated circuits can recognize. This form is binary. Binary form represents data as a group of *bi*nary digi*ts*, called bits. Each bit is either "off" or "on", as represented by the symbols "0" and "1".

BCD uses four bits to represent a decimal digit. For example, a decimal 5 appears as 0101. To convert the 0101 into a five, just multiply 8, 4, 2, and 1 by the appropriate bit in the set. For example

$$(0 * 8) + (1 * 4) + (0 * 2) + (1 * 1) = 5$$

The following table shows how digits 0 through 9 are represented in BCD:

Decimal	8421
0	0000
1	0001
2	0010
3	0011
4	0100
5	0101
6	0110
7	0111
8	1000
9	1001

As stated, each digit displayed is assembled from the 8421 binary number and is displayed independently of each other number on the display. Therefore, sixteen wires would be required for a number that has four digits (four wires per digit times four digits).

Each subsequent set of four wires represents a digit that is ten times smaller than the previous digit. For example, the number 1395 would be represented by

$$(1 * 10E3) + (3 * 10E2) + (9 * 10E1) + (5 * 10E0) = 1395$$

The lines that would represent this number in BCD would look like the following:

0001 0011 1001 0101

Some of the wires carry binary ones and others carry binary zeros; the specific combination of these ones and zeros represent a four-digit decimal number. Electrically, the wires with binary ones have voltages greater than 2.0 volts and wires with binary zeros have voltages less than 0.8 volts.

These binary signals can be used to obtain data from your instrument if the instrument does not have an RS-232 interface. That is, you can get information directly from the digital display. Alternatively, some instruments have connectors at their backs that output voltages directly from the display. (This connector is usually labeled "BCD".) In either case, you can convert voltages into binary zeros and ones with a device called a "Digital Input/Output" interface. You can then use a macro program to assemble the information into digits, the digits into numbers, and then place the numbers into your spreadsheet.

OTHER FORMS OF BCD

BCD uses systems other than 8421 to represent decimal digits. These systems are used much less frequently than 8421, but because of their existence, you need to determine which system your instrument uses so that you can develop an appropriate equation to convert the binary data to decimal.

For example 2421 and 5211 are two other weighted coding systems that are used in instruments. These systems work similarly to the 8421 system.

For example, in the 2421 system, a decimal 5 appears as 1011. To convert 1011 into a five, you multiply the 2, 4, 2, and 1 by the appropriate bit in the set. For example

$$(1 * 2) + (0 * 4) + (1 * 2) + (1 * 1) = 5$$

In the 5211 system, a five appears as 1000:

$$(1 * 5) + (0 * 2) + (0 * 1) + (0 * 1) = 5$$

The equation used in the last section to assemble the final decimal number (1395) does not change for either of these two systems. It remains

$$(1 * 10E3) + (3 * 10E2) + (9 * 10E1) + (5 * 10E0) = 1395$$

The lines that represent this number in BCD 2421 would look like the following:

0001 0011 1111 1011

The lines that represent this number in BCD 5211 would look like the following:

0001 0101 1111 1000

One other BCD code that is used is "XS3". This code is rarely used. It is similar to the 8421 code, except that three (0011) is added to the 8421 binary number. For example, a five would be 0101 + 0011, or 1000:

$$(1 * 8) + (0 * 4) + (0 * 2) + (0 * 1) = 8$$

$$1000 - 0011 = 8 - 3 = 0101 = 5$$

The lines that represent 1395 in XS3 would look like the following:

0100 0110 1100 1000

The following table allows you to compare the various BCD types:

Decimal	8421	2421	5211	XS3
0	0000	0000	0000	0011
1	0001	0001	0001	0100
2	0010	0010	0011	0101
3	0011	0011	0101	0110
4	0100	0100	0111	0111
5	0101	1011	1000	1000
6	0110	1100	1010	1001
7	0111	1101	1100	1010
8	1000	1110	1110	1011
9	1001	1111	1111	1100

Each of these codes can be represented by well defined equations. Thus, by determining which method your instrument uses, you can program accordingly. To determine your instrument's method, consult its User's Manual or Technical Reference manual, ask the manufacturer, or use trial and error.

DIGITAL INPUT/OUTPUT INTERFACES

A number of Digital I/O interfaces are available. Chapter 6 contained a list of manufacturers of stand-alone interfaces. These interfaces typically either have digital capabilities or these capabilities can be added. For example, the ACRO-400™ system has 16 digital lines. The ACRO-900™ system has a module (called the ACRO-913) that can be added to the system to give 60 digital lines. This number of lines is enough to run several BCD instruments.

Alternatively, if you have a plug-in or stand-alone A/D converter with enough lines, you can read voltages into your program and use {IF} tests to convert the voltages into binary zeros and ones. For example, you could use the following tests on the value of a parsed string for each BCD digit. The first {LET} gives an obviously wrong answer and allows you to spot errors.

```
{LET DIGIT,999999}
{IF VOLTAGE>2.0}{LET DIGIT,1}
{IF VOLTAGE<0.8}{LET DIGIT,0}
```

Review Chapters 6 and 9 for more details on how to use stand-alone and plug-in A/D boards, respectively. Although this alternative method will work, buying a dedicated digital I/O interface is usually much better.

In terms of plug-in digital I/O boards, at the time of this book's publication, Lotus Measure did not access this category of boards. However, you can inquire about the current release of Measure to see if the capability has been added to access parallel digital I/O interface boards (*NOT* parallel printer boards). These plug-in boards are dedicated to digital I/O and tend to be cost effective. If you find that this capability has been added to Measure, it is recommended for use, especially if you are under budgetary constraints. Chapter 6 has a list of plug-in manufacturers to contact if you decide to purchase a digital I/O interface board.

HOW TO CHOOSE A DIGITAL I/O INTERFACE

Choosing a digital I/O interface is relatively simple. You need four lines for each digit input. Some instruments may also require additional lines. For example, if your instrument displays a plus and/or minus sign, you need an extra line. Also, some instruments use a line as an indicator to signal that a reading has reached a steady state value. This line is often called a "strobe" and requires an extra line of input for your interface.

The interfaces most commonly available from the sources listed in Chapter 6 are able to accept input from as many as 24 digital lines. This capability translates to six digits, or five digits and a "strobe" line.

BRIEF DESCRIPTION OF THE EXAMPLE INSTRUMENT AND INTERFACE

The example instrument used in this chapter is a Gilford® Stasar III spectrophotometer from Gilford® Instrument Laboratories, 132 Artino St., Obelin, OH 44074.

The spectrophotometer has a flow-through cuvette system that allows sample introduction. After the absorbance of a sample is determined, the result is displayed on a four-digit *L*ight *E*mitting *D*iode (LED) read-out (e.g., #.###).

This instrument was manufactured prior to the personal computer revolution and therefore does not have RS-232 capability. However, it has a BCD output connector at the rear of the instrument. This output reflects the LED read-out and is intended for use by a BCD printer. The output also has a "strobe" line that indicates when a printer should print the results for the current sample.

The example connects the output of the BCD connector to an ACROSYSTEMS ACRO-913. This example Digital I/O interface is a module of the ACRO-900 system described in Chapter 6. The ACRO-400 works almost identically to the ACRO-913 and is a more cost effective alternative.

CONNECTING THE INSTRUMENT TO THE DIGITAL I/O INTERFACE

The method that you use to connect your instrument to the digital I/O interface will depend on your instrument. In the easiest case, a BCD connector will be found at the back of the instrument. In this case, create or purchase a cable and plug it in. Typically, the end of the cable will be a 24-pin Centronix type connector. That is, it will look like the connector for a parallel printer cable, but will have 24 pins.

Consult your User's Manual or Technical Reference Manual for the designation of each pin. Typically, these pins are arranged in groups of four sequential pins for each digit. Once you determine the groupings, decide which group of four lines represents the Least Significant Digit (LSD) and which group represents the Most Significant Digit (MSD) for the decimal number. You also need to determine the LSD and the MSD within each group of four lines.

The LSD of a number is the digit that has the least weight or "value". It is the digit to the extreme right of a binary number. For example, the LSD in the binary number 0001 is one. The MSD represents the opposite case. It is the digit with the greatest importance and is located at the extreme left of a group of digits. For example, the MSD in the binary number 1000 is one.

Your next task is to connect the lines of the cable to the digital I/O interface. Try to keep the lines organized so that all of the lines for a digit are in the order of LSD to MSD. It is also important to arrange decimal digits from LSD to MSD. Organization now will make your programming job easier later.

Figure 10-1 gives an idea of how the order should look. This figure shows the connector that would be used between the Stasar III and ACRO-913. Note that the strobe also has a line (pin 20). This line is used to signal the appropriate time for the computer to capture data. Also note the grounding pins used to communicate a common reference between the instrument and interface.

If your instrument does not have a BCD connector, you will need to tap into

ACRO-913	STASAR	
GND	GND	
GND	GND	
D1	A1	LSD
D2	B1	
D3	C1	
D4	D1	
GND	GND	
GND	GND	
D5	A2	
D6	B2	
D7	C2	
D8	D2	
D9	A3	
D10	B3	
D11	C3	
D12	D3	
D13	A4	MSD
D14	B4	
D15	C4	
D16	D4	
D17	STROBE	

Figure 10-1

the lines on the circuit board. This connection will require an electronics professional or someone with experience in electronics. If you do not have the required experience, contact your instrumentation maintenance shop, a consultant, or the Service Department of the instrument manufacturer.

The first step required to connect your instrument to a digital I/O interface is to identify the section of the circuit that gets BCD signals. Once identified, solder the wires onto the circuit board. Typically, these wires can be connected directly to the pins coming from the sockets that hold the integrated circuits supplying the information. If this case is similar to your set-up, it is extremely important that you remove the integrated circuits from the sockets (if possible) before you solder onto the socket pins. Most integrated circuits are heat sensitive and can be easily damaged if they get too hot.

Your next task is to connect the lines to the digital I/O interface. Group the lines as if they were coming from a BCD connector. Try to organize the lines so that all lines for a digit are in the order of LSD to MSD. It is also important to arrange the decimal digits from LSD to MSD.

Finish the task by fashioning a holding device to prevent the wires from being pulled off the circuit board.

THE MACRO PROGRAM

Figure 10–2 shows a program that contains all of the elements needed to get BCD data from an instrument. Portions of this program are similar to the ones used for the ACRO-900 examples. The ACRO-913 Digital I/O module works nearly identically to the ACRO-912 A/D Converter and ACRO-931 Thermocouple Interface.

OBTAINING BCD DATA AND CONVERTING TO DECIMAL

The macro in Figure 10–2 that gets data from the instrument is called GET_BCD. It begins with an example of how to monitor an instrument's output to see if it is the appropriate time to take a data reading.

If you recall, output line 17 is the "strobe" line that a Stasar III uses to indicate the proper time to take data from the display. The {WRITELN "DIN17"} and {READLN BUFFER} commands prompt for, and read in, the value of line number 17. The subsequent {IF} command checks the value read into BUFFER to see if it is a binary zero. If it is, the program knows that the data are not ready to be taken and the macro continues to monitor the line by {BRANCH}ing back to GET_BCD to read another value.

When a binary one appears in the BUFFER, the program is in a position to read data. However, a six-second delay in the reading is made to ensure that the reading is truly at a steady state. This requirement was reached experimentally; you may not need a pause.

```
---------N-------O--------P--------Q--------R--------S--------T--------U--------
 1      /*PROGRAM TO GET BCD DATA USING ACRO-913 DIGITAL I/O INTERFACE*/
 2
 3               /*AUTO-EXECUTING MACRO FOR LOTUS 1-2-3*/
 4 \0       {LET VERSION,0}   /*TELLS PROGRAM IT'S RUNNING UNDER 1-2-3*/
 5          {STASAR}                    /*CALL UP STASAR PROGRAM*/
 6
 7              /* AUTO-EXECUTING MACRO FOR LOTUS SYMPHONY*/
 8 AUTOEXEC{LET VERSION,1}   /*TELLS PROGRAM IT'S RUNNING UNDER SYMPHONY*/
 9          {IF @ISERR(@APP("DOS",""))}{HOOKUP} /*SEE IF DOS.APP ATTACHED*/
10          {SERVICES}AIDOS~MODE COM1:4800,E,7,2~          /*SET COMM PORT*/
11          {STASAR}                    /*CALL UP STASAR PROGRAM*/
12
13 HOOKUP   {SERVICES}AADOS~Q           /*ATTACHES DOS.APP*/
14
15          /*MACRO DRIVER FOR THE ACRO-913 DIGITAL I/O MODULE*/
16 STASAR   {LET DATE,@DATEVALUE(@NOW)}         /*DATE STAMP THE EXPT*/
17          {OPEN "COM1",M}   /*OPEN COM PORT AS DEVICE FILE NAME*/
18          {GETNUMBER "How Many Data Points?",POINTS}
19          {WINDOWSOFF}{PANELOFF}      /*FREEZE THE DISPLAY*/
20          {HOME}            /*SET CELL-POINTER TO CELL A1*/
21          {WRITELN "#1"}    /*INITIALIZE ACRO-913 (MODULE 1)*/
22          {FOR ROW,0,POINTS-1,1,GET_BCD}      /*GET THE DATA*/
23          {CLOSE}           /*CLOSE THE PORT*/
24          {CALC}            /*FORCE RECALCULATION OF THE SHEET*/
25
26                 /*GET BCD DATA FROM STASAR III*/
27 GET_BCD {WRITELN "DIN17"}            /*SEE IF READY TO PRINT*/
28          {READLN BUFFER}             /*0 = "NOT READY TO PRINT"*/
29          {IF @EXACT(@CLEAN(BUFFER),"0")}{BRANCH GET_BCD}
30          {WAIT @NOW+@TIME(0,0,6)}    /*WAIT FOR SAMPLE EQUILIBRATION*/
31          {WRITELN "DIN1-16"}        |*READ IN 16 PARALLEL LINES*|
32          {READLN BUFFER}            |*OF BCD DATA.              *|
33          {CONV_BCD}                  /*CONVERT BCD TO DECIMAL*/
34          {PUT RAWDATA,0,ROW,NUM}     /*PUT DATA INTO TEMPLATE*/
35          {WINDOWSON}{WINDOWSOFF}     /*UPDATE THE DISPLAY*/
36          {BEEP 10}                   /*SIGNAL SUCCESSFUL COMPLETION*/
37                 /*CONVERT 4 X 4 LINES OF BCD TO DECIMAL*/
38 CONV_BCD{LET NUM,0}                  /*INITIALIZE NUMBER TO ZERO*/
39          {FOR INC1,0,24,8,MK_NUM}    /*ASSEMBLE THE NUMBER*/
40          {LET NUM,NUM/1000}          /*SCALE READING TO #.### FOR STASAR*/
41
42 MK_NUM   {LET DIGIT,0}               /*INITIALIZE DIGIT TO ZERO*/
43          {FOR INC2,0,3,1,MK_DIGIT}   /*CALCULATE DIGIT FROM 4 LINES*/
44          {LET NUM,NUM+DIGIT*10^(INC1/8)}     /*ADD DIGIT TO NUMBER*/
45
46 MK_DIGIT{LET DIGIT,DIGIT+(2^INC2)*@VALUE(@MID(BUFFER,INC1+2*INC2,1))}
47
48          /*SCRATCHPAD OF NAMED RANGES*/
49 VERSION           /* 0=>LOTUS 1-2-3; 1=>SYMPHONY */
50 ROW               /*COUNTER FOR FIRST LOOP; ROW OFFSET FOR {PUT}*/
51 POINTS            /*USER DEFINED NUMBER OF DATA POINTS TO COLLECT*/
52 BUFFER            /*A BUFFER FOR INCOMING DATA*/
53 DIGIT             /*CURRENT VALUE OF THE DIGIT FROM BCD*/
54 NUM               /*NUMBER TO BE PLACED INTO TEMPLATE*/
55 INC1              /*LOOP COUNTER FOR DIGIT TO DECIMAL CONVERSION*/
56 INC2              /*LOOP COUNTER FOR BCD TO DIGIT CONVERSION*/
57
58
```

Figure 10-2

If you do not have a strobe line, use one of the snapshot programming techniques of Chapters 6 and 9 to have your program determine the appropriate time to take a reading.

After the {IF} test determines that a reading should be taken, the 16 lines carrying BCD data need to be read in. In this example, the {WRITELN "DIN1-16"} and {READLN BUFFER} commands are used to get the data. The "DIN1-16" instructs the ACRO-912 to read in digital inputs 1 through 16, make a string of the data and return it to the personal computer. The {READLN} command places the data into the BUFFER cell. In this example, the string contains a zero or a one for each digital line, separated by commas. For example

1,0,0,1,0,1,1,1,0,0,0,1,1,0,0,0

Your next step is to convert each packet of four bits in the string into a decimal digit and then assemble a decimal number from the decimal digits. In this program, this task is performed by the CONV_BCD, the MK_NUM, and the MK_DIGIT macros.

Begin your macro by setting the number that will be transferred to the template to a value of zero. Thereafter, each time a digit is calculated, add the digit to the number.

Next, use a {FOR} loop to calculate the value of each digit. The format of the {FOR} loop in this example is typical of the one you will need

```
{FOR INC1,0,24,8,MK_NUM}
```

This {FOR} loop's counter is set up to slide down eight characters of a string for each pass through the loop. Note that the string is separated by commas. These commas are considered characters too; by taking eight characters, you will get the offsets in the string for groupings of four binary digits (one digit in the decimal number).

The MK_NUM macro assembles a single decimal digit. It begins by setting the value of the digit to zero. It then uses a {FOR} loop to add the value of each of the four bits to DIGIT (starting at the offset specified by the first {FOR} loop). The format of the {FOR} loop in this example is typical of the one you will need

```
{FOR INC2,0,3,1,MK_NUM}
```

This {FOR} loop's counter is set up to take four sequential binary characters (starting at the offset specified by the first {FOR} loop) and assemble one of the decimal digits in the number.

To understand which characters are taken, look at the {LET} command in the MK_DIGIT macro. The offset in the string is

```
INC1+2*INC2
```

The first time through both loops, the offset is 0 (0+2*0). Then INC2 increments and the offset is 2 (0+2*1). The next offset is 4 (0+2*2) and the final offset is 8 (0+2*4). In the example string, this sequence gives 1001.

Each time through the MK_NUM loop, the DIGIT is increased in MK_DIGIT by the value specified by the bit. For the example string, the equations literally read the following on subsequent passes through the loop in MK_NUM.

DIGIT = DIGIT + (2^0)*1 = 0 + 1 = 1
DIGIT = DIGIT + (2^1)*0 = 1 + 0 = 1
DIGIT = DIGIT + (2^2)*0 = 1 + 0 = 1
DIGIT = DIGIT + (2^4)*1 = 1 + 8 = 9

The {FOR} loop in MK_NUM then terminates and the digit is calculated. In this case, INC1 remains zero and the equation literally reads

NUM = NUM + DIGIT * 10^(INC1/8)
NUM = 0 + 9 * 10^(0/8) = 0 + 9 * 10^0 = 9

Next, INC1 increments to 8 (the step size). This time, the offsets for the characters are

INC1	INC2	Offset Value
8	0	(8 + 2*0) = 8
8	1	(8 + 2*1) = 10
8	2	(8 + 2*2) = 12
8	3	(8 + 2*3) = 14

These characters are 0111 in the example string. For these characters, the equations literally read the following on subsequent passes through the loop in MK_NUM:

DIGIT = DIGIT + (2^0)*1 = 0 + 1 = 1
DIGIT = DIGIT + (2^1)*1 = 1 + 2 = 3
DIGIT = DIGIT + (2^2)*1 = 1 + 4 = 5
DIGIT = DIGIT + (2^4)*0 = 1 + 0 = 5

Next, the {FOR} loop in MK_NUM terminates and the digit is calculated. In this case, INC1 is eight and the equation literally reads

NUM = NUM + DIGIT * 10^(INC1/8)
NUM = 9 + 5 * 10^(8/8) = 9 + 5 * 10^1 = 59

Then INC1 increments to 16. The character bits at offsets 16, 18, 20, and 22 are 0001.

Work through the example to confirm that you must add 800 to 59 to get 859. The final value of INC1 is 24. The character bits at offsets 24, 26, 28, and 30 are 1000. This sequence means that 1000 is added to 859 to get a final number of 1859.

After a number is assembled from the digits, it is scaled with a {LET

NUM,NUM/1000} command. This command is necessary because the Stasar III is unable to signal a decimal point in its BCD output and the correct format is #.###. Thus, the final number placed into the template for this example is 1.859. You will probably need to scale the numbers that you get from CONV_BCD to the format of your particular instrument with a {LET} command like the {LET} command at the end of the CONV_BCD macro.

Take time now to confirm that you fully understand how this program is organized and how the loop counters affect the offset, the power of two in each digit and the power of ten in the final number. When you fully understand these concepts, you will be able to modify the program to suit your needs. Most BCD instruments, however, can use the program directly.

Also take the time to see how this program corresponds to the digital lines coming in from an instrument. That is, the first digit is the least significant bit of the least significant digit. The lines increase in significance in an orderly fashion until they end with the most significant bit of the most significant digit. Thus, by building your cable in a well organized fashion, you can make your programming easier.

COUNTING IN HEXADECIMAL

In all of the above discussions, each digit in a number was expressed as a group of four bits per digit. Electrically speaking, each group of four sequential wires coming into the digital I/O interface coded for one digit.

These digits were then converted into numbers using a system known as "positional weighting". The value of the final number was determined by weighting each of the digits in the number based on their positions within the number. In the decimal system, each subsequent digit is *ten times* larger than the previous digit. That is, they are base 10. The ten digits are 0, 1, 2, 3, 4, 5, 6, 7, 8, and 9. For example, a final number of 1859 can be formed with the following equation:

NUM = (9 * 10^0) + (5 * 10^1) + (8 * 10^2) + (1 * 10^3)

Because the base is 10, the following equation would normally be used to convert individual digits into a decimal number:

NUM = NUM + (DIGIT*10^INC1)

However, the loop in the CONV_BCD macro of Figure 10-2 increments by 8. Therefore, the following was used in the MK_NUM macro:

{LET NUM,NUM+DIGIT*10^(INC1/8)}

After the number was assembled from the digits, it was scaled with a {LET NUM,NUM/1000} command. This scaling was necessary because the Stasar III is unable to signal a decimal point in its BCD output and the correct format is #.###. The final number in this example, therefore, would have been 1.859.

Other instruments use other numbering systems. By far, the most common of these systems is hexadecimal. Hexadecimal (HEX, for short) is a numbering system that uses a base of 16. In base 10, the digits are zero through nine. Because each digit in HEX needs to be represented by a single unique symbol, alphanumeric values are used. The digits that represent the 16 hexadecimal values are 0, 1, 2, 3, 4, 5, 6, 7, 8, 9, A, B, C, D, E, and F.

The system of four bits (wires) per digit still applies. These bits can be readily converted to hexadecimal. For example, "0000" binary is a "0" HEX, while "1111" binary is an "F" HEX. The following table shows the relationship between binary, hexadecimal, and decimal:

Binary	Hexadecimal	Decimal
0000	0	0
0001	1	1
0010	2	2
0011	3	3
0100	4	4
0101	5	5
0110	6	6
0111	7	7
1000	8	8
1001	9	9
1010	A	10
1011	B	11
1100	C	12
1101	D	13
1110	E	14
1111	F	15

The MK_DIGIT macro shown in Figure 10–2 requires no changes to assemble hexadecimal digits from binary.

The hexadecimal numbering system is like the decimal and binary systems because it too is proportionally weighted. However, each subsequent digit in a final number is a power of 16 greater than the previous number. Therefore, you will need to use an equation different from NUM = NUM + (DIGIT*10^INC1), which converts decimal digits into decimal numbers. That is, you need to replace the equation in Figure 10–2 with one that converts the hexadecimal digits into decimal numbers. The decimal 1859 example would have been expressed as 0743 if your instrument reported its results in HEX. If the number 0743 HEX was indicated by the wires, the following equation would have been appropriate to convert the HEX number to decimal:

NUM = (3 * 16^0) + (4 * 16^1) + (7 * 16^2) + (0 * 16^3)
NUM = (3 * 1) + (4 * 16) + (7 * 256) + (0 * 4096)
NUM = 1859 (decimal)

The following equation can be used to convert a set of HEX digits to a decimal number:

NUM = NUM + (DIGIT*16^INC1)

Taking into account the fact that the loop in CONV_BCD of Figure 10-2 increments by 8, the following replacement equation should be used in MK_NUM if your instrument reports results in hexadecimal:

{LET NUM,NUM+DIGIT*16^INC1/8)}

With four hexadecimal digits you can have HEX numbers from 0000 to FFFF. However, you will never see these numbers in your spreadsheet. The equation in MK_NUM immediately converts the digits to decimal numbers and you will see numbers only from 0 to 65535. These numbers can be scaled to the format of your particular instrument with a {LET} command like the one used for the Stasar III at the end of the CONV_BCD macro.

WHAT'S NEXT?

The next chapter will give you a review of what you have learned in previous chapters and will give you some pointers on programming, building templates, setting up instruments, troubleshooting problems, etc.

Summary

This chapter is a summary of all the concepts you have learned in previous chapters. The concepts were assembled into a single chapter so that you could use the information as an outline whenever you wanted to interface a new instrument to Lotus®. The first section lists all of the integration tasks that you learned. The middle sections list some tips that you can use when designing your templates and macro programs. The final sections give you some tips on how to trouble-shoot communications and programming problems.

THIRTY EASY STEPS TO SUCCESSFUL INTEGRATION

This section outlines each step that you need to perform to get an instrument and computer to "talk". You have seen these steps in the many examples of the previous chapters.

1. Assemble the instrument and test it in manual mode.
2. Consult the User's Manual or the manufacturer of the instrument for the wiring configuration of the cable to use between the instrument and computer. If neither source helps, consult Appendix A at the end of this book.
3. Purchase or build the cable.
4. Connect the cable to the computer and the instrument. Secure the cable with screws.

5. Consult your instrument's User's Manual to determine how to set the Baud rate, number of character bits, parity check, and number of stop bits.
6. Set up the communications parameters in your instrument. Start with the slowest Baud rate and increase it after the program is running.
7. If your instrument is modular, give each module a unique address.
8. Make sure you turn your instrument off and then on before using it to reset the address and/or communications parameters to their new values.
9. Add a line to your autoexec.bat file to specify the path to the subdirectory containing the DOS Mode.com file. Alternatively, copy the MODE.com file into the subdirectory containing the Lotus 1-2-3® or Symphony® files.
10. If you are using Lotus 1-2-3, add a line to your autoexec.bat file to set the Baud rate to the appropriate speed. Reboot the computer.
11. If you are using Lotus Measure™, copy the files of the distribution diskettes into the subdirectory containing the Lotus 1-2-3 or Symphony files. If you are using Lotus 1-2-3, use the NEWLIB.exe program to add the appropriate driver to your driver set.
12. Spend some time designing the template and planning the macro program.
13. Change to the subdirectory containing your spreadsheet. Start the spreadsheet.
14. Enter the template into the spreadsheet.
15. "Activate" the template by creating named ranges and cells; format appropriate cells; set column widths; add equations; specify print ranges; etc.
16. Test the template.
17. Create and "activate" the auto-executing initialization macros. If you are using Symphony, specify the name of the auto-executing macro. (Issue the {SERVICES} Settings Auto-execute Set command.)
18. If you are using Lotus Measure, set up the communications parameters for the entire system. Use the Name option to save the parameters in a file. If you are using the DOS Device File Name method and Symphony, actuate the auto-executing macro to attach DOS.app and invoke it to issue a MODE command to set the communications parameters.
19. Enter a simple test program. This program should only try to establish minimal communications with your instrument. If your instrument allows you to write a string to its display or has a command that returns the instrument's software version number, start at that point.
20. "Activate" the test program by issuing the / Range Name Label Right command for each column of macro or scratch-pad names. Issue the / Range Name Create command for any named ranges.
21. Evaluate the test program. Modify it until it works correctly.
22. Expand the test program to try to retrieve some data. Once data can be retrieved, expand the program in small steps until you are able to get actual experimental data into the template. Test each program modification. This

step should culminate with a fully functional communications program that has optimal time coordination between the personal computer and the instrument.

23. Add the amenities and menus to the macro program. Be sure to use the name '\0 for the auto-executing macro if you have Lotus 1-2-3. Be sure that you include a DOS MODE command to set the communications parameters in a Symphony auto-executing macro. Also be sure to include comments.
24. "Activate" (or "re-activate") the macro program by issuing the / Range Name Label Right command for each column of macro or scratch-pad names. Issue the / Range Name Create command for any named ranges.
25. If using Symphony, create any "windows" that are needed.
26. Create a table of range names in an open portion of the spreadsheet. Review the table to ensure that the range names and their ranges are those expected. If not, make corrections.
27. Save the file under a descriptive filename.
28. Evaluate the program to ensure that it performs the way that you designed it to. (Symphony: make sure that you retrieve the file so that AUTOEXEC executes and sets the Baud rate.)
29. After the program is running correctly, increase the communications speeds (Baud rates) of both the instrument and the computer. Make sure that you turn the instrument off and then on again to reset the parameters. Make sure that you reset the communications parameters in the computer with the DOS MODE command. Re-test the program after each increase.
30. Save the spreadsheet (without data) under several names. Save back up copies on floppy disks. Also, print out the program for hard copy documentation. Save all of the items in several different, safe places.

PROGRAMMING TIPS

Trouble-free and efficient macro programs and templates require conscientious designs. Some of the tips listed in this section have been mentioned in previous chapters; others are new. Bringing these tips together in one section gives you a comprehensive checklist of topics to consider as you create your templates and macro programs.

Template Design Tips

Plan the template. Plan your worksheet application before entering your template. Remember, the template not only serves as a report, but it must also work in harmony with the macro that runs it. Plan ahead. The time that you spend planning your template will be more than offset by the time you will save getting it to run correctly. If you do not plan your worksheet, you will probably be required to make changes such as: inserting rows, deleting columns, moving data to different cells, etc. Re-designing templates and rewriting macro programs take a great amount of time and lead to errors.

Lay out the spreadsheet. When you design a template, determine the locations for all data. Also determine the locations for the macro programs. You do not want data overwriting program cells because the program will be destroyed. You will also want to take the time to organize the subsections of a template so that a macro can efficiently place data into them. For example, you may want an initialization section, a standardization section, a regression section, and a section for samples. To say the least, a well planned template leads to a more efficient macro program.

Allow for expansion. As projects progress, the amount of data that is collected in each experiment usually increases as new ideas are generated. This scenario is typical of most laboratories. You can plan your template for expansion from the start by reserving room on the spreadsheet for the extra rows and columns that will be required to hold the extra data.

Create "sectional" reports. A little planning will often permit you to design a template that fulfills multiple purposes. For example, you may be able to get day-to-day performance data for quality control purposes from the same template that holds your experimental data. Also, centralized assay-labs that perform analyses of samples from several other labs may need to issue individualized reports that contain only the data from a submitting lab. By laying out a template with foresight, you can use the /Worksheet Column Hide command (Lotus 1-2-3) or the /Width Hide command (Symphony) to hide certain data and print just the data needed.

Bigger is not necessarily better. Placing too many different types of experiments in one spreadsheet can make a spreadsheet too cumbersome and confusing. One way to lessen the confusion problem in Symphony is to create menus that switch between windows that display each section of a template.

Another problem arises if you have many @functions and formulae that tie big sections of a spreadsheet together. If your template is in this category, errors may become a serious problem unless you thoroughly test the template before you use it. That is, changing data in one section may have a detrimental effect on other sections.

Use easily understood labels. The labels that you use in a spreadsheet should be as descriptive and clear to understand as possible. Use as little jargon as you can. Bear in mind that some analysts that will use your spreadsheet may have minimal training and what may make sense to you, may not make sense to them. Using non-descriptive terms can confuse any user and lead to errors. This wastes time and can lead to misinterpretation of an experiment.

Minimize formulae. The more formulae that cells contain, the slower a spreadsheet recalculates. If you have several hundred formulae, every time you add a cell it could take what seems forever to update the spreadsheet. If you need to place data summaries in the spreadsheet, use a macro with {PUT} or {LET} commands. If you have formulae in a spreadsheet to remove, issue the / Range Values command. This command converts the formulae to their number equivalents and places the numbers into the cells.

Create and erase cells for maximum speed. The slowest process during data collection is the process of creating new cells in a template. Once created, the process of filling the cells with data is relatively quick. Likewise, overwriting old data in cells with new data is a relatively slow process. The fastest way to place data in cells is to have existing, empty cells. To implement this speed enhancement, force Lotus to create the cells at the same time that you make up your original template.

At creation, you fill the cells with "dummy" data and then erase the data. Erasing existing cells clears previous data, but leaves the cell's memory space and attributes intact. This tactic is faster than creating and filling new cells and is also faster than replacing existing data in cells. Therefore, if you need a fast data acquisition rate for an instrument, use the /Data Fill (Lotus 1-2-3) or the /Range Fill (Symphony) command to fill your template and scratchpad with numbers. Then, erase the numbers and use the empty cells to receive data.

Do not however become overzealous with this speed enhancement by creating all of the cells in a spreadsheet. Even though cells do not contain data, they still use up memory. Therefore, you are likely to run out of memory and you may actually slow the performance of your spreadsheet if you create too many cells. Created cells also mean that you will be storing bigger files on your disk. Therefore, only create cells that you plan to use in your template and scratch-pad.

Plan for memory usage. If you are going to be dealing with a *very large* amount of data, you may run out of computer memory. You may want to consider adding expanded memory (EMS) to your computer. You can use up to four megabytes of expanded memory from Lotus spreadsheets. This capacity translates to an increase from about 34000 cells to about 84000 cells if you are using Lotus 1-2-3 and an increase from about 23000 cells to about 68000 cells if you are using Symphony.

EMS memory is not nearly as fast as conventional memory. The use of EMS memory can profoundly affect the speed of your program and can thereby affect the communications between your instrument and spreadsheet. This effect is especially evident in Symphony, Release 2 because your programs execute approximately 35% slower when using expanded memory. This difference in speed is sometimes significant enough to have a considerable influence on the types of functions that you need to perform to successfully establish communications with your instruments.

Unless you anticipate collecting more than 23000 data points in your experiments, disabling the expanded memory is recommended. Your spreadsheet will run about 25% to 35% faster for nearly all operations. You can disable your expanded memory by removing the command lines that have been placed in your CONFIG.SYS file to specify the EMS device driver (i.e., the ".SYS" file) and then rebooting your computer.

If you are using Lotus with a co-processor accelerator board, such as the Orchid Technologies PCturbo 286e™, Lotus spreadsheets cannot access the EMS memory in accelerator mode. If you have added one of these boards to an old IBM PC/XT™ to try to improve its speed, you must choose between EMS memory and speed.

Program Design Tips

Use named cells and ranges. Use named ranges and cells whenever possible. This tip is especially important when using {PUT} and {LET} macro commands. Using named ranges will make your programs easier to read and follow. Having a name rather than a range or cell address becomes very important in large spreadsheets. Perhaps more importantly, if you change your template and have cell addresses in macro commands, you must go through your program and change each address. However, if you change the address specifications of a named range and have used just the range name in a {PUT} or {LET} command, you are not required to edit the command.

Use meaningful names for ranges and macros. The names that you use should be as meaningful as possible. Try to use names that relate to processes. Also try to use names that remind you of the macro's or the named range's purpose. Do not use names that look like cell addresses (e.g., A1); do not use any of the Lotus special keynames, keywords, or macro command names (e.g., PUT, QUIT); do not use spaces and/or symbols within a name; and do not exceed 15 characters. Sometimes you can break these rules, but most of the time you cannot.

Exercise caution when using range names with Measure. Specifying a range on a Measure Settings or Configuration sheet is different from specifying a range in Lotus 1-2-3 or Symphony. When you move a named cell or range in Lotus 1-2-3 or Symphony, the spreadsheet automatically makes the appropriate corrections. This update is usually not provided with Measure. Measure keeps a range designation the same regardless of what happens in the spreadsheet. Therefore, if you add or delete rows in your program or move your buffer, you MUST manually change the range designation in Measure's Settings sheets or your data will be placed into the wrong cell(s).

Comment! Thoroughly comment your programs. Comments make it easier for someone (including yourself) to read and understand your program. Commenting is such an essential part of every program that you should try to write a line of comments for every line of code that you write. Trying to remember what you were attempting to accomplish in a certain section of a program after a few weeks have passed can be difficult. If you have fully documented a program at the time that you wrote it, the program will be at its freshest in your mind and will make the greatest sense. It will also be more "readable" and easier to correct or modify the program, if necessary. You can use any commenting format that you want as long as the comments are not in a cell within the program. When Lotus executes a macro, it reads straight down a column until it detects an empty cell. If your comments are not within this "line of fire", they will be ignored and will not interfere with the program's execution.

Provide a program synopsis. Another important piece of documentation is a brief synopsis of the function of a program at the beginning of the program. Because macro execution begins at the cell that corresponds to the macro's name, any text above the cell is ignored. Your synopsis should contain the objectives of the program, instrument(s) controlled, type of testing performed, samples tested, special cable requirements, etc.

Add a development tracking table. Keeping track of a program's revision history can prove helpful when a problem arises. It is invaluable for debugging purposes. If you have a table that tracks the history of program/template changes that were made to a spreadsheet, you can easily keep track of additional pertinent changes. This table should be at the end of a macro program and should include the name of the person who developed the program, the date it was developed, the name of the individual(s) who modified it for each revision, the changes made, and the revision date(s).

Prepare a named range table. Prepare a named range table in your spreadsheet to ensure that you have created all of the ranges you need for your program. More importantly, carefully review the cell specifications for each range name to make certain they are correct. Incorrect cell specifications are a common source of program error.

Structure a program for readability and testing ease. Testing a macro program can require as much time as programming it. If you invest the time to create a readable and well organized program from the start, testing time will be minimized.

Use tables. Using tables from which your program selects instrument string commands presents many benefits. Tables make your programs more organized, more "readable", easier to follow, and easier to change, if necessary. These benefits present a powerful incentive to use tables routinely.

Program in sections. Macro programs should be constructed in sections, with each section being a macro subroutine. Do not try to make one large program. Large programs tend to be hard to follow, limited in their flexibility, hard to test, and hard to decipher when pinpointing the causes of problems. Moreover, if your program is modular you can usually call the modules from several places in the program, thereby decreasing the number of programming lines.

Test as you go. It is better to test macro subroutines as you create them rather than waiting until an entire program is complete. If you wait until the end, the many subroutines and program lines will make it difficult to isolate the errors. However, if you test as you build, you can verify that each subroutine works correctly before adding another layer of complexity and/or performing the final test that checks the entire program.

Test in sections. If your program has several macro subroutines and/or your template has several sections, it is preferable to test each subroutine and section independently. Then, when all are confirmed to work according to design, check to see that the parts interrelate correctly. It is easy to overlook discrepancies in a subroutine or section if you attempt to check too much all at once. It is also possible that compensating errors will escape detection when an entire worksheet and macro program is checked at once.

Thoroughly test the final program and template. A template and macro program that have not been thoroughly tested under actual experimental conditions should not be used to report results. Important decisions may be based on a report that looks formal, but contains erroneous results. You can compare the results of a macro program with your manual method, the instrument's display, etc. You should compare the results of the calculations in your template with the results that you obtain from a calculator, etc.

Analysis and documentation—the easy way. The Cambridge Spreadsheet Analyst™ (described at the end of Chapter 5) is a program that helps you document, test, validate, analyze, and troubleshoot your Lotus 1-2-3 and Symphony spreadsheet templates and macro programs. This relatively inexpensive program analyzes your spreadsheets for 25 different types of error conditions. The program makes it easy for you to avoid costly spreadsheet mistakes.

In addition to the above benefits, The Cambridge Spreadsheet Analyst can analyze your macro programs and graphically map a tree of the nesting, operation, and interaction of all macros so that you can better understand the commands that influence your spreadsheets.

These reports make nice additions to your documentation package.

Use auto-executing macros to initialize your program. It is important that you always start an experimental session with the communications parameters set to their correct values, all add-in programs attached, etc. Because you can never be sure of the configuration of a system when a spreadsheet is retrieved, use auto-executing macros to initialize the spreadsheet, your instrument, and your macro program. This safeguard ensures that all starting values within the system are set to the prescribed conditions needed by the system for operation.

Date-stamp the experiment. Date-stamp an experiment from the macro that collects the data. Use a {LET} command and the @NOW function. This measure ensures that an accurate record exists of when an experiment was actually performed. If several experiments are being performed on the same day, you may also want to time-stamp the experiment.

Use menus. Use custom menus whenever you can. These allow your users to interact with a program as easily as possible. Thus, your program can be used by personnel with a limited amount of specialized knowledge.

Minimize screen "flicker". "Flickering" is caused by the panel and spreadsheet updating during program execution. This occurrence can be quite annoying to a user and also slows execution of a program. Therefore, use a {WINDOWSOFF} {PANELOFF} combination to freeze the display.

Provide the user with status reports. It is important to inform your users of the status of a program as it executes. This feedback is especially important if your program takes a relatively long period of time to complete its tasks. Under these circumstances it may appear to the user that something has gone wrong. The most convenient way to provide a status report is with the {INDICATE string} {PANELON}{PANELOFF} set of commands.

Do not start with nested @functions and/or macro commands. It is better to start with individual simple formulae than lengthy formulae with many levels of nesting. Simplicity allows you to check out each component separately. Once you have individual formulae working correctly, you can combine them into a final (complex) formula.

Use {CALC} to prevent user confusion. The {CALC} command forces a recalculation of all formulae in a spreadsheet and updates (re-displays) the most current data. When a macro program executes, it does not necessarily recalculate the formulae in a template. Placing a {CALC} at the end of a macro avoids user confusion. More importantly, if a user forgets to recalculate a spreadsheet before printing it, the results may not be correct, because they may not reflect any new values that have been added to the template. To prevent the possibility of catastrophe, get into the habit of placing a {CALC} command at the end of all macro programs and, more importantly, before any automated printing is performed by your program.

Conserve disk space. Storing a program with data wastes disk memory storage space. For this reason, you should File eXtract just the template into a storage file. You can then review the new spreadsheet directly, combine it with other data spreadsheets, or even retrieve it back into the original spreadsheet containing the macro program for further analysis.

Spread your success around. Once you have a macro subroutine working correctly, you may want to use the File eXtract function to place a copy of it in another spreadsheet file. If you need the macro in the future, you can use the File Combine function to place it into another program, thus substantially decreasing your programming and testing time.

COMMUNICATIONS TIPS

Start with an instrument that you are sure is working correctly. Do not try to interface an instrument unless you are absolutely certain that it is working correctly. That is, check the instrument thoroughly in manual mode (if possible) before trying to communicate with it from your macro program. If your instrument is not working in manual mode, adding communications will only compound problems and make it very difficult to pinpoint the cause of a problem.

Plan ahead for your cable. Decide on the cable to use well in advance of when you want to start writing your program because you may need to purchase or make a cable.

Screw in the connectors. One of the most common communications problems is also the easiest to avoid. Connectors can easily become disconnected from personal computers and instruments. Adopt the policy that all connectors be secured with screws. This policy will save you hours of frantic work trying to pinpoint the cause of a seemingly "dead" system.

Start simple; then expand. Start with the simplest program that you can. If your instrument allows you to write a string to its display or has a command that returns the instrument's software version number, start at that point. These functions are usually the easiest ways to make sure that your personal computer and the instrument are talking properly. Next, try to get some data . . . any data . . . from your instrument into a buffer of your spreadsheet. Once you are able to prompt the instrument and/or get data back from it reliably, you can expand the control, extend the acquisition to retrieve data that is more useful, and add the amenities. Increase a program in small steps, writing only one functional unit at a time, testing it, and then working on the next functional unit. Remember, the larger the program, the harder it will be to find the cause of a problem if one occurs. Also, increasing the size of a program can sometimes affect its timing. By increasing a program's size in steps, you can monitor these timing changes, see how they affect performance and then optimize the program's speed according to need.

Start with a slow Baud rate; then increase it. Start your interfacing process with the slowest possible Baud rate. After you have your program running successfully, you can increase the Baud rate in a stepwise fashion until you no longer get reliable communications. Then, return to the last Baud rate that gave acceptable performance.

Be careful about how you change your instrument's settings. If you change the Baud rate, parity, etc., of your instrument, it is crucially important that you turn your instrument off momentarily and then turn it back on. Because most instruments check the values of their communications parameters only at startup, old values will be used unless an instrument is restarted again. Resetting the instrument is crucial to ensure that it is using new communications parameters.

Watch your computer's communications parameters. If you change a communications parameter in your instrument, make certain that you also change the parameter in the personal computer. Otherwise, you may be misled into concluding that the new parameter is not working correctly. It is very easy to forget to do this when you are increasing Baud rates to optimize performance; so be extra careful during this phase of program development.

Get the system's timing down. Timing is of utmost importance when you are dealing with an instrument that collects data. Optimize the {WAIT}s and any {IF} test conditions in *your* program for *your* instrument. Test the system with a series of samplings with both high and low values. Then, "fine tune" the numbers in the {WAIT} and {IF} tests to your precise application.

WORKSHEET PROTECTION: DESIGN TIPS

Once you have created your templates and macro programs and have begun generating data, your spreadsheets will represent a considerable time investment. You will want to protect this investment. The following measures are some suggestions to try that prevent devastating errors from ruining your work. You can also prevent unauthorized viewing of confidential information.

Restrict file access. You can assign passwords to worksheet files so that a password entry is required before a worksheet can be retrieved. That is, a user who wants to retrieve a spreadsheet will need to enter the correct password before Lotus will allow access to the spreadsheet. This protective measure permits you to restrict file use to only those people who are authorized to work with the file. A password will restrict access to a file, but once the file is retrieved the user will be able to make changes to the entire file.

To add a password to a file in Lotus 1-2-3, store the file with the / File Save filename P command (where filename is the name of the file that you want to save).

After the filename, type one or more spaces followed by a lower-case or upper-case P. In Lotus Symphony, use the {SERVICES} File Save filename P command. Lotus will ask you for the password before it saves the file.

Cell protection. You can ensure that less experienced users cannot accidentally ruin a worksheet by placing an entry in, erasing, or deleting the wrong cell(s). You can also ensure that a program will not be overwritten. Lotus 1-2-3 and Symphony both have protection features that allow you to choose specific cells that you want to protect. Once a cell is protected, it cannot be changed by a user or macro. Lotus establishes a protection default status of OFF for all cells in a new worksheet. This status allows changes to any cell. To use cell protection, you must first enable protection for the entire spreadsheet and then turn it off for the cells in which you will allow changes. In Lotus 1-2-3, issue the / Worksheet Global Protection Enable command. In Symphony, issue the {SERVICES} Global-protection Yes command.

These commands turn protection on for all cells. Next, issue the / Range Unprotect command in Lotus 1-2-3 or the / Range Protect Allow-changes command in Symphony and specify the range of cells in which you will allow changes. At a minimum, you should unprotect the data input sections of your template and the scratch-pad section of your macro program. As a general rule, you can leave cells that have formulae in them protected and they will be updated as usual. You MUST unprotect all cells with which macro commands such as {READLN}, {PUT}, and {LET} interact.

Hide confidential information and the macro program. If your template contains confidential information or if you do not want users to see your program, hide columns in the spreadsheet. In Lotus 1-2-3, use the /Worksheet Column Hide command. In Symphony, the command is /Width Hide. Symphony also has another means of hiding areas of a spreadsheet. It is the /Format Other Hidden command and it hides ranges that you specify. With this command, you are not limited to columns only.

Backup. Store your spreadsheet often during development. Then, if you have a power problem or a computer lockup, you will not lose the work that you have already completed. Plan for future disasters, too. If you have a large spreadsheet and/or a complicated macro program, store several copies without data on separate floppy disks and store the disks in separate locations. Also, print out the template and macro program and store the hard copies in safe places.

COMMUNICATIONS TROUBLESHOOTING GUIDE

Whenever you have a problem with communications, use the following tips to help pinpoint the cause of the problem.

Check for loose connectors. Loose connections are the source of most common communications problems and they are also the easiest to correct. If you have not secured your connectors with screws, do so and re-test the system to determine if the problem has been eliminated. If you have built your own cable, you may also want to make sure that the crimps on the wires are making sufficient contact. Perform a continuity check with an ohmmeter. The resistance between the pin at one end of the cable and its corresponding pin at the other end should be at some very low value. Also, look at the pins to ensure that they were not bent when you plugged them in.

Check the communications parameters. The next area to check is the Baud rate, parity, data bits, and stop bits of both your instrument and your personal computer. They must be identical. If they do not match, change them. Turn the instrument off and then on again before you use it. This task is required to reset the communications parameters to their new values.

Check addresses. If your instrument is modular, verify that each module is set to a unique address and that the address in your program is the correct one for the module from which you want to get data.

System errors. Many situations can cause Lotus to issue a System Error message. The most common cause is an incorrect match in the communications parameters of the instrument and the personal computer.

The next most common cause is speed. If you are using a Baud rate that is too fast or if your program executes too slowly to either start the reading process or clear the DOS buffer, you will get a System Error. The remedy is to slow down the Baud rate, freeze the display with {WINDOWSOFF}{PANELOFF}, remove @functions in cells and/or remove any unnecessary programming commands above {READLN} commands. This last strategy is especially important if you have back-to-back {READLN} commands or if you have a {READLN} command following a {WRITELN} command that prompts for data. If an instrument transmits a large amount of data at a fast Baud rate and you cannot get results from the instrument with the Device File Name method, try Lotus Measure, because it has greater data flow control.

The third most common cause of System Errors occurs immediately after an {OPEN} command is issued. This error usually exhibits itself every other time that you run a program. This error is harmless and you can usually remedy the problem by including {ONERROR} commands in your program. (See Chapter 8 for details.)

System errors caused by expanded memory. EMS memory is not nearly as fast as conventional memory. Thus, the use of EMS memory can profoundly affect the speed of a program and can, therefore, affect the communications between an instrument and spreadsheet. This affect is especially evident in Symphony, Release 2 because programs execute approximately 35% slower when using expanded memory. Unless you anticipate collecting more than 23000 data points in your experiments, disabling the expanded memory is recommended. Your spreadsheet will run about 25% to 35% faster for nearly all operations. You can disable your expanded memory by removing the command lines that were placed in your CONFIG.SYS file to specify the EMS device driver (i.e., the ".SYS" file) and then rebooting your computer. If you have developed an application and must add expanded memory, you may need to slow the Baud rate to compensate for the decline in your program's performance.

Try other cable configurations. Appendix A describes the use of the AutoCabler to determine cable configurations. This device can be invaluable if you are not getting ANY data at all. If such is the case, follow the method outlined in Appendix A to verify the cable. You will also want to try alternative switch settings to see if they remedy the problem. This entire process should take less than five minutes.

Use a terminal emulator program. Several terminal emulator programs are available from software supply houses. Two of these programs are PC/InterComm and Crosstalk. These programs are primarily designed to control modems. However, you can also use them to test the communications between your computer and instrument. That is, you can quickly switch between different Baud rates, parity, etc., to see how each parameter affects the data being exchanged with your instrument. In this way, you can send commands to the instrument, receive data back, and see exactly what is transmitted between the two. This tool will help you verify that an instrument is performing correctly and that the instrument's User's Manual has given you the correct DIP switch settings. These emulator programs will *not* tell you if your cable is working correctly because they control handshaking pins in a manner different to Lotus. Follow the instructions given in the emulator program's User's Manual when you use the program. When you configure the emulator program, use it in "half duplex" mode. In that mode, whatever you enter at the computer keyboard is displayed on the terminal, as well as being sent to the instrument. If you see each character that you type twice, switch to "full duplex" mode.

PROGRAM TROUBLESHOOTING TOOLS

Whenever you are confronted with a program problem, the first thing you need to do is to gather as much information about the problem as you can. Try to characterize when it happens, where in the program it happens, and the consequences of the problem. The more information that you have, the better your chances of correcting the problem.

A number of programming tools are available to increase the amount of information available to you. The following list contains some of the more useful tools.

Turn the windows and panel back on. The first action to take when you have a problem is place a {WINDOWSON}{PANELON} combination at the beginning of your program and temporarily remove all {WINDOWSOFF}{PANELOFF} commands. Executing these procedures will allow you to see exactly what is happening as your program runs. Also, place {CALC} commands throughout the problematical portion of the program. This measure ensures that the @functions recalculate and that the data you are viewing accurately reflect what is actually happening in the program.

Slow the program down with {WAIT} commands. Most macro programs execute very quickly. You can slow the execution of a program by placing {WAIT} commands throughout the program. You can then view the panel and the template or scratch-pad to try to get a clue of the problem. Again, use {CALC} commands to make sure that you are getting frequent updates of formulae. Alternatively, use the Lotus STEP mode to view the updates. (See below.) Be careful about placing {WAIT} commands before {READLN} commands, because {READLN} commands are very sensitive to timing. Thus, if you place a {WAIT} command in the wrong location, you may actually produce a potentially misleading System Error.

Single step through your program. Lotus 1-2-3 and Symphony both provide a STEP mode for macro testing. In Lotus 1-2-3, you can activate the STEP mode by pressing [ALT] F2. In Symphony, press [ALT] F7. Start the macro in the usual way (by pressing [ALT] and the macro's character name; or, if you have Symphony use F7 and the name of the macro). Thereafter, each time you press the SPACE BAR, another macro command is executed. The STEP method allows you to trace each step of a macro's execution and helps you to locate the instruction causing the problem. When you are finished press [ALT] F2 (Lotus 1-2-3) or [ALT] F7 (Symphony) again to return to normal operation.

Track the program. You can track the execution of your program at close to full speed by placing {LET} commands throughout the program. That is, you alternate your program lines with {LET} commands; each {LET} command places a unique number into a status cell. By viewing this status cell, you can determine your location in the program. This technique is useful if your program is locking up (most notably during the communications phase of your program, where quick timing is especially critical). You can use the information received from this technique to pinpoint exactly where a program has stopped executing. Backtrack from the number in the cell to the last {LET} command that executed when you were forced to press [CONTROL][BREAK] to abort the locked up macro.

Use {INDICATE}. Alternatively, you can use the status indicator box. By placing sequential numbers in {INDICATE} commands placed throughout a program, you can trace the execution of the program. If you use this technique, make sure that a {PANELOFF} command does not precede the {INDICATE}. The {INDICATE} command requires the panel to be "on" for it to update.

Update the named range table. One of the most frequent causes of program problems is ranges with incorrect cell specifications. A number of situations cause range specifications to change without your awareness. Therefore, any time you experience a problem, it is a good policy to update the Range Name Table and review it to ensure that it is correct.

WHAT'S NEXT?

Appendix A will show you a very simple method that you can use to determine the cable wiring configuration that your instrument and personal computer need to communicate properly. The appendix also includes instructions for building the cable.

You are now ready to use the power of a spreadsheet to analyze the data that you have acquired from your instrument. Lotus spreadsheets provide a wonderful tool for integration, differentiation, linear and curvilinear regression, graphics, etc.

Determining and Building RS-232 Cables

Because learning how to determine the configuration of a cable for an instrument and personal computer using conventional techniques is an arduous process, this appendix instead focuses on a device called the ''Auto-Cabler''. It is available from inmac Corporation (2465 Augustine Drive, Santa Clara, CA 95054-9977). (See Figure A–1.) It is also available under the names ''Quick Cabler-20'' from Black Box Corporation (P.O. Box 12800, Pittsburgh, PA 15241) and ''Easy Cabler''™ from E.I.L. Instruments, Inc. (10 Loveton Circle, Sparks, MD 21152-9989). A more enhanced version, the ''Easy CABLECHANGER''™, is available from Beckman Industrial (Instrumentation Products Division, 3883 Ruffin Road, San Diego, CA 92123-1898). (See Figure A–2).

The Auto-Cabler has been an accurate predictor for the cable configuration of every instrument that I have interfaced with several brands of personal computers and both Lotus spreadsheets. Furthermore, the Auto-Cabler can be left in line until a permanent cable can be made. With the device you can begin programming immediately. You can also test the system with the exact cable configuration that you will be using when you get a cable. Usually, several different cable configurations can be used for a particular instrument-computer combination. Using the Auto-Cabler in line will allow you to try all configurations under actual experimental conditions to find the configuration with the best performance.

To use the Auto-Cabler, read the remainder of this appendix. To learn more about RS-232, refer to the references at the end of this appendix.

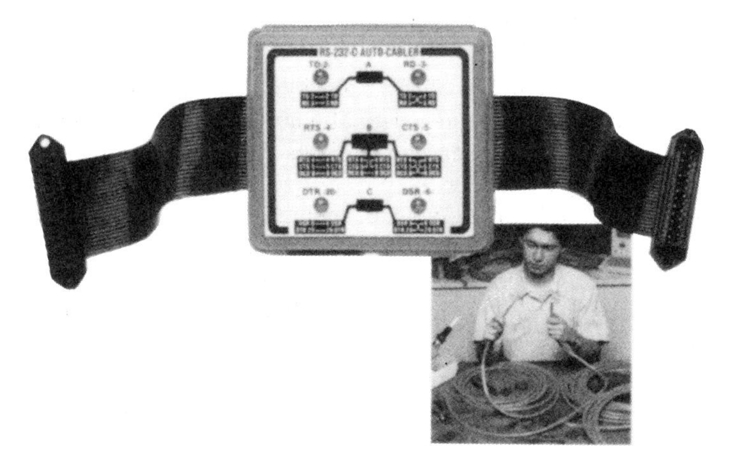

Figure A-1 (Courtesy of INMAC Corp.)

Figure A-2 (Courtesy of Beckman Industrial, Instrumentation Products Division)

BRIEF DESCRIPTION OF THE AUTO-CABLER

The Auto-Cabler consists of two flat ribbon cables and a controller box. One of the cables is plugged into your personal computer and the other is plugged into your instrument. The controller box contains three switches on the front panel. These switches swap the lines for the pins normally used in RS-232 communications (i.e., pins 2, 3, 4, 5, 6, 8, and 20). Thus, by moving these three switches, you can achieve and test all RS-232 configurations quickly and conveniently.

The Auto-Cabler controller box also has a pair of "tri-state" light emitting diodes (LEDs) for each switch. These LEDs provide information about the connection. They can determine if there is no voltage (not illuminated), a positive voltage (red), or a negative voltage (green) present at each of the connector pins. With this information, you can optimize your cable for best performance.

RULES FOR USING THE AUTO-CABLER

Your objective in using the Auto-Cabler is to position the switches so that they light the maximum number of LED lamps when an instrument is in an idle state. If an LED displays red in one switch position and green in another, green is preferred.

When the maximum number of LED lamps are lit, your next objective is to transfer data between the instrument and the computer (and vice versa, if appropriate). This transfer will be indicated by momentary transitions from green to red (flashing) on pin 2's and/or pin 3's LEDs.

USING THE AUTO-CABLER

The following checklist will help you use the Auto-Cabler to determine the configuration for your cable.

1. Connect the Auto-Cabler to the RS-232 connectors of your instrument and your personal computer. It is not important which end of the Auto-Cabler is connected to which device. If you need to extend the length of the cable, make sure that you use a 25-pin (wire) straight-through cable.
2. Move all of the switches completely to the left. (Note that switch "B" has three possible positions.)
3. Turn on the personal computer.
4. Set the Baud rate, parity, number of bits per character, and stop bits in both the instrument and the personal computer. If necessary, turn the instrument off and then on again to reset the parameters to the new ones. Use 300 Baud for this test.
5. Observe the six LEDs on the Auto-Cabler. The objective is to have all six LEDs lit, and preferably green.

6. If less than six LEDs are illuminated, change the switch associated with the extinguished LED(s) to see if the lamp can be lit. The switches are set properly when (in order of priority):
   a. The maximum number of lamps are green.
   b. The lamps are lit (either green or red) rather than extinguished.
7. When the maximum number of LED lamps are lit, write a simple Lotus macro program using the techniques in Chapter 4. Keep the program simple. At this point, it does not matter if the data are garbled. If your instrument allows you to write a string to its display or has a command that returns the instrument's software version, start at that point. These commands are usually the easiest ways to test a cable. If these commands will not test your system effectively, you will need to prompt for data or try to control one of the instrument's motors. At any rate, try to make your program as simple as the following example:

```
\R         {OPEN ''COM1'',M}
           {FOR INC1,0,50,1,TST_LOOP}
           {CLOSE}

TST_LOOP {WRITELN prompt}
           {READLN buffer}

INC1
BUFFER
```

Note that "prompt" is the character string that your instrument requires before it will send out a string of data. Also, the 50 in the loop counter may need to be increased or decreased, depending on your instrument. The objective is to get a flow of data that transmits a signal long enough for you to finish the Auto-Cabler test.

If your instrument cannot be prompted, you will need the following program:

```
\R         {OPEN ''COM1'',M}
           {FOR INC1,0,50,1,TST_LOOP}
           {CLOSE}

TST_LOOP {READLN buffer}

INC1
BUFFER
```

If you are using Lotus Measure,™ configure the Measure RS-232 module according to the instructions in Chapter 9. Substitute the Measure {RSEND} and {RRECEIVE} macro commands for the {READLN} and {WRITELN} commands in the above programs.

Activate the program by issuing /Range Name Label Right command and highlighting the names in the left-hand column (i.e., the column containing the \R). Also type a number into the cell to the right of BUFFER and erase the number using the /RANGE Erase command (Lotus 1-2-3) or /Erase command (Symphony). This action will create the cell and ensure maximum speed performance.

You may need to modify these programs somewhat for your particular instrument. For example, you may need to send some initialization strings to the instrument before you can obtain data. You may also need to use a trap loop or {ONERROR} commands (see Chapter 8). However, try to keep the program as small and simple as possible.

8. Start your program by pressing [ALT] R and attempt to transfer data between the instrument and personal computer. Data transfer will be indicated by a momentary green to red color transition on the TD (LED number 2) and/or RD (LED number 3) lamps. This transition may be accompanied by a color transition on the RTS (LED number 4) and CTS (LED number 5) lamps. Under no circumstances should the data transmission attempt cause an illuminated LED to be extinguished. If a lamp is extinguished, change the position of the switch that corresponds to the LED. If this condition persists regardless of the switch position, it indicates a major wiring error in the RS-232 interface of the instrument or personal computer.
9. If you cannot achieve data transmission, move the switches until you get data transfer. A maximum of 12 combinations (data transfer attempts) should be all that is required to interface an instrument with a personal computer.
10. When the instrument and personal computer are transferring data correctly, the data and control signals will cause the TD (LED number 2) and/or RD (LED number 3) lamps to flash both red and green. (If your instrument is not prompted for data from the computer, only one of the LEDs will flash.)

You can leave the Auto-Cabler in place permanently, or you can make or purchase a custom cable. You may want to leave the Auto-Cabler in line until you have your macro program up and running and have confirmed that everything is working correctly.

To switch to a permanent cable is easy. Write down the drawings in the boxes on the Auto-Cabler that correspond to the switch settings. These drawings will serve as the specification for your cable. You can now obtain the cable you will use permanently. Two other pins (pins one and seven) are required for your cable. They are electrical grounds and are *ALWAYS* straight-through.

HOW TO GET A CABLE

You can obtain a cable from a number of sources. You can

- have your instrumentation department or a consultant make it
- buy a custom cable from your local computer shop or an outlet that specializes in cables (such as inmac®)

- build your own cable from scratch
- buy a straight-through cable and build an adapter.

MAKING AN ADAPTER

The easiest way to get a cable for your instrument is to purchase a cable with wires that go "straight-through" from one end to the other and a kit from which you can build an adapter to convert the wiring to the one that your instrument needs. (See Figure A-3.) Adapter kits are available at your local computer shop or from one of the outlets given at the beginning of this Appendix. These kits usually consist of the following:

- Two connectors (one male and one female, two females, or two males)
- A snap together, two-headed hood cover (a plastic case that holds both connectors)
- A number of two-inch wires with pre-crimped pins on both ends to match your choice of connector.
- Male and/or female mounting screws
- "Y" jumper wires with pre-crimped pins
- An insertion/extraction tool (sometimes optional)

Make sure that you get a cable and connectors that are complementary to each other and the "sex" of your instrument and personal computer. For example, if your instrument is male, then you will need an adapter with a female connector.

Once you have purchased all the components for your cable, use the insertion

Figure A-3 (Courtesy of INMAC Corp.)

tool to push the pins into the appropriate holes of the shell specified by your Auto-Cabler. To accomplish this procedure:

1. Choose a wire to insert. Make sure to use a female socket for a female connector and a male pin for a male connector.
2. Position the wire in the insertion tool. Use the end of the tool that is not crimped. It is usually copper-colored. (The silver-colored end with the crimped end is for extraction.) The wire will lay in the tube of the connector. Pull back slightly so that the widened portion of the pin or socket (the "ring") rests against the end of the insertion tool.
3. Push the pin or socket into the appropriate hole of the connector. Make certain that you have the correct hole. Connectors are usually marked on both sides. You will feel a slight snap when the pin is completely pushed in.
4. Remove the insertion tool and repeat the procedure for the remaining pins or sockets. Push all of the pins or sockets into the holes of the connector that have been specified by your Auto-Cabler. Remember, pins one and seven MUST be used and must be straight-through.
5. Repeat the procedure described in steps 1 through 4 for the other connector.
6. If you make a mistake, remove the pin with the crimped (silver) end of the insertion/extraction tool. Thread the wire of the appropriate pin through the crimped portion of the tool, slide the tool into the hole and push until you feel a snap. The connector contains an "O" ring that needs to be pushed aside. Next, rotate the insertion tool and withdraw the pin by pulling outward. You may have to repeat this process until the pin is extracted. After you have extracted a couple of pins, you will get a "feel" for how far to rotate the tool before you pull the pin out.

Next, assemble the hood. The hood is usually assembled from two plastic covers. These covers snap together.

Begin by positioning the two connectors into the detents in the two covers. Then snap the covers together. Place the attachment screws and washers into the holes of the connector. Female mounting screws correspond with female connectors and male mounting screws correspond with male connectors.

After assembling the adapter, complete one final and important task—clearly label the adapter with a drawing of its pin configuration. It can be frustrating trying to recall what the configuration of an unlabeled adapter is; or worse yet, using the adapter for the wrong instrument.

MAKING A CABLE

Making your own RS-232 cable is not difficult if you have the proper tools. The following list of tools can usually be purchased from an electronics store or from one of the suppliers noted at the beginning of this appendix:

Small Flat Blade Screwdriver
Phillips Head Screwdriver
Small Needle Nose Pliers
Side Cutter
Wire Cutter/Stripper
Crimping Tool
Pin Insertion/Extraction Tool

The following supplies will also be needed and can be purchased from a computer store or from one of the suppliers noted at the beginning of this appendix:

RS-232 Cable
Crimp Pins or Sockets (as appropriate)
2 Connector Hoods (plastic cases)
Mounting Screws
2 Connectors (male/female, as appropriate)

When you purchase the RS-232 cable and pins, buy enough wires and pins to cover the amount specified by the Auto-Cabler. Most (if not all) RS-232 cables require less than 25 wires. Typically, you will only need wires for pins 1,2,3,4,5,6,7,8, and 20. Ordering just enough wires will not only save you added expense, but will make it easier to fit the wires into the connectors. Once again, remember that pins 1 and 7 are needed in every cable that you make.

You can typically purchase components for connectors as kits or "connector packs". Make sure to get connectors that are the opposite "sex"of your instrument and personal computer. For example, if your instrument is male, your cable must have a female connector.

A cable can be assembled using one of two techniques. The first is by soldering and the second is called "pin insertion". The pin insertion method is easier and produces more consistent results for beginners. The pin insertion cable will also be easier to service if a problem arises.

Start by using wire cutters to carefully cut back the outer sheath on the cable. (See Figure A-4). Expose about 1 to 1.5 inches of the inner wires. Be careful not to cut the insulation on the inner wires. Next, trim each internal wire 1/8 to 3/16 inches from its end.

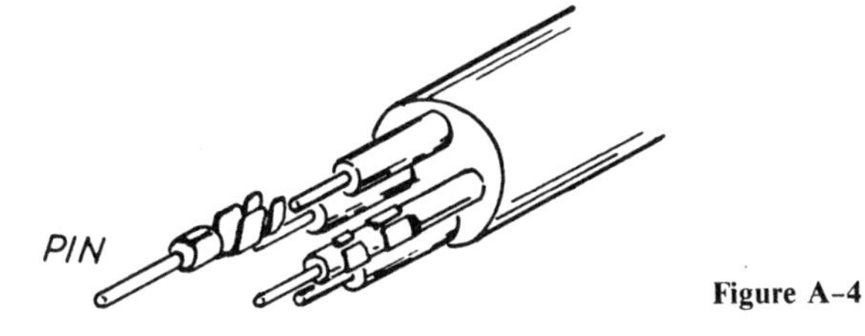

Figure A-4

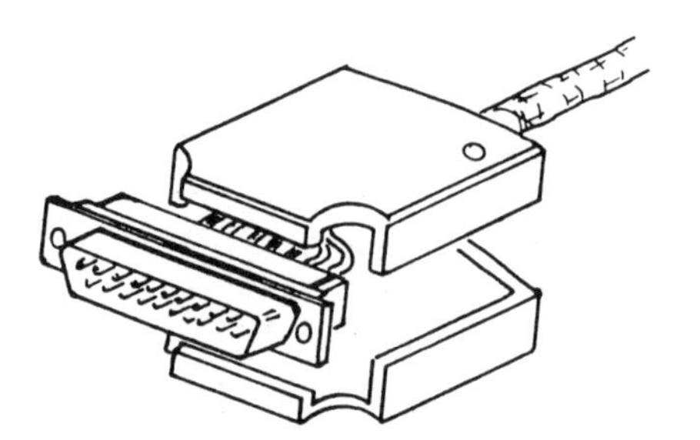

Figure A-5

Crimp an appropriate pin/socket onto each of the wires (i.e., a male or a female) using the crimping tool. If your Auto-Cabler specified more than one wire per socket, crimp all of the appropriate wires together into the pin.

Each pin has two crimping locations. Only the first crimp (the one nearest the pin) should touch the wire. The second crimping location should be on the insulation. This crimp provides strain relief.

Once all necessary crimps have been made, use the insertion tool to push the pins into the appropriate holes of the shell. That is, push the pins into the places specified by your Auto-Cabler. The procedure is the same as for making an adapter (see previous section).

Finally, assemble the hood. The hood is usually assembled from two plastic covers. (See Figure A-5.) These covers either snap together or are held together by screws. Begin by positioning the connector and cable between the two covers. Then snap or screw the covers together. If the hood has a strain relief screw, tighten it also. Then, place the attachment screws and washers into the holes of the connector. Female mounting screws correspond with female connectors and male mounting screws correspond with male connectors.

After assembling the cable, complete one final, important task—clearly label the cable with a drawing of its pin configuration. As mentioned before, it is frustrating trying to recall the configuration of an unlabeled cable or using a cable for the wrong instrument.

REFERENCES

CAMPBELL, JOE, *The RS-232 Solution.* Alameda, CA.: Sybex, Inc., 1984.

SEYER, MARTIN, *RS-232 Made Easy.* Englewood Cliffs, NJ.: Prentice-Hall, Inc., 1984.

SEYER, MARTIN, *Complete Guide To RS-232 and Parallel Connections.* Englewood Cliffs, NJ.: Prentice Hall, Inc., 1988.

WIDEMAN, GRAHAM, *Computer Connection Mysteries Solved.* Indianapolis, IN.: Howard W. Sams & Co., 1986.

B

Table of ASCII Characters

All instruments and personal computers that communicate through RS-232 connectors transmit one data bit at a time down a single wire of the cable. The receiving unit assembles a specific quantity of these bits into a number that represents one of the 128 numerals, punctuation marks, letters, or special control characters.

The conversion that occurs is based on a standard, called the *A*merican *S*tandard *C*ode for *I*nformation *I*nterchange (ASCII, pronounced "askee"). An ASCII code is a seven-bit binary code whose decimal values are between 0 and 127 and is used to represent alphanumeric characters.

For example, an alphabetic "A" is represented as a 65, a "1" is a 49, and a space is a 32. The ASCII standard is almost universally used as "the" system for representing characters in computers. The following table contains the 128 ASCII conversions:

Decimal	Hexadecimal	Binary	ASCII	Key Equivalent
0	00	0000000	NUL	CTRL/1
1	01	0000001	SOH	CTRL/A
2	02	0000010	STX	CTRL/B
3	03	0000011	ETX	CTRL/C
4	04	0000100	EOT	CTRL/D
5	05	0000101	ENQ	CTRL/E
6	06	0000110	ACK	CTRL/F
7	07	0000111	BEL	CTRL/G
8	08	0001000	BS	CTRL/H, BACKSPACE
9	09	0001001	HT	CTRL/I, TAB

(continued)

Decimal	Hexadecimal	Binary	ACSII	Key Equivalent
10	0A	0001010	LF	CTRL/J, LINE FEED
11	0B	0001011	VT	CTRL/K
12	0C	0001100	FF	CTRL/L
13	0D	0001101	CR	CTRL/M, RETURN
14	0E	0001110	SO	CTRL/N
15	0F	0001111	SI	CTRL/O
16	10	0010000	DLE	CTRL/P
17	11	0010001	DC1	CTRL/Q
18	12	0010010	DC2	CTRL/R
19	13	0010011	DC3	CTRL/S
20	14	0010100	DC4	CTRL/T
21	15	0010101	NAK	CTRL/U
22	16	0010110	SYN	CTRL/V
23	17	0010111	ETB	CTRL/W
24	18	0011000	CAN	CTRL/X
25	19	0011001	EM	CTRL/Y
26	1A	0011010	SUB	CTRL/Z
27	1B	0011011	ESC	ESCAPE
28	1C	0011100	FS	CTRL <
29	1D	0011101	GS	CTRL /
30	1E	0011110	RS	CTRL/=
31	1F	0011111	US	CTRL/−
32	20	0100000	SP	SPACEBAR
33	21	0100001	!	!
34	22	0100010	"	"
35	23	0100011	#	#
36	24	0100100	$	$
37	25	0100101	%	%
38	26	0100110	&	&
39	27	0100111	'	'
40	28	0101000	(	(
41	29	0101001	)	)
42	2A	0101010	*	*
43	2B	0101011	+	+
44	2C	0101100	'	'
45	2D	0101101	−	−
46	2E	0101110	.	.
47	2F	0101111	/	/
48	30	0110000	0	0
49	31	0110001	1	1
50	32	0110010	2	2
51	33	0110011	3	3
52	34	0110100	4	4
53	35	0110101	5	5
54	36	0110110	6	6
55	37	0110111	7	7
56	38	0111000	8	8
57	39	0111001	9	9
58	3A	0111010	:	:
59	3B	0111011	;	;
60	3C	0111100	<	<

Decimal	Hexadecimal	Binary	ASCII	Key Equivalent
61	3D	0111101	=	=
62	3E	0111110	>	>
63	3F	0111111	?	?
64	40	1000000	@	@
65	41	1000001	A	A
66	42	1000010	B	B
67	43	1000011	C	C
68	44	1000100	D	D
69	45	1000101	E	E
70	46	1000110	F	F
71	47	1000111	G	G
72	48	1001000	H	H
73	49	1001001	I	I
74	4A	1001010	J	J
75	4B	1001011	K	K
76	4C	1001100	L	L
77	4D	1001101	M	M
78	4E	1001110	N	N
79	4F	1001111	O	O
80	50	1010000	P	P
81	51	1010001	Q	Q
82	52	1010010	R	R
83	53	1010011	S	S
84	54	1010100	T	T
85	55	1010101	U	U
86	56	1010110	V	V
87	57	1010111	W	W
88	58	1011000	X	X
89	59	1011001	Y	Y
90	5A	1011010	Z	Z
91	5B	1011011	[	[
92	5C	1011100	\	\
93	5D	1011101	]	]
94	5E	1011110	^	^
95	5F	1011111	_	_
96	60	1100000	`	`
97	61	1100001	a	a
98	62	1100010	b	b
99	63	1100011	c	c
100	64	1100100	d	d
101	65	1100101	e	e
102	66	1100110	f	f
103	67	1100111	g	g
104	68	1101000	h	h
105	69	1101001	i	i
106	6A	1101010	j	j
107	6B	1101011	k	k
108	6C	1101100	l	l
109	6D	1101101	m	m
110	6E	1101110	n	n

(continued)

Decimal	Hexadecimal	Binary	ASCII	Key Equivalent
111	6F	1101111	o	o
112	70	1110000	p	p
113	71	1110001	q	q
114	72	1110010	r	r
115	73	1110011	s	s
116	74	1110100	t	t
117	75	1110101	u	u
118	76	1110110	v	v
119	77	1110111	w	w
120	78	1111000	x	x
121	79	1111001	y	y
122	7A	1111010	z	z
123	7B	1111011	{	{
124	7C	1111100	|	|
125	7D	1111101	}	}
126	7E	1111110	~	~
127	7F	1111111	DEL	DELETE

Lotus 1-2-3® and Symphony® use the Lotus International Character Set (LICS). LICS characters are represented by 256 decimal codes, 0 through 255. Codes 0 through 127 are the same control and alphanumeric characters as the ASCII codes. Codes 128 through 255 represent various international, scientific, and mathematical characters (e.g., British Pound, Japanese Yen, Dutch Guilder, German umlaut, Spanish tilde, pi, plus/minus, delta, degrees, etc.). With the exception of character 153 (the "unknown character"), codes 128 through 256 are rarely used for instrument communications. Please refer to your Lotus User's Manual for a list of LICS codes.

RS-232 Pinout Specifications

The latest version of RS-232, called RS-232C, was written by a committee of the Electronic Industries Association in 1969. The standard specifies the pins, connectors, control signals, timing signals, data signals, ground signals, etc., used in the intercommunications between devices.

Although the formalities of the standard are beyond the scope of this book, the nomenclature is not. This appendix gives an overview of the nomenclature.

First, each piece of equipment is referred to as either Data Terminal Equipment (DTE) or Data Communications Equipment (DCE). A DTE transmits its data on pin 2 and a DCE transmits its data on pin 3.

The following table describes the nomenclature used for each pin in the cable. The last column of the table denotes whether a pin is an input or an output for a DTE.

Pin	Abbrev	Name	To/From DTE
1	PG	Protective Chassis Ground	---
2	TxD	Transmitted Data	From
3	RxD	Received Data	To
4	RTS	Request To Send	From
5	CTS	Clear To Send	To
6	DSR	Data Set Ready	To
7	SG	Signal Ground	---
8	RLSD	Received Line Signal Detect	To

(continued)

Pin	Abbrev	Name	To/From DTE
9		Unassigned	
10		Unassigned	
11		Unassigned	
12	SRLSD	Secondary RLSD	To
13	SCTS	Secondary CTS	To
14	STxD	Secondary TxD	From
15	TSETDCE	Trans Sigl Element Timing DCE	To
16	SRxD	Secondary Received Data	To
17	RSETDCE	Recvd Sigl Element Timing DCE	To
18		Unassigned	
19	SRTS	Secondary RTS	From
20	DTR	Data Terminal Ready	From
21	SQD	Signal Quality Detect	To
22	RI	Ring Indicator	To
23	DSRSDTE	Data Signal Rate Select DTE	From
24	TSETDTE	Trans Sigl Element Timing DTE	From
25		Unassigned	

To summarize, pins 2, 4, and 20 are usually output pins for DTEs; pins 3, 5, and 6 are usually output pins for DCEs. Pins 1 and 7 are grouping pins. The remaining pins are rarely used.

Binary Coded Decimal and Hexadecimal Tables

One of the most common methods that scientific instruments use to display numbers is to have each digit of the display controlled by four wires. Each of the four wires for a digit represent a number, most commonly 8, 4, 2, and 1. The number displayed is the *sum* of the numbers represented by the wires that are switched on. This method of representing decimal numbers is called *B*inary *C*oded *D*ecimal (BCD).

Binary Coded Decimal can use systems other than 8421 to represent decimal digits. These systems are used much less frequently. For example, 2421 and 5211 are two other weighted coding systems that are used in instruments. These systems work in a manner that is analogous to the 8421 system.

One other BCD code that is used is ''XS3''. This code is used only rarely. It is similar to 8421, except that three (0011) is added to the 8421 binary number.

The following table compares the various BCD types:

Decimal	8421	2421	5211	XS3
0	0000	0000	0000	0011
1	0001	0001	0001	0100
2	0010	0010	0011	0101
3	0011	0011	0101	0110
4	0100	0100	0111	0111
5	0101	1011	1000	1000
6	0110	1100	1010	1001
7	0111	1101	1100	1010
8	1000	1110	1110	1011
9	1001	1111	1111	1100

Hexadecimal (HEX, for short) is a numbering system that uses a base of 16. In base 10, digits are zero through nine. The hexadecimal numbering system is like the decimal and binary systems because it is also proportionally weighted. However, each subsequent digit in the final number is a power of 16 greater than the previous number. Because each digit in HEX needs to be represented by a single unique symbol, alphanumeric values are used. That is, the digits that are used to represent the 16 hexadecimal values are 0, 1, 2, 3, 4, 5, 6, 7, 8, 9, A, B, C, D, E, and F.

The system of four bits per digit still applies. Bits can be readily converted to hexadecimal. For example, "0000" binary is a "0" HEX, while "1111" binary is an "F" HEX. The following table demonstrates the relationship between binary, hexadecimal, and decimal:

Binary	Hexadecimal	Decimal
0000	0	0
0001	1	1
0010	2	2
0011	3	3
0100	4	4
0101	5	5
0110	6	6
0111	7	7
1000	8	8
1001	9	9
1010	A	10
1011	B	11
1100	C	12
1101	D	13
1110	E	14
1111	F	15

Each of the codes in the table can be represented by well-defined equations. Thus, by determining which method your instrument uses, you can program accordingly. To determine the method used by an instrument, consult either your instrument's User's Manual or Technical Reference manual, ask the manufacturer of your instrument, or use trial and error.

E

Arithmetic, Relational, and Logical Operators

Lotus 1-2-3® and Symphony® use a set of "operators" to depict arithmetic operations. For example, the + operator causes the two values flanking it to be added together.

Lotus has a well-defined hierarchy that is used to determine the progression in which operators will be used. For example, consider the equation:

$$value = 10.0 + 35.0 * slope / factor$$

This equation contains addition, multiplication, and division. Clearly, the order of execution for the various operations can have a profound influence on the final value. To guard against chaos, and inconsistency, Lotus has a specified order of performance for its operators. This order is called "precedence" and each operator is assigned a precedence level.

The table shown below gives the precedence for each operator. In the table, division and multiplication have a higher precedence than addition and subtraction, so division and multiplication are performed first. If the precedence of two or more operators in a statement is the same, they will be executed according to the order in which they occur in the statement (left to right). If you are unsure of the precedence in an equation (or if you want to change it), use parentheses. Whatever is enclosed within parentheses is performed first. And within parentheses, standard mathematical rules apply.

Operator	Definition	Precedence
^	Exponentiation	7 (highest)
−	Negative	6
*	Multiplication	5
/	Division	5
+	Addition	4
−	Subtraction	4
=	Equal	3
&	String concatenation	1 (lowest)

In addition to arithmetic operators, Lotus has a category of operators that make comparisons. These operators are called "relational" operators. Each of the relational operators compares the value at its left to the value at its right. A relational expression evaluated from an operator and its two operands is assigned a value of 1 if the expression is true and a value of 0 if the expression is false.

Relational operators are most commonly used for testing the relationship of two expressions in an {IF} macro command or an @IF function. The following table provides a definition and precedence for each relational operator. The precedences are on the same scale as the arithmetic operators. Because the precedences for relational operators are lower than those for arithmetic operators, arithmetic operations occur before relational operations are evaluated.

Operator	Definition	Precedence
<	Less than	3
<=	Less than or equal	3
>	Greater than	3
>=	Greather than or equal	3
< >	Not equal	3

Lotus has a final category of operators called "logical" operators. Logical operators normally take relational expressions as operands. In this context, logical operators are useful for combining two or more relational expressions. The following table provides a definition and precedence for each logical operator. Because the precedences for logical operators are lower than the precedences for both arithmetic and relational operators, logical operators are evaluated last in an equation.

Operator	Definition	Precedence
#NOT#	Logical NOT	2
#AND#	Logical AND	1
#OR#	Logical OR	1

A logical expression evaluated from an operator and its operands is assigned a value of 1 if the expression is true and a value of 0 if the expression is false. The following explanations briefly describe each logical operator:

#NOT#: The expression is true if the operand is false, and vice versa.

#AND#: The combined expression is true if *both* operands are true, and false otherwise.

#OR#: The combined expression is true if one *or* both operands are true, and false otherwise.

Logical expressions are evaluated from left to right. An evaluation stops as soon as an expression is discovered to be false. The following examples illustrate the use of logical operands:

6>2#AND#3>1	is true
#NOT#(6>1#AND#10>4)	is false
6>7#OR#6>1	is true

Macro Command Summary

A set of instructions in a format and language that Lotus can understand is called a Lotus "macro" program. The language that is used is called the "Lotus Command Language" and the instructions are called "macro commands".

Lotus macro commands fall into six categories. They are

- System commands, which control the screen display and the computer's speaker.
- Interaction commands, which create interactive macros that pause for a user to enter data from the keyboard.
- Program flow commands, which include branching and looping in a program.
- Cell commands, which transfer data between specified cells and/or change the values of cells.
- File commands, which work with data in DOS files or communications ports.
- Menu commands, which allow you to design and manage menus.

The following is a list, by category, of the macro commands used in this book. The definitions are tailored to the primary purpose of this book—interfacing instruments to Lotus spreadsheets.

SYSTEM COMMANDS

{BEEP tone}

Sounds one of four different tones (1–4).

{INDICATE string}

Replaces the standard mode indicator message at the upper right corner of the display with the characters specified by *string*. Omitting *string* restores control of the mode indicator to Lotus.

{PANELOFF}

Suppresses updating of the control panel during macro execution.

{PANELON}

Restores standard updating of panel, cancels a {PANELOFF} command.

{WINDOWSOFF}

Suppresses updating of the display during macro execution.

{WINDOWSON}

Restores standard updating of display, cancels a {WINDOWSOFF} command.

INTERACTION COMMANDS

{?}

Halts macro execution temporarily, allowing a user to enter data, move the cell-pointer, and access Lotus menus. Macro execution resumes when Return is pressed.

{GETLABEL prompt,cell}

Halts macro execution, displays a *prompt* for a user to enter a line of characters, and stores the response as a label in a specified *cell.*

{GETNUMBER prompt, cell}

Halts macro execution, displays a *prompt* for user to enter a number or numeric expression, and stores the response as a number in a specified *cell.*

PROGRAM FLOW COMMANDS

{macro}

Calls ''*macro*'' as a subroutine.

{\R}

Calls *\R* as a subroutine. Same as pressing [ALT]-R.

{BRANCH macro}

Permanently transfers control of a program to the location called "*macro*".

{IF condition}

Conditionally executes the commands that follow the IF command in the same cell if *condition* evaluates to true.

{ONERROR macro,messagecell}

Branches to *macro* if an otherwise fatal error occurs. Optionally stores the error message that Lotus would have displayed in *messagecell.*

{FOR counter,start,stop,stepsize,subroutine}

Repeatedly executes the commands, beginning at the cell called "*subroutine*", as many times as indicated by the values of *start, stop,* and *stepsize.*

{QUIT}

Immediately terminates macro processing, returns to manual keyboard control.

{WAIT time}

Causes macro execution to pause until the computer's clock time matches or exceeds *time.*

CELL COMMANDS

{BLANK range}

Erases the cells within a specified *range.*

{LET cell,number}

Places *number* into a specified *cell.*

{LET cell,string}

Places *string* into a specified *cell.*

{PUT range,column,row,number}

Places *number* into a cell in *range* represented by offset coordinates (*column,row*).

{PUT range,column,row,string}

Places *string* into a cell in *range* represented by offset coordinates (*column,row*).

{RECALC range}

Recalculates the formulae in *range,* proceeding column by column.

FILE COMMANDS

{CLOSE}

Closes a disk file or, more commonly, a communications port.

{OPEN filename,mode}

Opens *filename* (disk file or communications port) for read only (R), write only (W), append (A), or both read/write (M).

{READ bytes,buffer}

Reads the number of characters specified by *bytes* from a file or communications port and stores the characters as a string in the cell specified by *buffer.*

{READLN buffer}

Reads a line of characters (until a carriage return is encountered) from a file or communications port and stores the characters as a string in the cell specified by *buffer.*

{WRITE string}

Writes a string of characters (without a carriage return) to a file or communications port.

{WRITELN string}

Writes a line of characters (appending a carriage return) to a file or communications port.

MENU COMMANDS

{MENUBRANCH menu}

Halts macro execution temporarily, branches to a customized menu whose instructions are found starting at the cell called "*menu*", prompts a user to make a choice, then continues execution based on the choice.

{MENUCALL menu}

Similar to {MENUBRANCH}; however, processes *menu* and its associated steps as a subroutine.

MACRO COMMANDS FOR SPECIAL KEYS

In addition to the above macro commands, commands exist which represent special Lotus keys. Cursor control commands allow you to specify the number of times a macro should "press" a named key. For example, {RIGHT 5} is equivalent to manually pressing the right arrow key five times. The following is a table of these commands:

{EDIT}	{PGUP}
{GOTO} location	{PGDN}
{WINDOW}	{END}
{CALC}	{ESC}
{UP}	{BACKSPACE}
{DOWN}	{BIGLEFT}
{LEFT}	{BIGRIGHT}
{RIGHT}	{MENU}
{HOME}	~(CARRIAGE RETURN)
{SERVICES} or {S} (SYMPHONY ONLY)	

@Function Summary

Special-purpose Lotus formulae are called "@functions". Each @function extends your calculating power beyond simple arithmetic and text handling operations. Lotus @functions fall into several categories: mathematical, statistical, string, special, logical, and date/time.

The following is a list of @functions used in this book and some others that you may find useful. The definitions are tailored to the primary purpose of this book—interfacing instruments to Lotus spreadsheets that have powerful data reduction templates.

MATHEMATICAL @FUNCTIONS

@ABS(number)

Absolute value of *number.*

@ACOS(number)

Arc cosine of *number.*

@ASIN(number)

Arc sine of *number.*

@ATAN(number)

Two-quadrant arc tangent of *number.*

@ATAN2(number1,number2)

Four-quadrant arc tangent of *number2/number1*.

@COS(number)

Cosine of *number.*

@EXP(number)

The number e raised to the *number* power.

@INT(number)

Integer part of *number.*

@LN(number)

Log of *number,* base e.

@LOG(number)

Log of *number,* base 10.

@PI

The number pi (approximately 3.14159).

@SIN(number)

Sine of *number.*

@SQRT(number)

Positive square foot of *number.*

@TAN(number)

Tangent of *number.*

STATISTICAL @FUNCTIONS

@AVG(list)

Arithmetic average of values in *list.*

@COUNT(list)

Number of cells containing values in *list.*

@MAX(list)

The maximum value in *list.*

@MIN(list)

The minimum value in *list.*

@STD(list)

Population standard deviation of values in *list.*

@SUM (list)

Sum of values in *list.*

@VAR(list)

Population variance of values in *list.*

STRING @FUNCTIONS

@CHAR(number)

Returns the ASCII/LICS character represented by *number.*

@CLEAN(string)

Removes control characters from *string.*

@CODE(string)

Returns the ASCII/LICS code of first character in *string.*

@LEFT(string,number)

Leftmost *number* of characters in *string.*

@LENGTH(string)

Total number of characters in *string.*

@LOWER(string)

Converts all uppercase letters in *string* to lowercase.

@MID(string,startingposition,numberofchars)

Extracts *numberofchars* characters from *string,* beginning at offset *starting-position.*

@N(cell)

Numeric value of *cell.*

@RIGHT(string,number)

Rightmost *number* of characters in *string.*

@S(cell)

String value of *cell.*

@STRING(cell,places)

Converts the number in *cell* into a string with *places* decimal places.

@TRIM(string)

Removes the leading and trailing spaces from *string* and compresses multiple spaces within *string* into single spaces.

@UPPER(string)

Converts all lowercase letters in *string* to uppercase.

@VALUE(string)

Converts a *string* that looks like a number into that number.

SPECIAL @FUNCTIONS

@CELLPOINTER(spec)

Attribute of the cell currently highlighted by the cell pointer.

@CHOOSE(selector,va11,va12,va13,...)

Selects value based on its position in the list.

@HLOOKUP(selector,rowrange,offsetnumber)

Table lookup, comparing *selector* value with a row of values. The value of *selector* may be either a string or a number.

@INDEX(range,coloffset,rowoffset)

Lookup based on position in *range*.

@VLOOKUP(selector,colrange,offsetnumber)

Table lookup, comparing *selector* value with a column of values. The value of *selector* may be either a string or a number.

LOGICAL @FUNCTIONS

@EXACT(string1,string2)

Tests whether *string1* and *string2* have exactly the same characters.

@IF(criterion,value1,value2)

If criterion is non-zero, *value1* is returned; if criterion is zero, *value2* is returned.

@ISERR(value)

If *value* has the value ERR, 1 (TRUE) is returned; otherwise 0 (FALSE).

@ISNUMBER(value)

If *value* has a numeric value, 1 (TRUE) is returned; otherwise 0 (FALSE).

@ISSTRING(value)

If *value* is a string, 1 (TRUE) is returned; otherwise 0 (FALSE).

DATE AND TIME @FUNCTIONS

@DATE(year,month,day)

Serial number of specified date (Jan 1, 1900 = 1).

@DATEVALUE(datestring)

Serial number of date specified by *datestring*.

@NOW

Serial number of current moment.

@TIME(hour,minute,second)

Serial number of specified time of day.

@TIMEVALUE(timestring)

Serial number of time of day specified by *timestring.*

@DAY(serialnumber)

Day number of *serialnumber.*

@HOUR(serialnumber)

Hour number of *serialnumber.*

@MINUTE(serialnumber)

Minute number of *serialnumber.*

@MONTH(serialnumber)

Month number of *serialnumber.*

@SECOND(serialnumber)

Second number of *serialnumber.*

@YEAR(serialnumber)

Year number of *serialnumber.*

@END(book)

The end of this book. (Not a real @function.)

Index

Accelerator boards, 4, 6, 273
Address:
  base, 208-10, 217, 232-33
  GPIB, 229-30, 244
  instrument, 14, 53, 116, 280
  I/O, 208-10, 217
  port, 209
  primary, 229-30, 234, 238, 244
  secondary, 229-30, 234, 244
Ampersand (&) (*see* Concatenation, string)
Analog:
  instruments/experiments, 90-91
  meters, 9-10, 13, 89-90, 102
  output, 9-10, 13, 89-91, 101-6, 128
  voltages, 2, 89, 116
Analog-to-digital converters, 9, 13, 89-101
  accuracy, 94
  background mode, 93, 111-17, 127, 207
  benefits of, 91
  choosing, 93-95
  conversion rate, 108-10, 114, 116, 124, 217, 222
  Data Acquisition Systems, 92
  definition, 90
  dual slope integration, 222
  electrical noise tolerance, 94-95, 127, 224
  full scale voltage range, 101, 108-10, 213
  integrating, 95
  interfaces, 92
  noise rejection, 94-95, 127, 224
  plug-in, 10, 92-93, 127, 189, 194, 207-26
  resolution, 91, 94, 108-10, 116, 127, 213-14, 219, 223
  sample and hold, 95, 117-19
  sampling rates, 95, 222
  sensitivity, 223
  speed, 94-95
  stand-alone, 91-128, 207-8, 238-40, 243-50
  steps, 213-14, 219
  successive approximation, 95, 222
Analog-to-digital data, obtaining, 107
@APP, 41
{APP} command, 195, 202-3
Arithmetic Operators (*table*), 301
ASCII (*see* Characters, ASCII)
Asterisks, column width, 34
ATN line, 231, 246
Attributes (cell) (*see* @CELL-POINTER)
Auto-Cabler, 284-88
Auto-executing macros, 37-42, 73-75, 139, 202, 221, 275
  differences in, 38
  Lotus 1-2-3, 39
  Symphony, 40
@AVG, 32, 125

Background mode (*see* Analog-to-digital converters)
Backing up, 35, 88
Barrier zones, 122, 206
Baud rate, 16-21, 25-26, 86, 132, 164, 168, 178, 187, 194, 278
Binary Coded Decimal, 2, 255-65, 299-300
Binary digits (bits), 15, 255-58
Bipolar input mode (*see* Input modes, bipolar)
Bits (*see* Binary digits)
{BRANCH} command, 75-76, 120
Buffer, 48, 61, 64, 116, 178, 205, 225
  overrunning DOS, 20, 168-69, 187, 189, 278, 280
  pointers, 116
Bus, 15, 227, 229, 231, 241

Cables, 21-24, 132, 162, 190, 268, 277
  adapters, 289

"crossed," 22-23
determination of (*see* Pin-out)
flow management, 21
GPIB, 227, 231
"jumped," 23
making, 290-92
maximum length, 24, 227, 234
null-modem, 21-23, 163-64
planning, 11, 198, 277
ribbon, 15
securing, 25, 277
straight-through, 21-23, 132, 162, 190, 286, 289
{CALC} command, 61, 63
Cell-pointer, 31, 54, 56, 76, 148, 181
finding (*see* @CELL-POINTER)
@CELLPOINTER, 149-51
Channel, 210, 219, 225
@CHAR, 54
Character bits, 17-18, 25, 132, 164
Characters:
alphanumeric, 15, 141, 152, 174-76
ASCII, 15-16, 54, 58, 176, 187, 204, 255, 293
non-printing, 60
unknown, 152, 174-76
Chart recorder output (*see* Analog, output)
@CHOOSE, 57, 86
@CLEAN, 58, 153, 222
{CLOSE} command, 63
@CODE, 176
Commenting, 43, 72, 87, 273
Communications:
establishing, 49-51, 81, 138, 277, 287
increasing the speed of, 187
parameters, 2, 15, 18, 132, 269, 280
serial, 15, 231, 255
setting with DOS MODE.com, 25-26, 38-40
troubleshooting, 279-81
Communications ports, 5, 8, 24, 39, 194, 222
closing, 53, 63
opening, 52-54, 172
Personal Computer-RS232, 5
Compatibility, computers, 4
Concatenation, string, 55, 114-16, 302
Concentration formulae, 137
Connecting, digital I/O interface, 259
Connector, GPIB, 227, 231
Controller, GPIB, 228, 240
Conventional memory, 5-6
Coprocessor accelerator boards (*see* Accelerator boards)

Daisy-chain, GPIB, 227
Data Acquisition Systems (*see* Analog-to-digital converters)
Data bits (*see* Character bits)
Data Communications Equipment (DCE), 22, 132, 162, 297
Data Set Ready (DSR), 22, 297
Data Terminal Equipment (DTE), 22, 132, 162, 297
Data Terminal Ready (DTR), 22, 162, 297
Date, 34, 66, 68, 79, 276
Debugging, 87, 281-83
tools, 87, 133, 281-83
Decimal numbering system, 261, 265
Device File Names, 46, 53, 164, 177, 189-93, 197, 202, 207, 269
Differential input (*see* Input modes, differential)
Digital displays, 2, 9-10, 89, 255-67
Digital Input/Output Interfaces, 257
Digital signals, 2, 10, 90-93, 128, 255-67
Direct Memory Access, 232, 234, 241
Documentation, 43, 72, 87, 275
{DOWN} command, 56
Drivers, 193-94, 196
Dual In-Line Package (DIP) switches, 18-19, 132, 198

Element Separator, 237, 246
ELISA assays, 160
End-Or-Identify (EOI), 232, 246
Erasing data, 145
ERR, from @functions, 32-33, 41, 87, 146, 175, 200
Error codes, 133, 142, 151
Errors:
bus, 241
GPIB, 247
handling, 33, 172-74, 176, 203, 206
messages, 41, 172, 177-78, 183, 206
system, 20, 169, 172-74, 178, 189, 206, 280
ETX, 54, 293
Expanded memory, 5-6, 272, 280 (*see also* Errors, system)
Extended talker/listeners, 228-30, 234

File, Extraction, 180
File manipulation commands (*see* Device File Names)
Flicker (*see* Screen flicker)
Flow control (*see* Handshaking)
{FOR} loops, 56, 63, 116, 120, 179 (*see also* Looping; Nesting)
Format:
date, 34, 167
fixed digits, 34-35, 167
Formulae, 29, 137, 213, 301-3
minimizing, 30, 272
Freezing the display (*see* Screen flicker)
@Functions:
general description, 7
mathematical, 111
speed considerations, 183, 272
summary, 309

Game Control Adaptor (GCA), 193
{GETLABEL} command, 77
{GETNUMBER} command, 77
{GOTO} command, 76, 170
GPIB (*see* IEEE-488)

Handshaking, flow control, 21, 24, 164, 190, 193, 231
Help windows (*see* Windows)
Hexadecimal, 2, 209, 229, 265, 300
Hiding columns, 30
@HLOOKUP, 57
HPLC, 94, 98, 240
HP-IB (*see* IEEE-488)

Idle, 120, 152, 159, 171
IEEE-488, 2, 10, 189, 194, 226-54
@IF, 302
{IF} tests, 41, 73, 176
@INDEX, 57, 86, 142, 179
{INDICATE} command (*see* Status, reports)

Initialization:
processes, 38
testing for successful, 142
Initialization macros, 38, 47, 52, 139-43
Lotus 1-2-3, 39
Symphony, 40
Initialize, GPIB, 240
Input (*see* User input)
Input configuration, 210
Input modes, 212, 215
bipolar, 208, 212-14
differential, 208, 210-11, 215
single-ended, 208, 210-11, 215
unipolar, 208, 212-14
Input-EOS, 236
Installation, instrument, 14, 49, 131, 161
Installing, Lotus Measure, 194
Integrating instruments, 1
Integration, dual slope, 222
Integration signals, 102
Interface, 1
Interrupt lines, 232, 234
I/O addresses (*see* Address, I/O)
@ISERR, 41

Jumpers, 18, 198

Least Significant Digit (LSD), 260
@LEFT, 58, 153, 180
@LENGTH, 153
{LET} command, 48, 67, 114
and cell-pointer, 150
using named ranges in, 85
Light Emitting Diodes, 259, 286-88
Linear regression (*see* Regression)
Linear rotation position transducer (LVDT/RVDT), 13
LIMs memory (*see* Expanded memory)
Listeners, 228-30
Logical Operators (*table*), 301
Lookup functions, 57, 312-13
Looping, 56, 62-63, 87, 142, 174-76 (*see also* {FOR} loops; Nesting, {FOR} loops)
definition, 48
re-adjusting, 150
Lotus Command Language, 7, 35-36, 67, 304
Lotus International Character Set, 176, 296
Lotus Measure, 2, 24, 46-47, 92, 189-254, 269, 273, 280, 287

Macro:
activating, 37-42, 51
commands, 7, 36, 304-8
creating, 36, 49-51
execution, 37, 42, 67
naming, 39-42, 44, 67
programs, 7, 35-38, 49
self-modifying, 146, 169, 180
Main programs, 47, 52, 69-70
Management lines, 231
Matrix, 31, 86, 165
@MAX, 32, 125
Memory (*see* Address, I/O)
Memory location, GPIB I/O, 229
{MENUBRANCH} command, 74-76
{MENUCALL} command, 74-76
Menus, 74-77, 276
linked, 169
programming techniques, 76
Meters (*see* Analog, meters)
@MID, 58, 153
@MIN, 32
Modular instruments, 13-14, 228
Most Significant Digit (MSD), 260

Named cells and ranges (*see* Range name)
Named range tables, 72, 274, 283
Nesting, 41, 56
{FOR} loops, 145, 179
@functions and macro commands, 64, 86, 276
limits, 74, 120
Non-volatile memory, 161, 164
@NOW, 79
Null-modem cables (*see* Cables, null-modem)
Numbers, method of transmission, 15-16

{ONERROR} command, 172-74, 177, 206
{OPEN} command, 53-54, 172

{PANELOFF} command, 79
{PANELON} {PANELOFF}, 78
Parallel transmission, 231
Parity, 17-18, 132, 164
Parsing, 57-64, 85, 122, 175, 178, 199, 237
formulae, 59-61
multiple data from single string, 151, 178-80, 200, 237, 246
variable length strings, 152
PATH command, 25
Pin-out, 21-23, 132, 284-88, 297-98
Planning (*see* Templates, planning)
Plug-in board (*see* Analog-to-digital converters)
GPIB, 10, 226-54
Polling, 249-52
parallel, 231, 250
serial, 232, 251
Portability, program, 52, 67, 73, 83, 148, 221
Ports (*see* Communications ports)
Position sensors, 13, 93
Precedence, 301
Primary address (*see* Address, primary)
Printing, compressed mode, 186
Process control, 128
Protection, worksheet, 278
{PUT} command, 48, 51, 62-64, 167, 180, 207
and cell-pointer, 150
conversion formulae and, 110
math @functions and, 111
using named ranges in, 85

{QUIT} command, 37, 77

Range name, 39, 42, 44, 137-38, 146, 273
creating, 37, 39, 42, 51
(*tables*), 72, 274, 283
Rate, of data collection, 217
{READLN} command, 57, 61, 152, 174, 183
{RECALC} command, 61
{RECALCCOL} command, 61
Recalculations, forcing, 61, 63, 86, 151, 276
Recorder signal (*see* Analog, output)
Regression, 135-36, 145
Relational Operators (*table*), 301-2
Reports, printing, 82, 186
Request To Send (RTS), 23, 162

Resistance Temperature Detectors (RTDs), 13, 93
Resolution (*see* Analog-to-digital converters)
Return, 41, 78, 84, 146, 182
Ribbon cables (*see* Cables, ribbon)
@RIGHT, 58, 153, 180
Robotics, 9, 129
RS-232 communications, 21, 90, 193-94, 197-206, 286
RS-232 pinout (*see* Pin-out)
RS-232 port (*see* Communications ports)

Sample and hold, 117, 119
Sampling rates, 222
Saving, files, 83, 180
Scratch-pad, 48, 63-64, 154, 183
  creating, 71
Screen flicker, 68, 78-79, 203
Self-modifying macros (*see* Macro, self-modifying)
Sensors, 9-10, 90, 93, 101, 210
Serial transmission, 15, 231, 255
Service Request, 231, 241-43, 252-54
SET driver files (*see* Drivers)
Signal (*see* Input modes)
Significant decimal digits (*see* Format, fixed digits)
Single-ended input (*see* Input modes, single-ended)
Snapshots, 117-27, 206-7, 222-24, 263
Speed considerations (*see* Baud rate)
SRQ line, 231
Stand-alone A/D converters (*see* Analog-to-digital converters)
Standard curves, 143
Standardization macros, 143
Star configuration, 227
Status, reports, 66, 68, 78, 282-83
{STATUSON} command, 247
Steady states (*see* Snapshots)
STEP mode, 282
Steps (*see* Analog-to-digital converters)
Stop Bits, 17-18, 132, 164
Straight-through cables (*see* Cables, straight-through)
Strain gauges, 13, 93
String, 16, 175, 247 (*see also* Concatenation, string; Parsing)
  commands, 54, 108, 114, 116, 122, 130, 138, 140-42, 154
  @functions, 58-61
  status, 140
@STRING, 55, 244
Strobe lines, 259, 261
STX, 54, 293
Subroutine, 47, 120
  calling, 41, 75-76
  levels, 74, 120
{subroutine}, 41, 76
Successive approximation (*see* Analog-to-digital converters)
System Error messages (*see* Errors, system)

Talkers, 228-30
Templates, 27-35
  assay, 133
  configuration section, 133
  creating, 30-35, 166
  design, 270
  planning, 29, 270
  regression section, 135
  report, 165
  sample data section, 136
  sections of, 29, 133-37
  standard curve section, 134
  testing, 65, 138, 275
Terminal emulator programs, 281
Testing, 65, 87, 138, 224, 275
Thermocouples, 9, 13, 27, 93
Tilde (*see* Return)
Time, 79-82
  delays, 67, 80-82, 117, 204
  elapsed, 68, 79-80
@TIME, 81
Time-out, 193, 235, 237-38, 246-49
@TIMEVALUE, 79
Timing, 81, 117-27, 142-45, 203, 222, 278
Transducers, 9-10, 90, 93, 101, 210
Trap loops, 174
Triggering data collection (*see* Snapshots)
@TRIM, 59
Troubleshooting tools, 281-83

Unipolar input mode (*see* Input modes, unipolar)
Unknown character (*see* Characters, unknown)
Unlisten command, 229, 244
Untalk command, 229, 244
User friendly programs, definition, 66-70
User input, 68, 77, 181

@VALUE, 58, 64, 153, 180, 200, 245
@VLOOKUP, 57
Voltage:
  configuration formulae, 213 (*see also* Analog, output)
  ranges, 101
  spikes, 126

{WAIT} command, 69, 77, 80-82, 117, 127, 143, 145, 149, 154, 222-24, 282
Waiting for data (*see* Snapshots)
{WAITSRQ} command, 238, 252-53
Windows, 147
  creating, 157, 185-86
  definition, 148
{WINDOWSOFF}, 79
{WINDOWSON} {PANELON}, 282
{WRITE} command, 55
{WRITELN} command, 54-55

XON/XOFF, 193
XS3, 257-58, 299

< > **definition,** 142

^**definition,** 154

{?} **command,** 77, 84, 181

Thank you for purchasing this book.

The author would like to offer you the opportunity to obtain the applications and examples in this book on diskette.

To save keyboard input time, all of the applications and examples in this book are available in the two disk set. Each application has the original template(s), named ranges, program(s), etc., described in the book. These applications will work with Lotus 1-2-3®, versions 2.0/2.01 and Lotus Symphony®, versions 1.1, 1.2, and 2.0.

To send for the *Laboratory Lotus®: A Complete Guide To Instrument Interfacing* disks, just photocopy this page, fill out the information completely and return to the address listed below. Thank you.

Date Purchased: ___________

Name: ____
Title: ____
School/Company: ___________________________________
Department: ______________________________________
Street Address: ___________________________________
City: __________________________ State: ___________ Zip: _________
Telephone: (   )____________________________________

Please enclose a check for $12 (US $dollars), payable to: Louis M. Mezei.

Louis M. Mezei
40815 Ondina Ct.
Fremont, CA 94539